Problem Structuring

Problem Structuring

Methodology in Practice

Mike Yearworth
Exeter, UK

WILEY

Published by John Wiley & Sons, Inc., Hoboken, New Jersey.
Published simultaneously in Canada.

Library of Congress Cataloging-in-Publication Data applied for

Hardback ISBN: 9781119744825

Cover Design: Wiley
Cover Image: © SEAN GL ADWELL /Getty Images

Set in 9.5/12.5pt STIXTwoText by Straive, Chennai, India

For Gemma, who, with love and considerable patience, helped me unlearn all the things that were getting in the way of understanding what really matters.

Contents

Preface

> *'We are not students of some subject matter, but students of problems. And problems may cut right across the borders of any subject matter or discipline.'*
> (Popper, 1963, p. 67)

The discipline of Operational Research (OR) emerged in the period between roughly the mid-1930s and the years encompassing the Second World War. The fact that an entire applied scientific endeavour focused on addressing operational issues came into existence and was crafted for the singular purpose of delivering interventions in support of winning a war is striking. The notion of addressing operational issues can be understood simply as conducting the research necessary for achieving the greatest effect from the limited resources available. The purpose was clearly self-evident and provided ample motivation. Much of the post-war diffusion of OR practices into government and industry reflected this single-minded purpose, leading to a great deal of success for the methods, tools and techniques at its core. Goodeve (1948a,b,c) provides an early account of the emergence of OR in the United Kingdom, and Kittel (1947) relates a similar story for the United States. Keys (1995, pp. 1–76) also provides a detailed account of these early years. As is the nature of knowledge gained from empirical work, OR became codified as a discipline and subject to formal teaching and research (for example, as presented in Churchman, Ackoff and Arnoff (1957)), leading to what Keys (1995, pp. 77–148) refers to as the period of growth and stability in the years 1945–1970.

Why go back over the emergence of a discipline that was initially shaped by fighting a war? What is the relevance? The answer is simple and is located in the challenges arising from new existential threats such as climate change, water shortages, flooding, resource depletion, pandemics of disease and diet, housing, waste accumulation, etc.[1] These are perhaps not as self-evident to all as a global

1 According to Frankopan (2023) these are not new threats, humanity has always faced them.

conflict and some still remain contested by certain constituencies – that are either isolated from the effects, choose not to look, or have some other motivation for denialism – but they are ample motivation for those who do choose to act. The contribution that OR can make today is the same as when it emerged – conducting the *operational* research necessary for achieving the greatest effect for its clients using the limited resources available. However, today, it must make itself relevant to problem contexts that are not so singular in nature, where there are different, often contested, viewpoints and perspectives on the problems and the means to address them; such problems are *wicked* (Rittel & Webber, 1973). In short, it needs to demonstrate that it can also operate as an applied *social*-science, an endeavour capable of moderating its positivism into accommodating a *softer*, more humanist framing. Thankfully, much of the groundwork has already been done by a small group of practitioners/scholars who emerged from the mainstream of OR in the 1970s and 1980s to pioneer a field colloquially known as Soft OR and centred on a methodological framing of techniques known as problem structuring methods (PSMs) (Rosenhead, 1989c).

This book is grounded in the history of the emergence of Soft OR and its development over time as a distinctively new endeavour. The origins of OR in the years leading up to and including the Second World War had already shown how knowledge of the '*hard*' physical sciences and mathematics could be applied to pressing operational issues. The notion of '*soft*' was intended to convey a further inclusion of ideas from the applied social sciences, broadening the scope of OR to deal with the more complex issues of transforming, strategising and planning – rather than just optimising. Although rooted in OR, the ideas presented here, together with practical methods and supporting case studies, are suitable for engineering students, OR practitioners that are not aware of *soft* methods and early notions of *problem formulation*, systems thinkers and, as is argued later, to applied social scientists. The first part of Chapter 7 is devoted to an account of how the evolution of Soft OR can be *read into* engineering practice leading to an equivalent notion of '*soft engineering*', which would have been a highly useful label had it not already been taken. The uptake of PSMs can also be found in areas of project management, environmental management, sustainability transitions and energy policy amongst many others. Chapter 13 surveys some of these applications and also discusses similarities between PSMs and applied techniques in the social sciences that have emerged from a completely different research tradition. The fact that communities of researchers and practitioners have traversed disciplinary landscapes and arrived at doing similar work should be no surprise to those aware of the *chaos of disciplines* (Abbott, 2001).

The content addresses the knowledge requirement to *select and apply, a range of problem structuring methods to understand complex problems* across a range of duties for an OR practitioner in the recently launched Operational

Research Specialist Degree Apprenticeship specification by the Institute for Apprenticeships & Technical Education (2021). It is is also suitable for engineering students at UK Level M/7 and above, i.e. final-year MEng, MSc, EngD/PhD and their equivalents as well as to support CPD for engineering practitioners. Students at schools or faculties signed up to the Council of Engineering Systems Universities (CESUN) should find the material of interest, offering an interpretive modelling perspective complementary to their main reading lists. The historical account of the emergence of Soft OR (in Chapter 4) and of a problem structuring sensibility in an engineering context (in Chapters 7 and 8) provides some insight into what might be involved to address recent calls for changes in engineering education (Sorby, Fortenberry, & Bertoline, 2021). The material presented in Chapters 9 and 10 was, in part, developed and used in teaching such M-level modules in Civil Engineering Systems and Sustainable Systems, and for an EngD in Systems taught programme at the University of Bristol between 2009 and 2016. This material was further adapted to provide an accessible approach to structuring an operations strategy problem for MBA students at the University of Exeter Business School from 2017 to 2022.

The approach to thinking about problem structuring presented in this book is intended for broad application. Whilst many of the developments in the field were the result of learning from consulting engagements, it is not just consultants who should benefit from studying problem structuring methods. Perhaps there has been a de-skilling taking place across businesses and the public sector that has left them in need of the services of the management consultant to help them deal with complex problems. However, in Chapter 6, the argument is put forward that there is both an abundance of everyday problem structuring taking place in organisations that could become more formalised and rigorous. It is also possible that quite often the keys to unlocking the understanding of a complex organisational problem can unwittingly be given away to consultants in the '*liminal*' space before a contract for services is signed. Therefore, building up organisation capability in problem structuring skills is unlikely to be a waste of resource even if, at the end of the day, organisations only go as far as becoming more discerning purchasers of the consultants' services.

The necessity for delivering an MBA programme online during the COVID pandemic in 2020 and 2021 provided further empirical evidence for the use of group support systems (GSS) to enable geographically distributed teams to work together effectively using PSMs and contributed material for Chapter 14. Further developments of platforms to support problem structuring in same time/different places workshops are also discussed in Chapter 14, especially in the light of more dynamic working environments that will need to support hybrid working and reduced flying.

Within OR, there has been a long history of borrowing, overlap and intersection with Systems Thinking (Ackoff, 1974b; Keys, 1991; Mingers & White, 2010). The emergence of Soft OR can be viewed as a recognition that a broad systemicity was (and is) needed in the practice of OR. Soft systems methodology (SSM) is claimed by the Soft OR community to be a PSM, and I must acknowledge the significant influence that *Systems Thinking, Systems Practice* (Checkland, 1981) has had on my own work, generally, not just in this book – every now and then we come across works that have such a profound effect. The use of hierarchical process modelling (HPM) in problem structuring that is presented in Chapter 9 was developed by academic practitioners and colleagues working in Engineering Systems at the University of Bristol. The similarities and differences between HPM and the Purposeful Activity System models from SSM are also discussed within the wider context of using HPM in an interpretivist sense, largely inspired by the argument presented by Checkland and Holwell (2004). The book is therefore of relevance to students on Systems Thinking, Systems Engineering (or, indeed, Engineering Systems) programmes, who may have been taught SSM but want to explore other PSMs that would be considered to be applicable in similar problem contexts.

This is the first time that the core PSMs have been discussed collectively[2] since the second edition of *Rational Analysis for a Problematic World* (Rosenhead & Mingers, 2001b), but here, instead of an edited volume, a novel approach has been taken in their analysis and review. Rather than presenting a straightforward account of approaches together with supporting case studies, this has been left for the texts that already exist to meet this need. Instead, a methodological perspective has been adopted throughout. A review of Strategic Options Development and Analysis (SODA), the Strategic Choice Approach (SCA) as well as SSM and their subsequent developments are presented together as a coherent set of methods in Chapter 5. Therefore, in the case of all of the PSMs presented and discussed in this book, the focus is directed at understanding constitutive rules, epistemology and the affordances of the modelling approach. This provides a common basis for comparison between the core techniques and opens up the field to further experimentation, e.g. in the development of new methodologies that share a similar theoretical and philosophical grounding. It also enables the exploration of links between PSMs and other applied social science techniques that have come from completely different traditions and different times – such as the work of Callon (1980) and areas such as knowledge co-production with decision-making under uncertainty (Moallemi et al., 2023). There is the potential for much cross-learning. There is also the enticing project of using the constitutive rules for PSMs to bring some rigour into problem contexts that hitherto have

2 SSM and SODA are covered by Pidd (2009, pp. 83–183) in a review of interpretive modelling approaches in OR/MS, but then segues into quantitative modelling through System Dynamics and multi-criteria decision analysis (MCDA).

relied on purely '*common sense*' or ad hoc approaches, the domain of non-codified problem structuring first explored by Yearworth and White (2014).

The motivation for this book is primarily to address issues within the practice of OR. Whereas some related fields have advanced beyond methodology choice (e.g. see Ulrich (2003)), much of OR practice today still seems trapped in the same situational logics outlined 40 years ago by Checkland (1983) and Rosenhead (1986). Even worse, the remedy, in Soft OR/PSMs, is also in danger of stumbling into similar traps. The problem is so pernicious and so easy to fall into that I have had to resort to liberal use of footnotes to haul my own thinking out of the trap. It seemed fairer to the reader to leave evidence, the breadcrumbs so to speak, of critical reflection rather than a perfectly polished text. Meeting deadlines has also made a contribution to this practice.

The arguments and analyses presented in this book are supported by two theoretical strands. The first is pragmatism, the influence of which perhaps was more apparent to earlier generations of OR practitioners through the work of Ackoff and Churchman when they were active, but has recently been brought to prominence in the OR field by Ormerod (2006, 2020b). Here, an appeal to pragmatism is made to help the OR profession (both Hard and Soft) from being trapped in situational logics that reduce practice to the formulaic spotting of problem contexts and the mechanical (admittedly rational) application of algorithms and methods. OR was more than this when it first emerged as a profession, but this has been somewhat lost over the years. The emergence of Soft OR as an attempt to break out of the situational logic of Hard OR (e.g. see Checkland (1983)) did nothing more than create its own, albeit more complicated, situational logic through its dichotomous framing of messy problems (Rosenhead, 1989b). The second strand is practice theory. The necessity of a pragmatic framing of OR to dig us out of the problem of situational logics then requires a suitable theoretical lens to analyse the practices of OR that lead to knowledge as a form of action that will change existing realities. Practice theory (or social practice theory) provides us with the conceptual language and useful approaches to analyse what OR practitioners are doing when they are dealing with complex problems. They are brought to bear in the survey of PSM applications in Chapter 13.

Although starting with the clear idea that pragmatism and social theory were sufficient grounding for an analysis of Soft OR practice, what has emerged in writing the book was a realisation that process philosophy had been waiting in the wings, as it were, to make an appearance in the final act. Process philosophy had been there all along, ever since I sat-in on the first lecture with my new colleagues in the Systems Centre at the University of Bristol. Panta rhei – everything flows – as John Davis started his lectures when he introduced the class to the idea that *anything* could be thought of as a process – including the normal objects that were the creation of engineers like flood defences – and that the realisation of this thinking could be addressed using HPM (see Chapter 9). The practicalities of this process-based modelling approach were fashioned not from some deep philosophical wrangling but from the experience of working out the implications of process thinking through consulting engagements and project work. As the application of HPM evolved into an approach to problem structuring – the work of Davis, MacDonald and White (2010) marked a turning point – it was clear that we had a modelling approach that fulfilled a similar purpose to activity system modelling in SSM, but one that also benefitted from the strong process philosophy. HPM enabled the simultaneous modelling of conceptual activities (as processes) in combination with non-reified entities (as processes), together with an inbuilt notion of process performance supported by measurements and, if needed, quantitative modelling. In thinking of this book processually, in its *becoming*, the importance of this work in engineering has become clearer, especially the contribution it could make in helping to resolve the divides that have emerged between Hard and Soft OR practice. These ideas are discussed in Chapter 16.

I make no apology for dipping into the history of the ideas that form the core of this book. Without an appreciation of the messiness of the emergence of Soft OR/PSMs in the 1980s (roughly), there is the likelihood of similar unhelpful debates developing as other researchers and practitioners, working ostensibly in STEM disciplines, find themselves making similar moves. Whilst there is currently much happening in the promotion of interdisciplinarity to address societal grand challenges, e.g. in the way in which funding bodies and Universities are marketing their products to each other, these seem to pre-suppose collaboration across tidy discipline boundaries. Again, no apology is made for muddying the waters by promoting a view that movement between disciplines, in the sense of either whole discipline trajectories or individual researchers, is a messy reality – although perhaps this is much harder today as academic careers become more precarious and narrower.

For those keen to go straight to the primary literature, it is possible to gain an immediate and useful overview of PSMs by referring to a number of core works. For the most recent, and brief, summary of PSMs, see Petropoulos et al. (2024, pp. 465-467)[3]. Franco and Rouwette (2022) review PSMs for the *Palgrave Handbook of Operations Research*, and White (2025)[4] presents an Invited Review of Soft OR for the *European Journal of Operational Research* as part of EURO's 50th Anniversary celebration. For another recent review of the development of PSMs over roughly the past decade, see Gomes Júnior and Schramm (2021). For a review of their fundamental characteristics, see Yearworth and White (2014) and Smith

3 This is the reference to an 'encyclopaedic' article on Operational Research assembled and edited by Petropoulos in celebration of the 75th Anniversary of the *Journal of the Operational Research Society*. The review of Soft OR/PSMs section was written by Yearworth and White.
4 This was in an advanced state of preparation at the time of writing.

and Shaw (2019). Rosenhead (1989c) should be consulted as the primary textbook that introduced PSMs to a wider audience – *Rational Analysis for Problematic World: Problem Structuring Methods for Complexity, Uncertainty and Conflict* (RAfaPW). Georgiou and Heck (2021) present a comprehensive bibliographic atlas of the key literature leading up to this important milestone. Rosenhead and Mingers (2001b) updated the material in RAfaPW for the second edition and also conducted an early Invited Review of '*PSMs in action*' for the *European Journal of Operational Research* (Mingers & Rosenhead, 2004). Occasional Special Issues have been commissioned by journals with guest editors assembling invited submissions on the subject of PSMs. Shaw, Franco and Westcombe (2006) and Franco, Shaw and Westcombe (2007) edited a two-part Special Issue for the *Journal of the Operational Research Society* consisting of 28 research articles in total, and Ackermann et al. (2014) compiled a Special Issue for the *EURO Journal on Decision Processes* consisting of 7 original research articles. The *Journal of the Operational Research Society* is curating an online collection of Soft OR articles (Kunc, 2021).

In any work it is important to point out its limitations. I made a decision to avoid producing a training manual for specific methodologies. Adequate pointers are provided throughout to the key textbooks, which have also been collected together into a single '*Selected Bookshelf*'. There are also excellent training courses available, such as from the UK OR Society. This has also meant scant treatment of one of the key skills of using PSMs – that of facilitation. My argument for a methodological approach grounded in pragmatism is that there is also no better way of learning than by doing. This is where facilitation skills can be honed, although some pointers to practical matters are given in Chapter 6. I have also avoided a general coverage of systems, with the exception of SSM. It is a vast area and the reader is much better served by an encyclopaedic treatment, such as by Jackson (2019a). The use of causal loop diagrams (CLDs) and qualitative system dynamics has also received less attention than perhaps it should. Like SSM they sit firmly in the systems world yet also provide another way into structuring problems, especially ones of dynamic complexity. They are linked into the PSM narrative as the use of CLDs has changed over time. Originating as a useful tool to communicate the complex behaviour-over-time dynamics of a System Dynamics simulation study, their use has evolved to be one of the means of *eliciting* system structure at the start of a study. Consequently, they have found increasing use as a modelling approach to support problem structuring. Developments in group model building (GMB) by the System Dynamics community has provided useful learning into the Soft OR/PSM field.

To achieve the purpose of a methodological focus on PSMs, I have chosen to highlight what I have referred to as the '*core*' methods, which has meant that methodologies such as value-focussed thinking (VFT) and the drivers, pressures, state, impact and response model of intervention (DPSIR) have only been briefly covered in Chapter 13. Other methods covered in the two editions of RAfaPW have been skipped entirely – hypergames, robustness analysis and drama theory. I am also conscious of the overlaps with the whole subject of *Group Decision and Negotiation* (GDN) with its associated journal and conference. Work from the GDN community that intersects directly with Soft OR/PSMs has been frequently cited. The subject of the digital transformation of PSMs through the use of GSS, firmly grounded in GDN, is covered in Chapter 14. Readers interested in finding out more about this field should consult the second edition of the *Handbook of Group Decision and Negotiation* edited by Kilgour and Eden (2020).

In sum, this book is likely to be of value to any student of problems who is looking for approaches to problem structuring that have a successful track record in application. As explained in Chapter 1, problem structuring is quite distinct from the general, common sense[5], notion of problem-solving by first taking a look at the co-dependency of the idea of problem formulation with the idea of categorisation of types of problems, which then shows why problem structuring is necessary. Regardless of whether your original discipline is in STEM, business and management, or in the social sciences, this book is likely to provide a useful resource for dealing with the sorts of problems you are facing in professional practice.

May 2024

Professor Mike Yearworth
Exeter, UK

---

5 Common sense is a spongy term to try and define or debate. Unfortunately, talking about complex problems seems to generate outbreaks of 'common sense-ism', which I try to dispel by focusing on the *processes* used to engage with complex problem contexts.

Acknowledgements

I start by thanking all my colleagues at Hewlett Packard's European Research Laboratories (HP Labs, Bristol) with a special mention to Martin Sadler, Ian Cole, Giuliano Di Vitantonio, Chris Whitney, Richard Taylor, Chris Tofts, Brian Monahan, David Pym, Chris Preist and Claudio Bartolini. Managing the Model Based Analysis team led me to join the MBA programme at the University of Bath where I thank my fellow students and superb faculty, with particular acknowledgement of the help received from Mike Lewis (Operations Management) and Neil Allen (MBA project supervision) in setting me on this path. Studying for my MBA convinced me that there was a need to develop a better understanding of the '*softer*' end of operational research (OR) and I am grateful for the opportunity to join the Faculty of Engineering at the University of Bristol to continue this work. In making the move I was fortunate to work with (the same) Chris Preist, Dan Schien and Paul Shabajee and make inroads into working on energy-related projects as exemplars of wicked problems.

I thank all the staff and students on the Engineering Doctorate in Systems programme at the University of Bristol that contributed to the development and review of modules in Research Methods, Advanced Research Methods, Civil Engineering Systems, Sustainable Systems, Systemic Problem Structuring and Research Methods and Advanced Systems. Many are co-authors and our work is extensively cited in this book. My specific thanks are due to Patrick Godfrey, John Davis, Colin Taylor, Oksana Kasyutich, Katharina Burger, Ges Rosenberg, David Lowe, Sarah Cornell, Neil Carhart, Gordon Edwards, Adrian Terry, Jacqui Wilmshurst, Rachel Freeman, Patrick Tully, Steven Barr, Dawn Gilbert, Tom Walworth, Alasdair MacDonald, Theo Tryfonas, Andy Hale and Bengt Cousins-Jenvey. I would also like to thank Sarah Tauwhare and Sophie Causon-Wood for their dedication in the day-to-day running of the Industrial Doctorate Centre (IDC) in Systems. I also appreciate the contribution of Alexander Coles, Oliver Jefferies, Luke Potter and Anna Tolmarsh for permission to use their model in Chapter 10.

Beyond the IDC in Systems, I thank Daniel Black for collaboration in the Healthy Resilient Cities project in Bristol, Dawei Han for the RESIST project, Brian Collins and many others in the International Centre for Infrastructure Futures project and Bronwen Lewis for the assembly of the 2012 Olympics data set and discussions on aporia. My fellow participants in the IFSR conversation in Linz and preparatory meeting in Cómpeta – Janet Willis Singer, Michael Singer, Jim Kijima, Duane Hybertson, Rick Adcock and Gerhard Chroust. I give my thanks to my many collaborators in the EU-funded STEEP and REPLICATE projects who contributed a great deal by way of feedback and support in the further development of Hierarchical Process Modelling as a Problem Structuring Method. I specifically thank Chiara Tavella for her process modelling work with the Comune di Firenze, the many participants of the workshops organised by Fomento San Sebastián and the participants of the workshop held in Nilüfer, Bursa. As External Examiner for the Master's degree programmes in the Systems Group at the University of Hull, I became better acquainted with the work of the Centre for Systems Studies through Gerald Midgley, Mike Jackson (whom I thank for the invitation to contribute the viewpoint piece that appears as Chapter 8), Amanda Gregory, Angela Espinosa and Giles Hindle.

At the University of Exeter Business School, I thank Richard Owen (now Bristol) and Alex Gerbasi, and colleagues in the Centre for Simulation Analysis and Modelling: Steve Disney, Nav Mustafee, Alison Harper and Andi Smart. I also thank all the staff and students on the MBA programme who contributed to the development and review of my core Operations Management course, especially Jackie Bagnall for her leadership and encouragement, and co-convenors and students on the Advanced Interdisciplinary Research Design module for the South West Doctoral Training Partnership, especially Maria Fannin at the University of Bristol and Graham Room at the University of Bath. I also acknowledge the grant of Emeritus status and access to the university's library resources.

In the Soft OR/PSM field, I acknowledge all the scholars and practitioners who have contributed an enormous amount towards my learning. Leroy White deserves a special mention here as my long-standing co-author and bridge between the engineering systems world, the problem structuring community in OR, and more recently to management scholarship generally. Conversations with Leroy inevitably lead to a flow of new ideas. Richard Ormerod has been a patient guide through his vast reading of social practice theory and the American pragmatists, and his long contribution to theorising OR practice. I thank Colin Eden, Fran Ackermann, Robert Roithmayr and Anita Reinbacher for providing access to an early version of Strategyfinder during the COVID pandemic and for ongoing discussions about Group Support Systems and problem structuring methodology. I also acknowledge the help I have received over the years from Alberto Franco, Gilberto Montibeller, Jonathan Rosenhead, David Lane, Isabella

Lami, Frances O'Brien and John Friend; Eleanor Reynolds for her meticulously detailed PhD work that cast further light on the generic constitutive definition of PSMs; Ine Steenmans for discussions about evaluation and policy relevance; Graham A. Forbes for his observations and answers to questions about process philosophy and PSMs; and Dave Friend for permission to use screenshots from the STRAD software. For those working in Soft OR/PSMs that I have not specifically mentioned, I hope that I have suitably acknowledged the weight of their contribution to the field through the citation of their work in these pages. I also acknowledge the contribution made through presentations and discussions at the UK OR Society PSM SIG meetings (thanks to Martin Parr and Christine Philips for organising), and the Soft OR and Problem Structuring Methods streams at the UK OR Society Annual Conference (thanks to Adrian Small) and the EURO Annual Conference (thanks to Chris Smith, Ashley Carreras and Irene Pluchinotta).

Finally, I recognise the work of Robert Dyson for his many years of service as Co-Editor-in-Chief of the *European Journal of Operational Research* and passing on the baton to me so that I can continue the work with my fellow Co-Editors-in-Chief Roman Słowiński, Emanuele Borgonovo, José (Zé) Fernando Oliveira, Ruud Teunter and Steffen Rebennack.

At Wiley I would like to thank Vishal Paduchuru and Brett Kurzman for their patience and support as this project developed.

Glossary

AHP	Analytic Hierarchy Process
ANT	Actor-Network Theory
AQA	Augmented Qualitative Analysis
CAQDAS	Computer-Aided Qualitative Data Analysis Software
CDT	Centre for Doctoral Training
CLD	Causal Loop Diagram
CM	Cognitive Mapping
COPE	COgnitive Policy Evaluation
CR/E/A	Constitutive Rules/Epistemology/Affordance of Modelling Approach
CSH	Critical System Heuristics
CSP	Critical Systems Practice
CST	Critical Systems Thinking
DPSIR	Drivers–Pressures–State–Impact–Response
EDEN	Evidential Discourse for Engineering
EJOR	European Journal of Operational Research
ÉLECTRE	ÉLimination Et Choix Traduisant la REalité
EPSRC	Engineering and Physical Science Research Council
ESRC	Economic and Social Research Council
EURO	Association of European Operational Research Societies
GCD	Generic Constitutive Definition
GDN	Group Decision and Negotiation
GMB	Group Model Building
GSS	Group Support System
HPM	Hierarchical Process Model/Modelling
IBIS	Issue-Based Information System
IDC	Industrial Doctorate Centre
IFORS	International Federation of Operational Research Societies
IFSR	International Federation for Systems Research

INCOSE	International Council on Systems Engineering
IOR	Institute for Operational Research
ISM	Interpretive Structural Modelling
ISSS	International Society for the Systems Sciences
JORS	Journal of the Operational Research Society
LLM	Large Language Model
MCDA	Multi-Criteria Decision Analysis
MCDM	Multi-Criteria Decision-Making
NERC	Natural Environment Research Council
PSM	Problem Structuring Method
QSD	Qualitative System Dynamics
RAfaPW	Rational Analysis for a Problematic World
SCA	Strategic Choice Approach
SCOOP	Systems Engineering in a Context of Systemic Cooperation
SODA	Strategic Options Development and Analysis
SOSM	System of Systems Methodologies
SSM	Soft Systems Methodology
STRAD	STRategic Adviser
TCP	Transformational Competence Perspective
UKRI	UK Research and Innovation
VFT	Value Focussed Thinking
VSM	Viable System Model

Section I

The Problem of Problem Formulation

The first section of this book delves into the problem of problem formulation. This conundrum sits at the heart of operational research (OR) – basically deciding what exactly is the problem that motivates the research. An orientation to problems is what makes OR distinctive as an applied science. The framing also positions OR as a technology, as a means of achieving practical ends and the reason why we should be interested in understanding how it is used in context.

Chapter 1 starts with a justification of why a methodological approach has been adopted. The differences between problem-solving and problem structuring are explored from a number of perspectives. The problem of problem formulation is then grounded in the type of problems that OR can address, the multiplicity of problematisations facing a researcher and the need for methods to structure complex problems. This chapter also surfaces the first critiques of mainstream OR and emergence of '*Soft OR*' that culminated in Rosenhead's grouping together of the new problem structuring methods (PSMs) as a means for keeping OR relevant to wider social issues, i.e. dealing with '*wicked*' problems rather than ones that have already been '*tamed*'.

Chapter 2 reviews the characteristics of wicked, messy and swampy problems. These definitions articulate the realm of Soft OR and PSM-based interventions. Some of the ontological frameworks used to categorise complexity and orient practice are introduced. This provides the basis for considering the means of avoiding category mistakes and paradigm incommensurabilities in practice.

Problem Structuring: Methodology in Practice, First Edition. Mike Yearworth.

Chapter 3 then directly confronts the resultant split that now exists between Soft and Hard OR arising from the dichotomous framing of both problems and methodological approaches to intervention. Pragmatic philosophy is discussed through notions of enquiring systems, phenomenal regularities and that action *is* knowledge as a potential means of framing the processes of OR, generally, that cut across this Hard/Soft split.

1 Origins

'Operational research is committed, not to the unravelling of knowledge about the universe, but to practical impact on the functioning of organized human activities. Evidently, then, it is best considered as a technology rather than as a science. So we should not be surprised that the particular circumstances of O.R.'s employment have shaped the forms of its tools, and in so doing have limited the scope of their use. Yet many O.R. people do resist this perception, asserting for our subject both timelessness and universality - claims which are dubious even for science, but preposterous for technology'. (Rosenhead, 1986, p. 338)

'What is the real problem? How do you know you are working on the right problem? How do you decide what to do about the problem?'
(Woolley & Pidd, 1981, p. 197)

1.1 Why Methodology?

I start by declaring my viewpoint as a methodologist and asking a performative question: What is it that operational researchers[1] are *doing* when they are dealing with a complex problem? A justification for a methodological approach is that it avoids trying to get inside the mind of a practitioner as the primary means of developing an understanding of Operational Research (OR) practice and instead focuses on the observable *actions* taken by practitioners when they engage in dealing with a client problem. The observational and methodological viewpoints are closely linked by way of process. The methodology can be observed in action as a process and the means of codifying and exploring OR practice as a methodology can also be taught.

1 Or engineers, or for that matter any '*expert*' practitioner.

Problem Structuring: Methodology in Practice, First Edition. Mike Yearworth.

A methodological focus also avoids widespread notions of common sense in relation to dealing with complex problems. Observing an expert '*doing common sense*' would be challenging, perhaps nonsensical. On the other hand, having observed an expert dealing with a complex problem, one might be tempted to ascribe the application of common sense. However, I am arguing throughout this book for a methodological interpretation of action, i.e. an expert engaged in a sequence of observable actions that can be documented, analysed, and refined, ready for deployment to deal with the next complex problem. Of course, there is no reason for common sense to not continue to be explored and debated *philosophically*.

Following '*Messing about in Problems*' (Eden, Jones, & Sims, 1983), this book could also be thought of as a similar exploration, looking at the similarities and distinctiveness of the different means for engaging with complex problems through a common framework, mindful of the interpretation of OR as technology and therefore with an interest in how the techniques of OR work and how they are used in practice.

By concentrating our attention on the actions taken by practitioners when dealing with complex problems and making a methodological interpretation we can also investigate how moral choices surfaced and enacted when making decisions. I take the view that practitioners have agency when dealing with complex problem contexts. They are not subservient to the implementation of rigid methods or algorithms.

1.2 Why Not Problem-solving?

In lists of key employability skills, problem-solving usually appears at, or near, the top, and that makes sense. In running a business or managing an organisation, constant attention is required to solve endless problems. Entire industries have grown up marketing the virtues of problem-solving skills, techniques, and toolsets, and the solutions based on their embodiment. So why read a book on problem structuring? What is the difference between problem-solving and problem-structuring? Why do we not see problem-structuring in the lists? The key to answering these questions is not in a protracted wrangle about the differences between solving and structuring but in examining the nature of problems.

Since we all recognise problems when we encounter them, there seems to be little point in understanding their nature. There is a common-sense notion of what problems are that launches us into action to solve them without much thought. We have a felt sense of problems, we experience them, and this triggers action. However, to assume that we have a clear understanding of the nature of a problem before '*diving in*' is an error with significant consequences. Failure to understand what we are dealing with leads to what are known as category mistakes. A category mistake arises when we construct and then attempt to act on infelicitous propositions, i.e. that are not semantically and pragmatically coherent. They arise in the context of discussing problems when the ontological nature, the category, of the problem itself is misunderstood. If our skills, techniques, and toolsets are oriented towards problems that can solved and we then encounter a problem that has no solution, then it would be a category mistake to try and use them. The argument in this book is that such problems do exist and that the best we can do as practitioners is to structure them for intervening. What follows from this is largely an account of the methodological approaches the OR profession has developed to avoid making category mistakes in relation to the nature of problems.

I am not the first to try to disentangle the difference between problem solving and problem structuring in OR. For example, Smith (1988) developed a heuristic theory for problem structuring in management science. However, it is grounded in the notion that problems are solvable and that '*ill-structuredness*' rests with the individual's ability, or not, to find a way to a solution. Smith uses the well-known Tower of Hanoi *puzzle* as an example. The reference to solving puzzles is useful here; some people have the know-how (innate or learned) to get to a solution while others do not know how to start or get lost on the way. The fact that there may be a class of problems that no individual can solve is only considered from the perspective of goal indefiniteness leading to a lack of a stopping rule, but this is to highlight only one aspect of a '*difficult*' problem. However, prior to this attempt at a comprehensive theory, Churchman (1967) had already drawn attention to so-called wicked problems in an Editorial in *Management Science.* In a similar vein, Ackoff introduced the notion of a mess (Ackoff, 1974b, pp. 20–21), and Pidd (2009, pp. 43–46) built on Ackoff's definitions and defined the notion of a puzzle in an OR context to produce a simple taxonomy, which I will refer to throughout the book and is summarised in Table 1.1.

Pidd's taxonomy is a useful starting point for defining categories of problems for the purpose of avoiding category mistakes. '*Puzzles*' have a correct answer. There is a formulation of the problem that means a single solution can be found without

Table 1.1 Formulation and solution to puzzles, problems, and messes.

Category	Formulation	Solution
Puzzles	Agreed	Agreed
Problems	Agreed	Arguable
Messes	Arguable	Arguable

Source: Adapted from Pidd (2009, p. 44).

disagreement. Any ill-structuredness rests with an individual's approach (the skills, techniques and toolsets used) in reaching a solution, either by producing the wrong answer or failing in the process to arrive at any answer. On the other hand, '*Problems*' are a category that may be formulated such that they have a range of plausible solutions and means of reaching them, but none can be considered the single correct answer. The solution chosen to address the problem can only be resolved through a debate. Finally, as defined by Ackoff (1974b, pp. 20–21), '*Messes*' are a system of problems. Such systems of problems are also associated with the notion of wicked problems as defined by Rittel and Webber (1973) or a swamp (Schön, 1987, p. 3). Whilst the definitions of wicked/messy/swampy problems are discussed further in Chapter 2, for now, they stand-in for a category of problems where the process of formulation is debatable, and the nature of (any) solutions is contested.

The category mistake of crucial importance to the argument of this book is treating a problematic situation as a '*Puzzle*' or a '*Problem*', according to the definitions above, when it is in fact wicked/messy/swampy. However, a paradox lurks at the heart of this categorisation. Whilst it may seem to provide a useful orientation device, it presumes that there exists some means of knowing whether we are dealing with a puzzle, problem or mess. Unless the context provides a clear indication, as is usually the case with puzzles or what we might expect in professional practice working within a particular domain of known problems and solutions, when dealing with an unknown problematic situation we cannot assume anything. If our skills, techniques and toolsets as OR practitioners are not predicated on this uncertainty then we run the risk of endlessly making category mistakes.

Since wicked/messy/swampy problems defy solving, and if we do not know, a priori, the nature of the problem we are confronting, the process of deciding what those problems are and how they might be *addressed* is bound together. I use the labels '*addressed*' advisedly; whether we use addressing, intervening, dealing with, etc. all are reasonable labels for the process, and I have used all of them interchangeably in this book when making the point that these processes are not about solving problems. The important thing is to avoid the label *solving* since this propagates an unhelpful mindset. Smith (1988), in reference to Rittel and Webber (1973), refers rather disparagingly to this sort of insistence as a '*desiderata*'. However, I consider it essential to avoid making category mistakes. I develop this point in Section 3.2 to define a number of pathologies in OR practice arising from this failure.

The circularity of the notion that deciding what a problem is, and the process of dealing with it are bound together presents a challenge to a scientifically minded discipline such as OR. Whilst we may be able to codify and examine what we are doing when addressing such problems, without a specific solution how do we know whether we have succeeded? Again, the notion of '*success*' implies a solution, therefore in the absence of a stopping rule (a solution) the process of addressing must also generate within it the definition(s) of improvement and also a means of judging progress towards improvement, i.e. a measurement scheme as well. This process of problem *structuring* will, if carried out rigorously, provide the participants in the process with a way forward to *improve* the problem situation. Unfortunately, assessing the final value of the intervention is not possible until after it has been carried out, largely because the wickedness/messiness/swampiness of a problem means that value is itself a *co-produced* appreciation of what '*improve*' means, arising from the process of the intervention itself. Deciding what improvement means and how it is evaluated is part of the process of problem structuring; it is inherently open-ended, i.e. without precise stopping rules when started.

Given the open-ended nature of problem structuring without precise stopping rules, why would a customer pay for problem structuring? For that matter, why would a practitioner want to get involved in an open-ended process? It is arguable that many professionals are quite content with avoiding messes. Focusing on problems with definite solutions is, after all, how most experts get paid; *solutionism* pays the bills. Who wants or needs problem structuring if this is going to make the life of the expert harder? For me, the answer is reasonably straightforward, at least in principle. I believe we have a moral responsibility as practitioners in OR, which means we cannot ignore messes.[2]

1.3 From Puzzles, Problems, to Messes

Unlike dealing with messes, the outcomes from puzzles and problem-solving leading to definite solutions are usually easy to quantify, especially financially. For example, innovation-as-a-service platforms offer precise financial rewards for solving very specific problems. If we see the same problem again, then we can solve it in the same way. We do not need to work out the method from scratch. There is regularity in the world that we can rely on. Thus, we can build a repertoire of solutions to known problems, improving on them as new insights are gained. We can develop new ways of generating solutions that are '*better*' by some objective measure of performance. Best practices for professionals can be established, and the knowledge can be codified and reproduced in textbooks and other bodies of knowledge. The conundrum for OR is that we have become extremely successful at this process. We have a huge repertoire of solutions to known problems, and our process of problem formulation has largely become one of spotting the problem *type*.

2 I come back to the question of moral choices in Chapter 11, but for now I suggest the best reflective piece on the professional practice of OR, despite its age, can be found in Churchman (1970).

However, if we go back to early descriptions of the process of problem formulation, it was not always the case.

> *'Both the consumer's and the researchers' problem must be formulated. The research consumer is the person (or group) who controls the operations under study. (He is also referred to as the decision-maker.) In formulating the consumer's problem an analysis must be made of the system under his control, his objectives, and alternative courses of action. Others affected by the decisions under study must be identified and their pertinent objectives and courses of action must also be uncovered. What we have called the over-all viewpoint is closely connected with the attempt to define objectives. O.R. tries to take into account as broad a scope of objectives as possible. In most general terms, the research problem is to determine which alternative course of action is most effective relative to the set of pertinent objectives. Consequently, in formulating the research problem a measure of effectiveness must be specified and its suitability must be established'.*
>
> (Churchman, Ackoff, & Arnoff, 1957, p. 13)

As an observation about what is required from the process of problem formulation, it seems to be just as relevant to the practice of OR today but still does not seem to be normalised. Any serious attempt to engage with this process would, if the problem was sufficiently complex, have run into the reality of competing objectives or differing viewpoints on the problem. That is, almost any problem addressed in this way would reveal some element of messiness. Of course, this approach to problem formulation clearly spells out that there will be alternative courses of action and that the effectiveness of the action taken will be measured relative to a set of pertinent objectives, not a single one. It assumes that different courses of action *can* be identified and the '*best*' chosen. However, resolving the question of '*best for whom*?' i.e. deciding which viewpoint to privilege is a matter left to the owners of the research.

If we fast-forward to 1981, we get to Checkland's problem formulation for soft systems.

> *'The thinking embodied in "systems engineering" and "systems analysis" is essentially the same. Analysis of many different accounts of these activities shows that they all assume that problems can be formulated as the making of a choice between alternative means of achieving a known end. The belief that real-world problems can be formulated in this way is the distinguishing characteristic of all "hard" systems thinking'.*
>
> (Checkland, 1981, p. 15)

He was drawing attention to the deficiencies of systems engineering and systems analysis, as they were conceived in the 1960s and 1970s, when applied to the complex problems of business and management in the consulting engagements undertaken by the team at the University of Lancaster. However, his critique of systems engineering and systems analysis is relevant precisely because it echoes the problem formulation of Churchman, Ackoff and Arnoff, i.e. making choices between alternative means of achieving known ends. Whilst his argument was not directed at OR, it was nonetheless relevant in that systems engineering and systems analysis shared a similar objectivity from the viewpoint of OR. Checkland went on to say that

> *'[t]he success of systems engineering and systems analysis led to many attempts to use the concepts in problems of social systems, including those of formulating public policy. It was not always noticed that these problems could not necessarily be formulated as hard problems .... The results were on the whole disappointing, and this in turn led to polemic criticism of the whole idea of making the transfer'.*
>
> (p. 15)

Checkland's argument for the need for a soft systems problem formulation was a direct result of the failure of systems engineering and systems analysis to intervene in complex problem contexts encompassing social systems and dealing with public policy. The polemic he refers to here is one of criticism of these '*hard*' methods and their failure to work or, as he put it, more-kindly, their '*disappointing outcomes*' when used in such complex problem contexts. In effect, he is flagging here the effect of making a category mistake.

> *'The research described ... assumed that the concept of a human activity system would be relevant to tackling the "soft" ill-structured problems of the real world, those before which the methodology of natural science is impotent. The idea was to apply one of the versions of hard systems thinking to real-world situations in which the actors perceived they had problems, in order to find out whether, why, and how the hard methodology was inadequate. The intention was to find a systems methodology for tackling problems which defy formulation in the hard sense, and also to enrich the concept human activity system in order to understand better the "social systems" of the real world'.*
>
> (p. 15)

The remainder of Checkland's argument was to set out the development of a systems methodology for tackling problems that defy formulation in the hard sense – the means to tackle soft, '*ill-structured*' problems. The ontological framing of problems is very clear here; *the difficulty exists in structuring the problem.*

1.4 The Apparent Success of Operational Research

Coming back to the enormous success that OR has achieved through its accumulation of standard solutions to problems, here is Checkland commenting on a standard OR textbook of the time.[3]

> *'The typical textbook of O.R. emerged as a brief account of the war-time emergence of O.R., a bald statement of its methodology, and a series of chapters considering in turn the logical structure of the problem situations which recur: the investment portfolio problem; the queueing problem; the depot location problem etc. (A major university text in the late 1970s takes but 22 of its 1000 pages to discuss the origins, overall methodology and implementation of O.R.; the rest discuss the algorithms.)'*
>
> (Checkland, 1983, p. 665)

OR had become successful precisely because of its discovery of the '*logic of situations*' (p. 665) i.e. the identification of a recognisable set of ready structured problems[4] such as optimising, controlling, queueing, routing, scheduling, locating, bidding, packing, cutting, timetabling, gaming, forecasting, '*frontiering*'[5] and so on, which occur time and again in industrial and commercial operations and are eminently teachable and subject to continual research and development. That is, finding better algorithms for solving such problems, where better can be understood in the objective sense of being faster to execute on a computer or achieving numerical improvements in the outcomes, such as more efficient utilisation of resources. The crisis facing the OR profession, in Checkland's words, was that it was in danger of becoming '*practitioner-free*' (p. 667). There were two facets to this crisis. On the one hand, the necessity for the practitioner to formulate or structure the problem was disappearing because the solutions to ready structured problems were now well known and easy to apply. On the other, there was the recognition that such well-structured problems were rarely found in dealing with real-world

3 In this case Wagner (1975).

4 The problem of situational logics is not unique to Operational Research. Rowlinson (2023) has commented on the practice in management scholarship. For example, logics like '*if a motivation problem, then Maslow's hierarchy of needs (HON)*' or '*if a change management problem, then Lewin's change-as-three-steps (CATS)*' without a full understanding of the original theory. Cummings et al. (2017) set out an agenda for re-problematising management scholarship in relationship to how the original theories have been historically misrepresented over time. In a similar vein, Zyphur and Pierides (2020) argue that the logic-then-ethics approach of traditional statistics is only really useful for simple/complicated problems – i.e. the use and/or creation of scientific knowledge but not intervening in complex problem contexts.

5 The non-parametric analysis of production frontiers across decision-making units (see Charnes, Cooper and Rhodes (1978)).

messes (p. 667). The ideals and ambitions for the wide application of OR in its early days were being lost and it was Ackoff (1979a, b) and Churchman (1970) that were signalling their concerns. Ackoff (1977) commented on the propensity for OR practitioners to '*opt out*' of addressing the messy problem at hand and fall back on solving the structured part of the problem.[6]

Looking at this from the perspective of technical rationality, Schön observed that

> '*...professional practice is a process of problem solving. Problems of choice or decision are solved through the selection, from available means, of the one best suited to established ends. But with this emphasis on problem solving, we ignore problem setting, the process by which we define the decision to be made, the ends to be achieved, the means which may be chosen. In real-world practice, problems do not present themselves to the practitioner as givens. They must be constructed from the materials of problematic situations which are puzzling, troubling, and uncertain*'.
>
> (Schön, 1983)

The key question emerging from these observations and vesting them as a '*crisis*' facing OR was the means for OR practitioners to recognise the existence of a messy problem and find out something about it – the idea of *problematisation* as a process.

1.5 The Process of Problematisation

The pre-eminence of situational logic as the driver of problem-*solving* activities in OR is strongly coupled with the idea of professional best practice. However, the existence of wicked/messy/swampy problems requires a rigorous problematisation as the initial activity when conducting an investigation into a complex problem context. The idea of problematisation in OR is thus a development of the original idea of problem formulation, but with an awareness of the inherent circularity involved when the nature of a problematic situation is constructed from the process of dealing with it.

There are various places to look for suitable theories of problematisation. Schmidt (2011) provides a very general philosophical overview of the question '*what is a problem*?' Bird (1998, pp. 215–235) looks at the question from the perspective of a reliabilist epistemology, and Schutz (1967, pp. 86–96) presents a phenomenological perspective. Landry (1995) delves into an organisational viewpoint and provides a useful taxonomy that identifies objectivist, subjectivist and

6 Prominently, as the first article in the very first issue of the new European Journal of Operational Research.

constructivist perspectives. However, what seems to be missing in these accounts of the processes of problematisation is any reference to the quality of experience of the properties of a situation that makes it problematic and specifically problematic for one person and not another.[7] Looking at the issue from a different perspective, that of management research, Alvesson and Sandberg (2011) argue for the use of problematisation to generate research questions in an effort to challenge the hegemony of gap-spotting, which can be likened to the situational logics of OR.[8] They recognise the origins of problematisation in, amongst other theoretical strands, Pragmatism – through the work of Dewey,[9] and Actor-Network Theory (ANT) – through the work of Callon.[10]

Callon (1980) proposes a general structure of problematisations as a frontier that emerges between what is *chosen* to be analysed and what is suppressed from the analysis, forming a closed domain with its own coherence and logic, a division between the property of the scientist (analyst) and what is left for outsiders. A second frontier then emerges between what is intangible, taken for granted, and what is problematised or unknown.

> *'Problematisation culminates in configurations characterised by their relative singularity. There is not one single way of defining problems, identifying and organising what is certain, repressing what cannot be analysed'.*
>
> (Callon, 1980)

Each problematisation leads to a problematic situation with specific demarcations (i) the un-analysed, (ii) the network of certainties, and (iii) the area of suspicion. What Callon is articulating here is a '*[d]escription of mechanisms through which reality is problematised*' which he then goes on to describe as the '*[f]orces of problematisation*' and the '*[a]nalysis of the relationships various forces of problematisation*'. This switches the focus from problem categorisation to observation and analysis of the forces of problematisation, i.e. the '*problematic situation is a dual process of construction and de-construction*'. This is a crucial point, Callon presents a theoretical description of the process of problematisation that goes beyond problem categorisation, i.e. practising according to situational logics.

---

7 Such that when exposed to the same problem context one person feels sufficiently motivated by their problematising to take action, and another person to not act at all.

8 In management research this is the filling-in of knowledge gaps as opposed to generating new management theories. Of course, management research and operational research have different aims. The former is looking for generally applicable theories, whereas operational research is (or should be) addressing the specific problem context that motivated the research. Nonetheless, gap-spotting and the following of situational logics are similar.

9 As discussed in Chapter 3.

10 The contribution that Actor-Network Theory (ANT) could make to the analysis of the practice of OR was brought to the attention of the OR community by Keys (2006) and then by White (2009).

The usefulness of Callon's work on problematisation is that it draws attention to the choices faced by an operational researcher when investigating a problem, and that there is not a single right answer to what is problematised. Whilst this supports the claims of Churchman, Ackoff and Checkland that these choices exist and that operational researchers are making conscious decisions about what to work on (and therefore what to ignore), it does not provide an answer to the problem of OR becoming practitioner-free, i.e. the problem of dealing with situational logics. The answer is found in the emergence of problem structuring methods (PSMs).

1.6 Problem Structuring Methods

The fundamental assertion made in the articulation of a need for problem *structuring* is that there exists a category of problems that are *not* solvable by general problem-solving skills, techniques, and toolsets – i.e. they have no solutions and that a different approach is required. However, rather than starting with setting out another situational logic, that of spotting wicked/messy problems, Rosenhead (1989b, p. 12) set out to define an alternative *paradigm* to the dominant or mainstream approach of OR at the time. This was constructed empirically by defining a set of characteristics of practice that were literally the opposite to '*normal*' OR practice, i.e.

1. not based on optimisation and acceptable without trade-offs,
2. not totally based on analysing data but also included social judgements,
3. simple and transparent in use,
4. conceptualised people as active,
5. enabling of bottom-up planning, and
6. accepting uncertainty and delayed resolution of options.

Over time, these characteristics of a methodological approach for a new '*Soft*' OR were refined to provide a canonical definition of PSMs. For example, in the following

> *'The methods (or methodologies, see below) are not mathematical, but they are nevertheless structured and rigorous. They are based on qualitative and often diagrammatic modeling procedures.*[11] *Obviously numerical information may be included, but not complex equations.*

---

11 The fundamental analytical object in operational research is the model. Models are abstractions, abstract representations of a problem, see Pidd (2009, pp. 3–20).

They allow a range of distinctive views to be expressed and explored, and embrace multiple and conflicting objectives without collapsing them into a single, often financial, measure.
They encourage the active participation of stakeholders in the modeling process often through facilitated workshops of those affected by the problem. In order to encourage participation, models should be transparent to participants. This is aided by the first point that they are generally non-mathematical.
Significant uncertainty is expected and tolerated as is a lack of reliable quantitative data. They aim for exploration, learning and commitment rather than optimization'.

(Mingers, 2011, p. 731)

In setting out the rationale for PSMs as a methodological approach to intervention in complex problem contexts, Rosenhead (1989b, p. 10) focused on the *dichotomous* framing that was inherent in the original definitions of wicked/messy/swamp problems, i.e. Rittel and Webber's wicked versus tame problems, Ackoff's articulation of messes versus problems, Schön's swamp versus the high ground, and Checkland's problem formulation that sets out the difference between hard systems thinking and soft systems thinking. The implication from this positioning in RAfaPW was that wicked, messy, swampy, soft… were all of a similar character, all related, and that if the hard analytical methods of OR, and by association systems engineering and systems analysis, were used in these contexts it would lead to failure. Unfortunately, this assertion and the dichotomous framing of problems has become entrenched in the field of Soft OR/PSMs to the detriment of the original ideals associated with problem structuring. In effect a new situational logic has emerged '*If a wicked/messy problem then use a problem structuring method*'. Whilst this is not incorrect, and mostly a perfectly reasonable decision, it has had the unintended consequence of driving a wedge between what are now two distinct branches of OR. Hard OR, characterised by the original situational logic diagnosis set out by Checkland and others,[12] and Soft OR, following the initial prescriptions of Rosenhead.[13]

---

12 For example, Ackoff (1987) characterised the OR profession as having '*left him*' because it seemed to abandon a *market* orientation – i.e. focussed on dealing with actual customer problems to one that was *output* oriented and just dealing with recurring types of problem where '*human choice and purposeful behavior had virtually no role*'.
13 Of course, this is not a completely clear-cut distinction. However, the normal social processes of academic publishing and organisation of conferences lends support to non-overlapping streams, non-intersecting interests, and in some circumstances questions of legitimacy and scope. The reasons are discussed in more depth in Chapter 16, drawing on Kuhn (1962) and Abbott (2001).

1.7 Summary

The origins of problem structuring in OR have been traced through their emergence in criticisms of practice in the 1980s and is mapped in further detail in Chapter 4 as an historical account. At the same time, Callon's study of the work of scientists and his articulation of the mechanism of problematisation provides an account of a process by which decisions are made. Of crucial importance is the idea of a dual process of construction and deconstruction in problematisation, which was not sufficiently emphasised in the original articulation of PSMs in RAfaPW. Whilst the resolution of this is tackled in Chapter 3, which sets out an approach based on the philosophy of pragmatism, Chapter 2 reviews some of the characteristics of wicked/messy problems that were first introduced in the rationale and motivation for the emergence of PSMs.

2

Wicked and Messy Problems

'The picture which has emerged, I believe, is of a dichotomy of problem situations, for which operational research has deployed a unitary paradigm appropriate to only one of the two classes. This dichotomy has been presented here as that between tactical and strategic problems. Others have coined more vivid terminology to dramatize the difference. Ackoff distinguishes "problems" from "messes", and Rittel and Webber characterize "tame" versus "wicked" problems.'

(Rosenhead, 1989b, p. 10)

2.1 On Dichotomies

Abbott (2001, p. 10) observes in the *Chaos of Disciplines* that dichotomous framings occur throughout social science as an emergent expression of the organisation of fields of study and that these dichotomies can concern either (i) the object of study, (ii) the aspect of the phenomenon taken as problematic or (iii) differentiation in methodological styles. This categorisation helps make sense of Rosenhead's original motivation for flagging a particular set of dichotomies as the domain of relevance for problem structuring methods (PSMs), i.e. wicked/messy/strategic problems, and the claim that existing Operational Research (OR) practices were not appropriate for addressing them. Soft OR has been bound to this dichotomous framing of problem type ever since. However, while dichotomies such as pure and applied research, and research methodologies broadly aligned with quantitative and qualitative methods and philosophies are fairly intuitive, the Hard/Soft dichotomy in OR is a more complicated construct as it embodies all three aspects of Abbott's taxonomy. Interpreting these three aspects as aligned roughly with ontological, epistemological, and methodological viewpoints provides a convenient way of looking at how the practice of OR

Problem Structuring: Methodology in Practice, First Edition. Mike Yearworth.

'*dichotomises*' phenomena, but there is also the problem of circularity. Do we take the existence of problems as tame/wicked, problems/messes, and tactical/strategic as existing somehow independently of our investigations, or are they a product arising from them? Clearly, in the accounts of the researchers that have coined these labels, the phenomena are a *co-production* of their investigations, but for the purposes of presenting a linear account, it is easier to start with the definitions '*fully formed*', so to speak, and then deal with the circularities. It is also worthwhile to briefly cover a more ontological framing of complex problems based on a characterisation of the systems in the world. Some of these frameworks provide useful orientation devices.

2.2 Wicked[1] and Tame

The work of Rittel and Webber (1973) was first brought to the attention of the OR/MS community by Churchman (1967). The quote from Rosenhead highlights its centrality in defining the scope of Soft OR – PSMs are designed as a practical means for addressing the challenges of wicked problems.

Rittel and Webber (1973, pp. 161–167) characterise wicked problems as follows[2]:

1. There is no definitive formulation – formulating a wicked problem *is* the problem. It is not possible to approach a wicked problem with preconceived notions of how it might be addressed, the only way forward is to start a process of enquiry and develop ways forward from there.
2. There are no stopping rules. The process of intervening is also the same as understanding the nature of the problem – the intervention is '*good enough*' or the best that can be achieved within other limitations external to the process (e.g. time, budget, patience, etc.)
3. Interventions are not right or wrong, there are no formal decision rules for defining correctness, they can only be viewed as making things better or worse for certain interests. Judgments will depend on personal interests and values.

1 The label '*wicked*' seems anachronistic. However, since its use is pervasive in the OR literature that refers to this original definition, I have stuck with it. It is worth repeating the original intention behind the label '*As you will see, we are calling them "wicked" not because these properties are themselves ethically deplorable. We use the term "wicked" in a meaning akin to that of "malignant" (in contrast to "benign") or "vicious" (like a circle) or "tricky" (like a leprechaun) or "aggressive" (like a lion, in contrast to the docility of a lamb). We do not mean to personify these properties of social systems by implying malicious intent. But then, you may agree that it becomes morally objectionable for the planner to treat a wicked problem as though it were a tame one, or to tame a wicked problem prematurely, or to refuse to recognize the inherent wickedness of social problems.*' (Rittel & Webber, 1973, pp. 160–161).

2 I have modified Rittel and Webber's definition slightly to remove references to '*solution*' in preference for '*intervention*'.

4. There is no immediate or ultimate test for an intervention. Interventions will generate '*waves of consequences over an extended – virtually an unbounded – period of time*'. The consequences of an intervention are thus difficult to evaluate because the consequences will be continually changing.
5. Interventions are '*one-shot operations*', and experiments are difficult to conduct. Every intervention is consequential and effectively irreversible. Interventions are essentially unique in nature – we cannot intervene in the same problem context twice as our interventions change the problem context.
6. There is no enumerable, exhaustively describable set of possible interventions. There are no criteria that enable us to judge whether we have found all of the interventions that are possible in a given problem context. It is a matter of realistic judgment about how expansive the process of enquiry should be.
7. Every wicked problem is essentially unique. '*Essentially*' implies that aspects may be common, but to think in terms of categories or classes of wicked problems with common '*solutions*' is misleading.
8. Wicked problems can be considered symptoms of other problems, i.e. there is inherent systemicity in the world. '*The level at which a problem is settled depends upon the self-confidence of the analyst and cannot be decided on logical grounds. There is nothing like a natural level of a wicked problem. Of course, the higher the level of a problem's formulation, the broader and more general it becomes: and the more difficult it becomes to do something about it.*'
9. Can be contested at the level of explanation; there is likely to be conflicting evidence or data. It is not possible to rigorously test hypotheses about interventions due to their unique circumstances.
10. Whereas scientific progress arises as a consequence of refuted hypotheses (in a sense, being wrong is good), in the area of policy and planning, decisions that have negative consequences are not tolerated (being wrong is not good).

Similar to Ackoff's motivation in arguing for the social responsibility of OR (Ackoff, 1974b), Rittel and Webber's work is grounded in a justifiable response to the challenge of addressing prevailing social issues in the USA in the 1970s. Their objective was to repudiate a '*scientific*', and hence objective and definitively positivist, approach to social policy. They critiqued the prevailing '*RAND-style*' of systems analysis as

> '*… the professional style of the systems analysts, who were commonly seen as forebearers of the universal problem-solvers. With arrogant confidence, the early systems analysts pronounced themselves ready to take on anyone's perceived problem, diagnostically to discover its hidden character, and then, having exposed its true nature, skillfully to excise its root causes. Two decades of experience have worn the self-assurances thin. These analysts are coming to*

> *realize how valid their model really is, for they themselves have been caught by the very same diagnostic difficulties that troubled their clients.'*
>
> (Rittel & Webber, 1973, p. 159)

This is in much the same vein as Checkland's observations[3] quoted in Chapter 1. The practice of OR does not their escape their criticism either

> *'The methods of Operations Research play a prominent role in the systems-approach of the first generation; they become operational, however, only after the most important decisions have already been made, i.e. after the problem has already been tamed.'*
>
> (ibid, p. 162)

The association that Churchman and then Rosenhead made with wicked problems is that OR *should* be policy relevant (Churchman's moral imperative) and that Soft OR/PSMs provide OR practitioners with the *means* to be policy relevant (Rosenhead's instrumental imperative). Circularity is again evident from the definition, in that the process of formulating a wicked problem *is* the problem, and therefore a problematising and structuring approach is required for *any* uncertain problem context.

In articulating the characteristics of wicked problems as the defining issue of planning and their relevance to the issues of deciding social policy, Rittel and Webber also provide observations about the implications for *any* process of enquiry that will lead to interventions. It is possible to see how, in Chapter 4, the newly emerging PSMs are constructed methodologically to address such problems. However, there is no lack of work (outside of Soft OR/PSMs) that builds on this definitional framework and sets out similarly practical approaches to processes of enquiry. For example, Alford and Head (2017) construct a new typology from the original Rittel and Webber definitions as a means of constructing different types of collaborative approach. The notion of wicked and tame problems has made a broad impact across a range of disciplines in addition to planning. For example, Head (2022) presents a recent and comprehensive work on the policy implications of wicked problems, and Lazarus (2009) introduces a legal perspective on the notion of '*super wicked*' problems in the context of climate change. Their work has also provided inspiration for sociological perspectives on future studies (Tutton, 2017).

To convey something of that breadth, the distribution of 9835 citations of their 1973 paper by subject area is shown in Figure 2.1 and a keyword co-occurrence analysis of the 2000 highest cited of these articles is shown in Figure 2.2. The relevance of this broader work here lies in posing the question, '*if (some of) the*

3 Checkland (1981) constructs his '*soft*' systems method of enquiry empirically through an analysis of the failure of hard systems approaches to deal with complex problems of business and management. His work surfaces a similar critique of '*RAND-style*' systems analysis as Rittel and Webber.

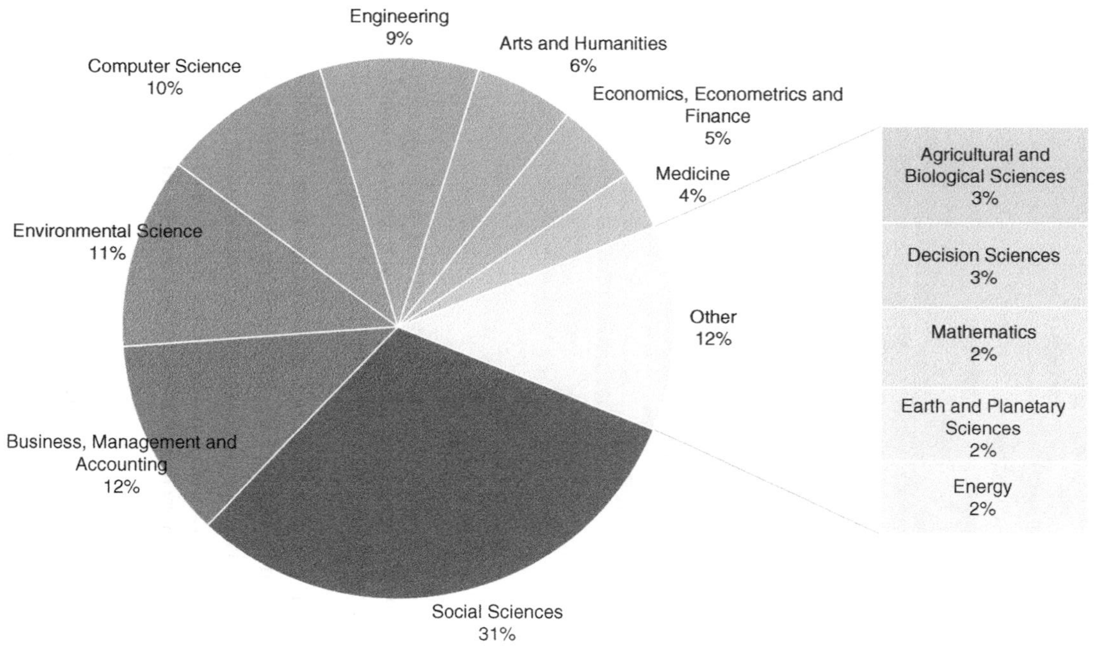

Figure 2.1 Distribution of citations of Rittel and Webber (1973) across subject areas. Data from Scopus.

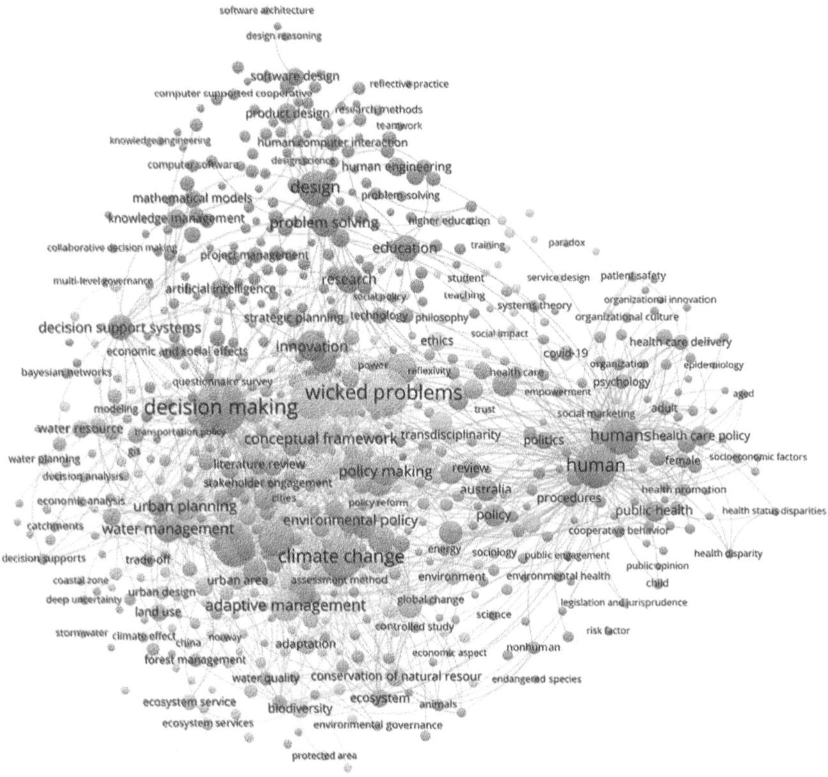

Figure 2.2 Keyword co-occurrence analysis of the 2000 highest cited articles that cite (Rittel & Webber, 1973). An interactive online version of this figure is available at https://problemstructuring.com/Book/Chapter2/Figure-2.2.html.

OR profession is interested in the notion of wicked problems, who else is?' I return to this question again in Section 13.7, when looking at the overlaps between PSM applications and practices from other traditions, and in Section 16.1 when looking at problem structuring practice, generally, as an applied social science and potentially where new disciplinary linkages need to be made by the Soft OR community.

2.3 Swamps and High Ground

In his inaugural lecture at the London School of Economics (LSE), Rosenhead (1992) again draws attention to the relevance of the OR profession to social issues and matters of policy.

> *'Many, perhaps most, of its practitioners have seen operational research as apolitical in nature, or even positively technocratic - with a mission*

to substitute technical expertise for "irrational" political decision making processes. Equally, and at least partially in consequence, operational research has made little contribution to those processes; its visibility to those active in public-policy making has been low. Its tools and techniques, methodology and approach have been applied elsewhere.'

(Rosenhead, 1992, p. 293)

In providing further historical detail concerning the emergence of OR as a discipline, Rosenhead reveals insight into attitudes towards OR by policymakers and planners resulting from the failures of technocratic policy approaches. In effect, these failures had led to the dismissal of the relevance of OR by association. However, finding ready acceptance of the mainstream techniques of OR in industry, focus in the discipline switched accordingly. In drawing attention back to matters of social policy, Rosenhead made use of Schön's dichotomous framing of the swamp and high ground. In *Educating the Reflective Practitioner*, Schön states that

'[i]n the varied topography of professional practice, there is a high, hard ground overlooking a swamp. On the high ground, manageable problems lend themselves to solutions through the use of research-based theory and technique. In the swampy lowland, messy, confusing problems defy technical solution. The irony of this situation is that the problems of the high ground tend to be relatively unimportant to individuals or society at large, however great their technical interest may be, while in the swamp lie the problems of greatest human concern. The practitioner must choose. Shall he remain on the high ground where he can solve relatively unimportant problems according prevailing standards of rigor, or shall he descend to the swamp of important problems and non rigorous enquiry?

This dilemma has two sources: first, the prevailing idea of rigorous professional knowledge, based on technical rationality, and second, awareness of indeterminate, swampy zones of practice that lie beyond its canons...Depending on our disciplinary backgrounds, organizational roles, past histories, interests, and political/economic perspectives we frame problematic situations in different ways.'

(Schön, 1987, p. 3)[4]

4 This is a slight variation to the quote in Schön (1983, p. 42), where '*non rigorous enquiry*' was originally stated as '*forsake technical rigour*'. Either way, the seeds are sown for the association of rigour with the formal mathematical and engineering approaches of technical experts, and lack of rigour with addressing swampy problems. In fact, Schön uses OR as an example of a rigorous profession because it uses formal (mathematical) modelling, but then notes its failure to '*yield effective results in the more complex, less clearly defined problems of business management, housing policy, or criminal justice*'. (Schön, 1983, p. 44). The whole of Chapter 5 is devoted to an argument, supported by the definition of the CR/E/A framework for specific instances of an enquiry, that PSMs are in fact a *rigorous* means of addressing swampy (wicked/messy) problems.

In using this definition of swampy problems, Rosenhead makes the case for reclaiming rationality in decision making from a '*narrow, calculative and predictive interpretation as the elimination of the need for judgement*' and positioned PSMs as the means for intervention that instead '*do not have the aim of finding optimal solutions to well-formulated problems. Their function, rather, is of helping a group with disparate perspectives to agree nevertheless on a common problem focus and a shared commitment to action*'. While PSMs offer analytic help in the swamp of important problems by being transparent and participative, the abandonment of the high ground comes at a price.

> *'This price is an abandonment of much of the mathematical formalism (continuous variables, algorithms, probabilities, trend extrapolation) which has, until recently, been the principal stock-in-trade of operational research. It is a bargain.'*
>
> (Rosenhead, 1992, p. 302)

The bargain, as such, is bluntly stated in his lecture, and no doubt to achieve a dramatic conclusion. However, as argued later, this only reinforces ideas of dichotomies and splits in OR. There is still the need for mathematical formalism; however, it is not the first thing to look for when in a swamp. Again, testing the ground – is it swamp or high ground – requires some way of finding out.

2.4 Problems and Messes

Ackoff meets the challenge of circularity head-on in his articulation of messes as a system of problems. In *Redesigning the Future*, he states

> *'Management involves decision making and decision making involves problem solving whenever the decision maker is in doubt about the choice to make. Therefore, problem solving has traditionally been taken to be an essential function of management. Through systems thinking, however, we have come to doubt the existence of problems and solutions to them. This doubt, and the sense in which "existence" is a part of it, requires explanation. In the Machine Age problems were thought of as "out there," as purely objective states of affairs. But John Dewey, the great American philosopher, challenged this notion and argued that decision makers have to extract problems from the situations in which they find themselves. They do so, he said, by analyzing the situation. Hence problems are products of thought*

acting on environments; they are elements of problematic situations that are abstracted from these situations by analysis. What we experience, therefore, are problematic situations, not problems which, like atoms and cells, are conceptual constructs. We have also come to realize that no problem ever exists in complete isolation. Every problem interacts with other problems and is therefore part of a set of interrelated problems, a system of problems. ... English does not contain a suitable word for "system of problems." Therefore, I have had to coin one. I choose to call such a system a mess'.

(Ackoff, 1974b, pp. 20–21)

Here we have a succinct reframing of the notion of a problematic situation as a product of thought (analysis) acting in an environment. Like Schön, Ackoff's work is grounded in the philosophy of the American pragmatists, and I will come back to the influence of Dewey and pragmatism in Chapter 3.

2.5 An Ontological Viewpoint

By way of contrast to these three different characterisations of complex problems as co-productions of processes of enquiry, a number of attempts have been made to build taxonomies of complexity theoretically, from the ground up, so to speak, and are hence ontological in nature, even if the labelling of regions and intervention strategies are themselves empirically informed. They start with a decomposition of the space of complex problems into two, presumed orthogonal,[5] dimensions.

(i) The Cynefin framework – constructed along dimensions of [1] degree of central direction (of the system), and [2] strength of connections among constituent components (Kurtz & Snowden, 2003; Snowden & Boone, 2007)
(ii) '*Wickedness and anatomy of complexity*' – constructed along dimensions of [1] complicatedness, and [2] complexity (Andersson & Törnberg, 2018)
(iii) System of systems methodologies (SOSMs) – constructed along dimensions of [1] degree of divergence in values/interests, and [2] complexity (Jackson & Keys, 1984; Jackson, 1993)

Unlike the declarative/empirical strands above, these approaches either offer relative quantification along each dimension or divide the dimensions into distinct categories such that the resultant space can be classified as a number of identifiable

5 Categorical dimensions chosen such that they are independent.

regions. Their main purpose is as an aid to orientation and sense-making in order to make decisions about choice of intervention methods and/or strategies. The ingoing assumption is that it is possible for the consultant to orient themselves in these spaces before any process of problematisation.

2.5.1 Cynefin

The Cynefin framework is a sense-making device for appreciating the complexities of problem situations and how to deal with them (Kurtz & Snowden, 2003; Snowden & Boone, 2007). The framework, shown in Figure 2.3, was developed from the use of narrative theory and complexity theory in consulting practice and emerging questions about whether rational choice held in all problem contexts (Jackson, 2019a, pp. 160–162). The framework is constructed along dimensions of the strength of the central direction of the system and the strength of connections among its constituent components. The two axes are continuous, not categorical, but the complexity space so defined is then partitioned into two regions – the ordered domain, in turn partitioned into simple and complicated problems, and the unordered domain, itself partitioned into complex and chaotic problems. The problem types thus range from simple, complicated, and complex to chaotic. A fifth domain is labelled disorder and corresponds to complete uncertainty about uncertainty.

Each of the four problem types of the Cynefin framework has an associated strategy for intervention. The ordered domain is the realm of '*known unknowns*' and interventions can be based on the use of situational logics and best practices i.e. the realm of problem contexts where rational choice makes sense. The strategies for dealing with the unordered domain of complex problems bear some relationship to a problem-structuring first approach, whereas the chaotic domain puts action before anything else. Jackson (2019a, pp. 160–162) offers a detailed critique of the framework and flags weaknesses compared to the SOSM (discussed in Section 2.5.3).

2.5.2 Wickedness and Anatomy of Complexity

Andersson and Törnberg (2018) offer another sense-making framework that describes a more fine-grained decomposition than Cynefin through a '*meta-ontological*' map of system types. Their framework is constructed by defining the space of complex problems along dimensions of complicatedness and complexity. Their ambition is to discriminate system '*types*' by attempting to separate out the notion of complicatedness to reduce its conflation in the colloquial reference to complex systems, or as they call it the '*overwhelmingness*' of complex systems. They provide the example of NASA's Space Shuttle as an entity that is

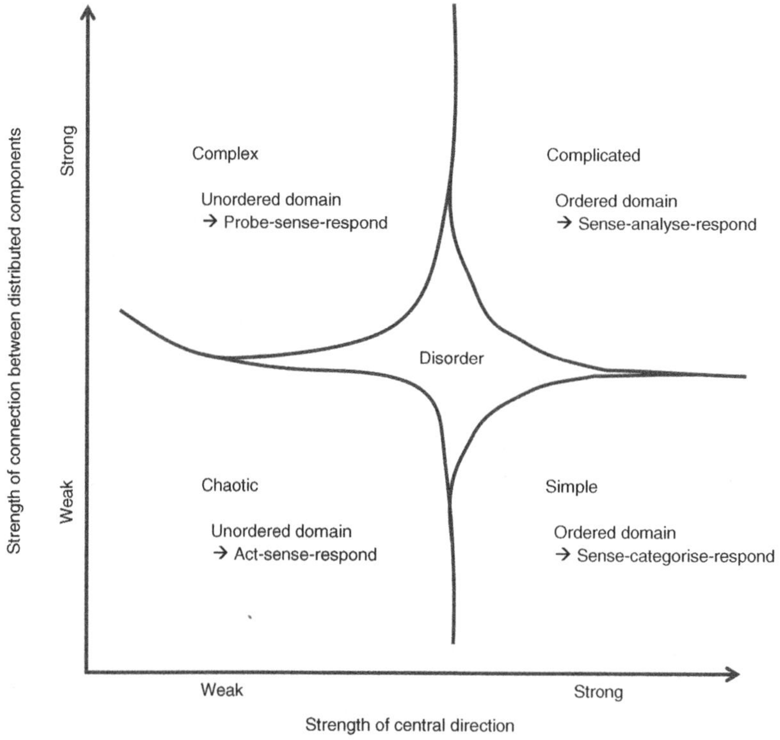

Figure 2.3 Cynefin sense-making framework. Source: Adapted from Kurtz and Snowden (2003) and Snowden and Boone (2007) with additional refinement following Jackson (2019a, p. 160).

colloquially referred to as a complex system, but on analysis using their taxonomy, it is seen to be merely highly complicated with low complexity. Entities that remain both highly complex and highly complicated on this separation are, by definition, wicked. Entities that are neither are classified as simple systems. The taxonomy is further refined by decomposing highly complicated and highly complex systems into a further two-by-two classification with trans-complex, trans-complicated, sub-wicked and wicked categories. Their conclusions from the development of this meta-ontological map are summarised into three main themes, which provide a sense-making framework in relation to dealing with complex problems.

The Cynefin and the '*anatomy of complexity*' frameworks are both empirically derived from practice but are not theoretically informed in terms of the meaning of the categories of the regions so defined. The Cynefin framework at least offers well-reasoned arguments for the associated strategies for dealing with different complexities, whereas the '*anatomy of complexity*' framework offers more by way of an imaginative/communicative device for discussing complexity.

2.5.3 The System of Systems Methodologies

Following Ackoff, Jackson and Keys (1984) defined two dimensions of problem complexity; the first arising from the complications of having an increasing number of decision makers involved in the process and/or the degree with which they are in agreement or conflict, and the second from the complicatedness of the system itself. Jackson and Keys then positioned the four Burrell and Morgan paradigms (described below) within these two dimensions. The intention was to develop a means for *theoretically informed* methodology choice, the SOSM, as shown in Figure 2.4.

The SOSM has been an enduring feature of the systems landscape, and since its introduction, Jackson (2000, 2003, 2019a) has been using the SOSM as a structuring device to characterise the different systems approaches and offer assistance to practitioners in choosing an appropriate intervention methodology (Jackson, 2019a, pp. 519–522). The intention was to encourage plurality by emphasising the complementarity of the different systems approaches, viewing the diversity as an opportunity for increased competence rather than a paradigmatic crisis. However, the SOSM (as do other definitions based on ontology) implies the need for a meta-methodology to, in effect, *look down* on the framework and make a decision about methodological choice. Nonetheless, despite advocating

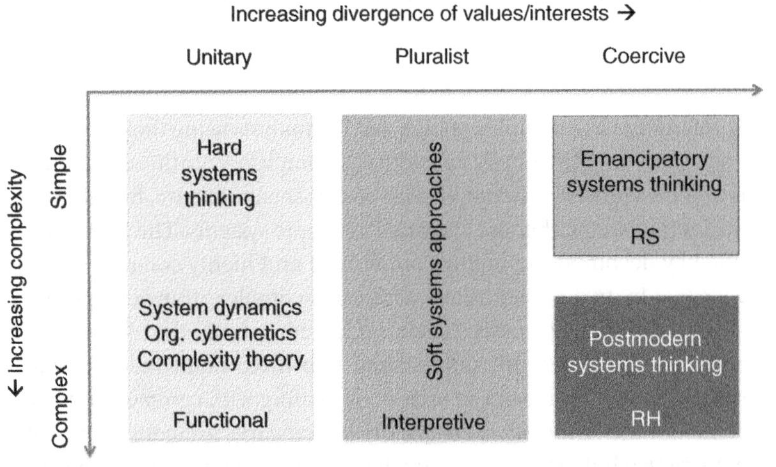

Figure 2.4 The system of systems methodologies (SOSM). Source: Adapted from Jackson (2000, p. 95) and Jackson (2019a, p. 158).

the position of plurality, and complementarity, the grounding in the Burrell and Morgan framework drew attention to the problem of paradigm incommensurability – the mis-application of a systems approach developed within one paradigm and applied to a problem context arising in another. For example, Jackson (2003, p. 81) offers a critique of group model building and the use of causal loop diagram (CLDs)/qualitative system dynamic (QSD) in problem structuring as just an '*undertheorized soft systems approach*'. The general argument is that why risk the effectiveness of a methodology used in the '*wrong*' problem context when methodologies already exist that are '*right*'. This raises an important question – what are more important to avoid, paradigm incommensurabilities or category mistakes? This is discussed further below.

2.6 The Burrell and Morgan Framework

The Burrell and Morgan classification of sociological paradigms (Burrell & Morgan, 1979, pp. 21–37) has been extensively used in the systems field to characterise assumptions made about the nature of social science and views about social order and how change happens when designing systems approaches. Their classification is set out as a two-by-two grid defined by two axes that range over (i) subjective to objective views of social science and (ii) regulation to radical change views of social order (Figure 2.5).

The two dimensions are further defined according to different strands of debate within the different social theories summarised in Table 2.1.

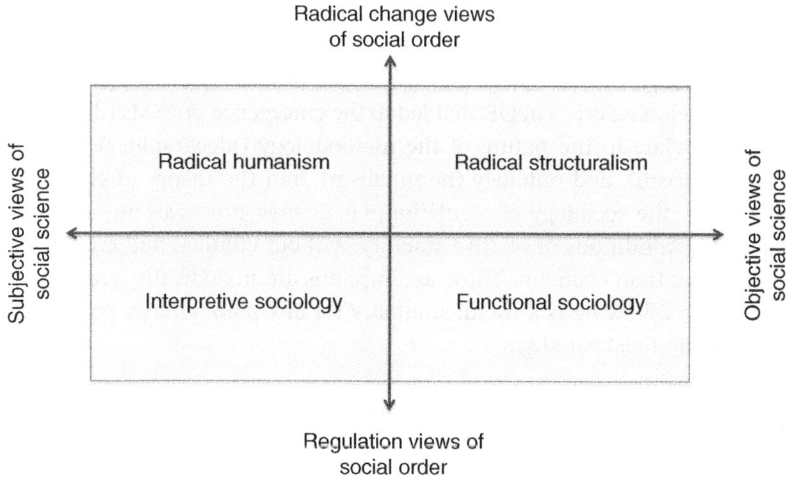

Figure 2.5 Four paradigms for the analysis of social theory. Source: Adapted from Burrell and Morgan (1979, p. 22).

Table 2.1 Social theory strands of debate (top) and social change strands of debate (below).

Subjective views		**Strands of debate**		**Objective views**
Nominalism	←	Ontology	→	Realism
Anti-positivism	←	Epistemology	→	Positivism
Voluntarism	←	Human nature	→	Deterministic
Ideographic	←	Methodology	→	Nomothetic
Regulation views		**Strands of debate**		**Radical change views**
Stability	←	Dynamism	→	Change
Integration	←	Order	→	Conflict
Functional coordination	←	Operating logic	→	Disintegration
Consensus	←	Social positioning	→	Coercion

Source: Adapted from Lane (1999) following Burrell and Morgan (1979, pp. 1–20).

The Burrell and Morgan classification schema was used by Checkland (1981, pp. 280–281) to highlight the differences between hard and soft systems thinking and to firmly position soft systems methodology (SSM) within the sociology of regulation/subjective views of social science quadrant of the typology, i.e. in the interpretivist paradigm. Although Checkland argued that SSM applications were '*limited by the desire to achieve action in the real world,*' i.e. culturally feasible applications, thus conservative, and why extension towards application in the radical change quadrant has not been seen.

By association, it is reasonable to posit that PSMs also sit in the interpretivist paradigm. Certainly, the strands of theory that are of most relevance to characterising the developing crisis in OR that led to the emergence of PSMs (described in Chapter 4) relate to the nature of the methodology (ideographic[6]), human nature (voluntarism), and ontology (nominalism), and the theory of change is firmly rooted in the sociology of regulation, i.e. change processes are assumed to take place in conditions of relative stability, without conflict, and employing consensus rather than coercion. These assumptions are periodically questioned, but for now, the following is a useful summary for any methodology positioned within an interpretivist paradigm.

> *'[s]oft Systems Methodology embodies a paradigm of learning…The notion of "a solution", whether it optimizes or satisfices, is inappropriate in a methodology which orchestrates a process of learning which, as a process, is*

6 White (2025) discusses the specific role of representation in Soft OR.

never-ending. To this extent, the methodology as a whole clearly articulates a phenomenological investigation into the meanings which actors in a situation attribute to the reality they perceive. And at a more detailed level, too, there are many parallels between the operations within the methodology and the philosophical/sociological tradition of interpretive social science'.

(Checkland, 1981, p. 279)

2.7 Category Mistakes and Paradigm Incommensurability

The ostensible purpose behind the frameworks that have been set out here is to avoid dealing with a problem context believing it is one thing when it is in fact another. Building on the notion of category mistakes introduced in Chapter 1 it seems perfectly reasonable to simply assert the existence of wicked/messy/swampy problems, despite the obvious circularities, and then just accept Rosenhead's original framing of PSMs as another sort of situational logic, i.e. '*if a wicked/messy/swampy problem then use PSMs*', and declare any other action as a category mistake. There are many positive consequences of this decision in terms of a number of strong assertions we can make. Some are straightforwardly bound up in the definitions. For example, a declaration or assertion that it is possible to solve a wicked problem is a simple category mistake. Within this framing there can be no '*solutions*' or '*cures*' to wicked problems. If a practitioner/expert[7] states they have a solution to a wicked problem, then they are likely to be making a claim about only an aspect of the problem, or only from a certain viewpoint; or their formulation is not that of a wicked problem i.e. they are talking about something '*tame*', or a problem that has already been tamed. As another example, assuming to possess an objective viewpoint and thus *know* what is e.g. optimal, best, right, smart, is another category mistake and arises from the rejection of the existence of a singular objective viewpoint in the definition. Similarly, to reject or ignore plurality is another category mistake. Applying the definitions of wicked/messy/swampy problems to any assertion that has been made means that we can test statements about such problems to see if they can be rendered false or meaningless. It is possible to cut through a lot of noise to discriminate between genuinely motivated interventions, grounded in methodologically appropriate processes of enquiry, and category mistakes. In the context of wicked/messy/swampy problems we should not expect to encounter

---

7 Making the distinction between a practitioner/expert, who should know that such claims do not make sense – i.e. they are making an infelicitous statement that is not semantically and pragmatically coherent – given the definition of a wicked problem, with others who perhaps do not know that wicked problems have such a definition.

statements about solving, solutions, optimal/best/right outcomes, singular statements about plurality i.e. '*The Public*', '*silver bullets*', '*conclusive proof*', and so on. Category mistakes are a warning sign for us to be sceptical of the claims that are being made; they suggest either misunderstanding, partiality, or (worse) deliberate deceit. On the other hand, whilst the identification of category mistakes would seem useful, relying on such dichotomous framings is suspect and does have negative consequences. It entrenches another form of situational logic, albeit more complex, and it is clear in opinions expressed both by scholars and practitioners that problems, and associated methodologies, should not be framed dichotomously (e.g. see French (2021)). We should, in fact, see problems as wholes that are not separable. The arguments for avoiding paradigm incommensurabilities seem of a similar kind, although the latter is far more theoretically informed. Pluralism and critical systems thinking, in the sense of putting the different systems approaches to work according to their relative strengths and weaknesses (Jackson, 2019a, pp. 519–524), are both arguments for treating problems as whole, and at the same time, rather clear when errors of practice are being made.

Conducting OR interventions with a view to avoiding category mistakes or running into paradigm incommensurabilities that weaken the effectiveness of our approaches would seem to be desirable if we are claiming to be practising in the best interests of our clients. However, as noted by Jackson, practice can be an exercise in being pragmatic[8] leading to making decisions about approaches that are judgements – what seems to be needed by the context. Between the two, my own pragmatic judgement is that the main difficulty facing an OR practitioner is the decision about whether a problem is wicked/messy/swampy and that paradigm incommensurabilities are less of a concern. The latter have more to do with reinforcing divisions between practitioners using different methodologies and within the OR discipline as a whole, i.e. in particular to entrench the Hard/Soft divide. Perhaps that reflects my own engineering background and its essential pragmatism. For example, I would regard using a CLD in a process of problem structuring (say) as less of an issue than using an optimisation algorithm when trying to develop a strategy. The former is eminently doable and seems to work, whereas the latter would seem to be nonsense.

Given that any problem context is likely to be wicked/messy/swampy, or that we just do not know before we start engaging with the problem, it seems reasonable to start any process of enquiry in OR with problem structuring.

8 In the everyday meaning of the word, not as pragmatic philosophy as discussed in Chapter 3.

3

A Pragmatic Framing

'Both Dewey and Foucault agreed that thinking arose in the context of problems. ... Foucault, like Dewey, asserted and affirmed that thinking arose in problematic situations; that it was about clarifying those situations, and that ultimately it was directed towards achieving a degree of resolution of what was problematic in the situation'.

(Rabinow, 2011, p. 12)

'In philosophy, we draw a distinction between "episteme" and "techne", with episteme being the theoretical kind of knowledge which traditionally is what philosophy is interested in, and techne being the mechanics' knowledge. Dewey thinks traditional philosophy has tried to exalt theoretical knowledge and downplay technical knowledge. Dewey thinks that's backwards. All knowledge begins with dirty hands, with doing things, with getting dirt under your nails. The theory stuff comes later.'

(Talisse, 2013)

'We don't act because we have ideas, we have ideas because we must act, and we act to achieve ends.'

(Menand, 2001, p. 364)

3.1 Dichotomies, Again

The dichotomous framing of problems set out in Rosenhead's critique of Operational Research (OR) has remained more or less unchallenged throughout the subsequent history of the development of problem structuring methods. However, recent work has questioned the damaging impact of this dichotomous framing on

Problem Structuring: Methodology in Practice, First Edition. Mike Yearworth.

the overall profession of OR by reinforcing a division between '*Hard*' and '*Soft*' OR (Yearworth, 2022). In a similar critique, Franco and Rouwette (2022) remind us that, in reviewing the state of problem structuring practice, Mingers and Rosenhead (2004) observed that '*…the very construction of the situation as being a problem of a particular type is a result of the process of problem structuring rather than being a given starting point*' (p. 532). There is a pragmatic feel to the circularity of this observation, and this is *not* a situational logic at play. Mingers and Rosenhead are reminding us that we need to understand that the notion of wicked/messy problems is *co-creational* with problematisation, i.e. the process of problematisation literally defines the messiness.

> *'It may sometimes be the case that, just as a well-defined problem clearly implies a particular solution technique, so a strongly organised problem situation may indicate the method by which it can fruitfully be tackled. Commonly, however, this will not be the case in the "messy" situations that PSMs are appropriate for. It may well be that only after the investigation is underway will the situation attain sufficient definiteness that a method can be selected. Furthermore, since the use of PSMs is a form of "organised finding out", it is quite possible that this process will change the understanding of the problem context.'*
>
> (Mingers & Rosenhead, 2004, p. 543)

The example problem framings from engineering on a dimension of '*messiness*' listed in Table 11.1 illustrate the point. The forces of problematisation, the dual process of construction and deconstruction (Callon, 1980), position interventions along this dimension of messiness. Conversely, an absence of any process of problematisation that is at least attuned to the notion of messiness is likely to '*opt out*' in Ackoff's words and fall into the '*practitioner free*' situational logic of Hard OR[1] (Checkland, 1983). Whilst Chapter 1 alerts us to the trap, the rational choice of methodological individualism, it does not help practitioners avoid it. Unfortunately, much of the problem formulation reported in the Soft OR literature falls into the original dichotomous framing of wicked/tame, soft/hard and so on. The pervasive nature of these dichotomies and the consequent emergence of fractal discipline landscapes were introduced in Chapter 2. Ideally, to avoid being trapped in these dichotomies we need to be mindful of the words of Mingers and Rosenhead, that the use of problem structuring methods (PSMs) is a form of *organised finding out*. However, disentangling Soft OR/PSMs from the ontological, epistemological, and methodological considerations that manufacture the divide

1 It is not without irony that Hard and Soft have become so entrenched as a polarising framing within the OR profession that it is difficult to avoid using the labels when trying to articulate a way of breaking out of the dichotomy.

from Hard OR is difficult; the seemingly inevitable dichotomous choice makes this a challenge (Abbott, 2001). Despite this, the ideas of pragmatism appear to offer some useful insight into the question of relevance to the practice of OR, not least the opportunity to perhaps break out of some of the framings defined in Chapter 2. The remainder of this chapter follows the path set out by Ormerod (2006, 2020b). It draws on Menand's (2001) historical account of the American Pragmatists and Baert's (2005) framing of the philosophy of the social sciences '*towards pragmatism*'.

3.2 Organised Finding Out

Pragmatism as a philosophy is associated with the ideas of Charles S. Peirce, William James, John Dewey, and Oliver Wendell Homes emerging from the turmoil of another war, this time the American Civil War. With so much philosophical and scientific development since then, it appears an odd choice to go back to the second half of the nineteenth century for ideas relevant to the practice of OR today. However, the motivation is driven, in part, by what Baert refers to as an impasse in the current framing of social research, an appeal for a new perspective for our investigations into the social world (pp. 1–9). Also, as Menand observes, the philosophy of the American pragmatists was an '*idea about ideas*' and that

> *'[t]hey all believed that ideas are not "out there" waiting to be discovered, but are tools – like forks and knives and microchips – that people devise to cope with the world in which they find themselves. They believed that ideas are produced not by individuals, but by groups of individuals – that ideas are social. They believed that ideas do not develop according to some inner logic of their own, but are entirely dependent, like germs, on their human carriers and the environment. And they believed that since ideas are provisional responses to particular and unreproducible circumstances, their survival depends not on their immutability but on their adaptability.'*
>
> (Menand, 2001, pp. xi–xii)

There is already something of the flavour of Soft OR in this description, and some of this must be due to the direct influence of the American pragmatists through Edgar A. Singer Jr, who was a pupil of William James and who, in turn, taught Churchman and Ackoff. Ormerod (2006) provides a useful overview of the ideas of pragmatism and their relevance to OR practice. Here, I focus specifically on those ideas that seem to address the key points of divergence in OR practices between what has become codified as Soft OR/PSMs and the solutions to well-structured problems in mainstream or '*Hard*' OR practice. Four strands appear to be of particular relevance: (i) the notion of an enquiring[2] system, (ii) the production of knowledge through collective experience, (iii) instrumentalism and (iv) phenomenal regularity. These come together such that the process of taking action *is* the process of producing knowledge.

At the heart of pragmatism is the idea that dealing with practical issues requires placing theory at the service of practice, accepting that all knowledge begins with doing things, getting stuck in to dealing with the messiness of the real world and foregoing the exalted status of theoretical knowledge. Theoretical knowledge might emerge from this process of enquiry, but it is not its primary purpose, the purpose is to take action. Dewey '*insisted on the centrality of practical problem solving to inquiry ... each issue has to be argued through an activity of doing and making which reshapes antecedent intellectual material which sets the problem of inquiry*' Ormerod (2006, p. 902). Ingoing propositions are challenged in the process and are not immutable. '*During the course of an inquiry, propositions will be scrutinized or examined not as to their truth or falsity, but rather as to their "relevancy and efficacy of their subject-matter with respect to the problem in hand." Everything depends on the problem, the tools, and the situation, not on a priori elements or fixed rules as components of inquiry*' Rabinow (2011, p. 14).

Whilst these ideas of enquiry were developed in the broader context of an overarching scientific method and creation of scientific knowledge, the principles are relevant to the scale of action research inherent to the processes of OR, within a single intervention or across a whole community of knowledge such as Soft OR.

> *'The term inquiry is taken to apply to both substantial investigations such as a consultant project, and to the myriad of small problems met during an investigation that need resolving. ...the totality of activities that stretch from project beginning to end and the mindset this involves is usually referred as the process of OR; the pragmatists would refer to the process of OR as the OR method reflecting their initial interest in the "scientific method".'*
>
> (Ormerod, 2020b, p. 1693)

> *'Inquiries have a beginning and an end. The end is reached when the initial doubt or puzzle is satisfactorily resolved in belief or a change of habits; in other words, when the investigatory activities (intelligent inquiry) have resolved the issue with sufficient clarity and certainty that the conclusions can be used as the basis for action with confidence.'*
>
> (Ormerod, 2020b, p. 1696)

---

2 Enquiry or inquiry? I prefer to use enquiry, unless I specifically mean Inquiry in the sense of a formal process e.g. the United Kingdom COVID-19 Inquiry.

Dewey '*defined inquiry as the transformation of a puzzling indeterminate situation into one that is sufficiently unified to warranted assertion or coherent action*' Ormerod (2006, pp. 900–901). The process of enquiry into a problematic issue will lead us to express our ideas about what is going on – in effect, the model (map, diagram, graph) of our structured problem. We want to do this rigorously, hence taking a methodological orientation with a focus on the specifics of constitutive rules, epistemology, and the affordances of the mapping approach for PSMs, as set out in Chapter 5. A problem structured is thus an expression of a collective understanding of the problem that is open to inspection, challenge, and ongoing investigation. The inherent systemicity of the world and its messiness mean that a process of enquiry into a problematic situation and taking action will not necessarily lead to a singular satisfactory outcome. Acting will almost certainly reveal deeper and/or wider issues. '*[S]uch a search for clarification and resolution, of course, was not assured success. Further, for Dewey, the resolution of one problem often led to the opening of another one: such was the nature of inquiry*' (Rabinow, 2011, p. 13). Our understandings are only contingent, and acting on them, the process of taking action to intervene in the problem situation, will inevitably lead to a circling back to our modelling as we observe the effects of our actions. The American pragmatists dispelled notions of solutions and ultimate goals, and instead seeded an awareness that the process of enquiry into the messiness of the world and its continually changing problematic situations is never ending.

Problem contexts are not singular in their perception. There is no single overarching viewpoint that defines them; '*knowing cannot be a matter of individual mind "mirroring" reality. Each mind reflects differently-even the same mind reflects differently at different moments-and in any case reality doesn't stand still long enough to be accurately mirrored. Peirce's conclusion was that knowledge must therefore be social*' (Ormerod, 2006, p. 898). The processes of enquiry that we construct must therefore reflect this plurality in our attempts to capture this social knowledge. Holmes held that '*experience is not individual but collective and consensual; it is social not psychological*' (Ormerod, 2006, p. 896).[3]

OR is intrinsically problem and client oriented (Eden, 1989a, p. 43). Our purpose as practitioners should be directed to intervention in the problem context that presents itself, not an excuse to address some other theoretical problem of interest. In this regard, OR shares a similar position to pragmatism. Dewey '*shifted the emphasis on philosophy from the big "wholesale" questions of metaphysics and epistemology to the empirically specific "retail" problems specific individuals and groups*' (Ormerod, 2006, p. 902).

---

3 Which also supports the argument by Yearworth and White (2014) against methodological individualism. We need theories about what groups of people are trying to achieve through their actions, not trying to recover what they were thinking or reconstructing the rational choices behind their recourse to situational logics.

Debates about this interpretation about the purpose of OR as a profession were all part of the emergence of Soft OR as a distinctive strand in OR (see Chapter 4). Ormerod (2006, p. 905) refers to an exchange of letters in the Journal of the Operational Research Society on this point and draws attention to the observation by Raitt (1979) that '*[t]he distinctive feature of O.R. in my view is its use of models and analogies. A model is not a theory, it has no direct substantive implications. No-one expects the history of O.R. to show an accumulation of models of increasing power, precision or generality. A model is constructed for practical application in a particular situation. It is wholly instrumental*'. Rosenhead and Keys argue that OR should be treated as technology, which aligns with this view. Technologies have no independent role except as instrumental means to achieve ends.

> '*An idea has no greater metaphysical stature than, say, a fork. When your fork proves inadequate to the task of eating soup, it makes little sense to argue about whether there is something inherent in the nature of forks or something inherent in the nature of soup that accounts for the failure. You just reach for a spoon.*[4] *But philosophers have worried whether the mind is such that the world can be known by it, and they have produced all sorts of accounts of how the "fit" is supposed to work—how the mental represents the real.*'
>
> (Menand, 2001, p. 361)

This focus on means-ends is a defining aspect of pragmatism, '*abstract principles cannot determine actions because ends cannot be separated from means…[t]he answers might be different at different times and in different contexts*' (Ormerod, 2006, p. 902). The uncertainties associated with our understanding of the problem context, our expression of uncertainties in our modelling and measurement, and the uncertain effects of our actions in the world *are* the thesis of Rational Analysis for a Problematic World (RAfaPW) (Rosenhead, 1989c). Dewey '*called this whole process reconstruction…. "Reconstruction can be nothing less than the work of developing, of forming, of producing (in the literal sense of the word) the intellectual instrumentalities which will progressively direct inquiry into the deeply and inclusively human—that is to say, moral—facts of the present scene and situation"*' Rabinow (2011, p. 15). This comes closest to describing what we would expect from an approach to problem structuring in OR, one that is moral and humanist in its orientation (Churchman, 1967, pp. B-142).

It is the idea of phenomenal regularity that most starkly highlights the divergence between Hard and Soft OR. Ormerod (2006, p. 893) on Peirce notes that

---

4 I pick up on this again – first in Chapter 5 as the affordance of modelling approach in a PSM and then in Chapter 9 where an entirely processual interpretation of practice is introduced; i.e. what matters is the process '*eating soup*', the reification of the entities involved in enacting this purpose can be delayed.

'the central task of phenomenology is to bring order to the manifold of phenomena that appear before the mind'. This bringing of order first appears similar to the situational logic in OR. The Soft OR reasoning *'if messy then use a PSM'* is not in itself wrong, but does have the flavour of Checkland's original critique of mainstream OR. However, the pragmatic framing produces something close. It still preserves a subtlety that is important and incorporates the co-creational nature of wicked/messy problems in OR, as highlighted by Rosenhead and Mingers. With this perspective, we should not declare a problem context wicked/messy and then invoke an intervention using a PSM as a response. We should first declare the uncertainty in our understanding of the problem context and then embark on problem structuring to help decide. The importance of this point is difficult to stress and is not easily examined empirically. Despite the analysis presented in Chapter 6 on the practicalities of re-problematising a client's instructions/requirements, the actual transition from client problem context to offering a *'solution'* and thus bypassing any attempt at problem structuring might happen very quickly, literally nothing more than a momentary thought before moving on. This is where a moral and humanist orientation to practice should assert itself.[5] Phenomenal regularity is not then in any logic that flows from an uninformed assumption about the problem context, but in the consistency of our problematising response to this initial uncertainty. This difference is summarised in Table 3.1.

'Hard' OR has been incredibly successful precisely due to exploiting situational logic, leading to well-established sub-fields and continuous improvement in its solutions to well-known problems. The potential for making an impact through the first-time application of a solution, or by the substitution of a *'better'* solution,

Table 3.1 Comparison between situational logic and phenomenal regularity.

	Situational logic	**Phenomenal regularity**
Intent	The repeated encounter of a specific *type* of problem that leads to the development of standard methods for addressing the problem type. These methods become well known *solutions*, they become teachable *best practices*.	Phenomenology as a task to bring order to the investigation of the phenomena we encounter. Rather than starting with the expectation that there is some regularity in the problem context (which might turn out to be the case) we instead have regularity in the method of investigating the phenomenon and acting to resolve our uncertainty in understanding.

5 I return to this consideration in Section 11.2 where I introduce the notion of weak injunctive translations.

is reasonably assured, hence it is entirely logical to establish them as best practices to be followed.[6] This very success is likely to lead to objections that the interpretivist grounding of Soft OR and the process of (re-)problematising would seem like a recipe for *creating* problems rather than solving them. On the other hand, Soft OR has developed a repertoire of problem structuring methods that provide well-worked-out regular means for investigating phenomena. For example, starting with a declaration of uncertainty about a decision problem is fundamental to the strategic choice approach (SCA) presented in Chapter 5 and built-in to the hierarchical process modelling (HPM) introduced in Chapter 9. Any OR practice that is not methodologically equipped to investigate a messy/wicked problem context may miss the real problem facing a client and thus fall into making a category mistake. Therefore, the notion of category mistakes can be restated and extended as a misreading of the implications of phenomenal regularity and extended into the idea of practice that exhibits pathological behaviour. The idea is summarised in Table 3.2.

Whilst none of the rows in Table 3.2 should be seen as being totally independent, each casts a different perspective on the issue of pathological practice. Some cases may be unintentional due to inadequate education/training, and others may

Table 3.2 Pathological practice in OR.

Type	Description	A misreading of (amongst others)
Category mistakes	Declaring that it is possible to solve or cure a wicked problem – there are no '*solutions*', '*cures*'…	Rittel and Webber
	Declaring possession of an objective viewpoint and thus knowing what is e.g. optimal, best, right, smart… i.e. Deliberately misrepresenting or ignoring plurality	Checkland
Fallacy of composition	Believing that solving the solvable part of the problem solves the whole problem is mistaken	Ackoff
'Carving-off'/Taming wicked problems	Deliberately deceiving, being morally wrong	Churchman
Adopting coping strategies	Locking down/resisting, asserting solved, measuring, casting '*just-like*', following orders, declaring limited options/re-framing	Conklin
Avoiding the wicked/messy problem	Being '*practitioner-free*', situational logic	Checkland
	Solving the puzzle part	Pidd
	Opting-out	Ackoff

6 We would reasonably place such best-practices on the right-hand side of the Cynefin framework as shown in Figure 2.3, the realm of the simple/complicated, known and knowable problems.

be genuine coping strategies for satisfying a particular brief, task, or command. Churchman and Ackoff characterise some practices as deliberately deceiving and morally wrong. It is hoped that greater awareness of these pathologies filters into the general understanding of what constitutes normal practice[7] in OR.

3.3 Action Is Knowledge

In the pragmatism of Dewey, agency in a problematic situation is engaging in activity to do something about it. Knowledge arises from the process of finding out.

> *'Dewey…meant that knowledge is inseparably united with doing. Education at the Dewey school was based on the idea that knowledge is a by-product of activity: people do things in the world, and the doing results in learning something that, if deemed useful, gets carried along into the next activity. In the traditional method of education, in which the things worth knowing are handed down from teacher to pupil as disembodied information, knowledge is cut off from the activity in which it has meaning, and becomes a false abstraction. One of the consequences (besides boredom) is that an invidious distinction between knowing and doing – a distinction Dewey thought socially pernicious as well as philosophically erroneous – gets reinforced.'*
>
> (Menand, 2001, p. 322)

In this Dewey-inspired mode of pragmatic enquiry, the statement '*I know how to solve this*' in the context of a messy problem is a form of category error and seems hopelessly backwards. A genuine failure to grasp the implication of trying to deal with a messy problem is perhaps forgivable if education and professional exposure have constantly talked about problem-solving, but a wilful disregard is either hubristic or worse, a deceit.

There is no problem that exists independently of the stakeholders to problematise. There is no idealised perfect solution that exists independent of the problematisation. The whole enquiring system only exists due to the will of the '*affected*' to problematise. This reflects a pragmatic orientation to the nature of enquiry – not that we will, through effort and enough time, find the '*right*' problem to address or eventually reveal a metaphysical truth or pre-existing reality, but that our endeavours will generate knowledge as a form of action that will change existing realities.

> *'One way forward is to conceive of knowledge, not as representational, but as a form of action, as something active. As William James rightly pointed out, "the Pragmatic method … appears as an indication of the ways in which existing realities may be changed." … To take pragmatism seriously, therefore, is to avoid the "ontological fallacy", the mistaken assumption that methodological questions can be reduced to matters of ontology. Against this assumption, I*

7 That is, evidence of these pathologies should become indications of malpractice in OR.

argue that questions of method always entail questions of aims and, as such, that the method used depends at least in part on what the research wants to achieve. No reference to the ontology of the social can ever be sufficient to settle matters of social methodology; there is nothing essential about the social that compels the use of a particular method. However, this is not to say that methodology is simply an issue of aims. ... it is possible that a certain methodological path might lead the way towards a particular objective in one field of inquiry but possible that the nature of a particular object of study excludes the possibility of obtaining a particular objective altogether.'

(Baert, 2005, p. 154)

Consider, for example, a definition of OR as the '*scientific approach to the solution of problems in the management of complex systems that enables decision makers to make better decisions*' (The Operational Research Society, 2023). Better for whom? In what context? How is '*better*' defined? Can the problem be '*solved*'? And so on. There is no objectively defined goal that would set a target for our knowledge in OR. It is not directional; it is only an endless set of problematisations that face us as practitioners in a world full of contingencies. What we learn from every engagement will be *both* process and content (or representational) knowledge. The process knowledge is applicable every time and, over time, deepens our understanding of what it means to engage in the process of enquiry into the messy problems of the world.

3.4 Ways Forward

Menand makes it clear that the ideas of pragmatism are not ideologies. These ideas are tools for thinking about problems in particular circumstances. They help us *cope* with the world in which we find ourselves. Merely coping seems like a lowly ambition compared to solving problems, but part of what makes the philosophy of pragmatism so appealing is the lack of hubris. The relevance to OR practice generally seems remarkably clear.

Ormerod's review of pragmatism (2006) makes a useful summary of the subject[8], and his presentation of 12 reasons '*for thinking that pragmatism could serve OR practitioners well*' (pp. 905–907) provides guidance for incorporating the ideas of pragmatism into OR practice. These reasons have been re-interpreted as a hierarchical process model[9,10] in Figure 3.1 to operationalise the purpose – putting pragmatism to the service of operational research.

8 Also see Simpson and den Hond (2022) for a relevant overview of pragmatism for studying organising, generally.
9 See Chapter 9 for the technique used.
10 Element 11, on biological approaches and their future relevance to OR seemed too speculative to model in this context. However, see the discussion in Section 10.8 on the relevance of a processual re-interpretation of biology.

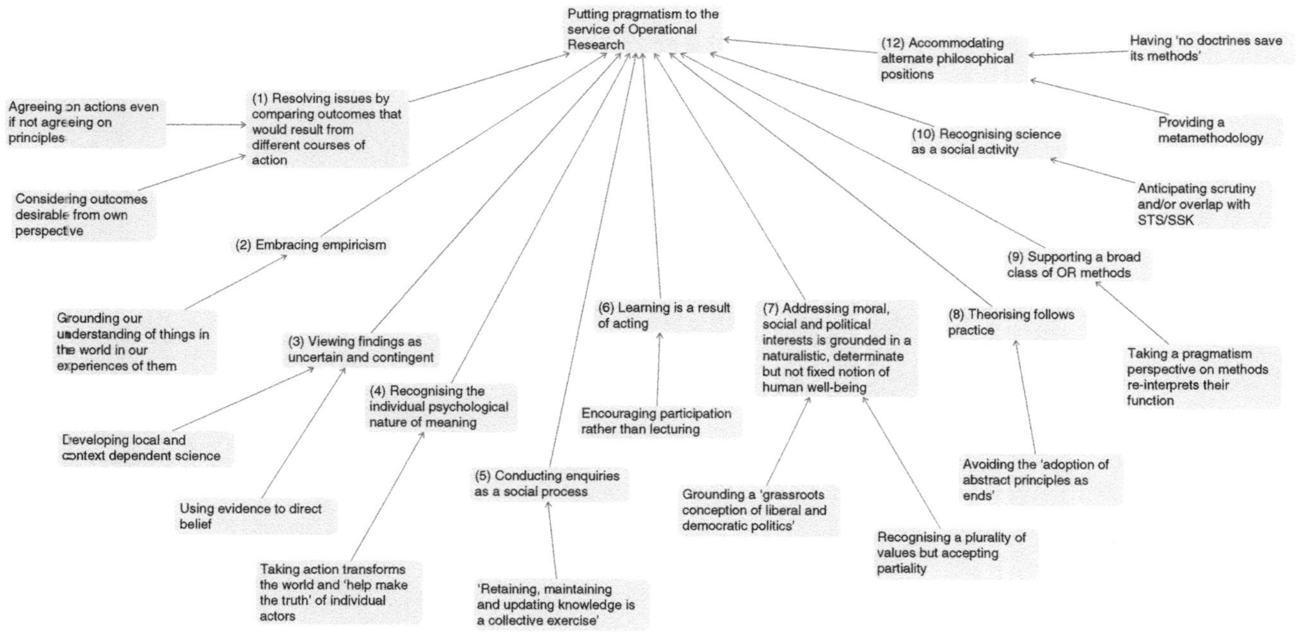

Figure 3.1 Putting pragmatism to the service of OR as a process model. SSK, Science of Scientific Knowledge and STS, Science and Technology Studies. Source: Adapted from Ormerod (2006, pp. 905–907). Numbers refer to the original items in Ormerod's list and text in quotes is from the original list.

Section II

Problem Structuring Methods

The second section of the book reviews the core methods of problem structuring – soft systems methodology (SSM), strategic options development and analysis (SODA – and its subsequent developments) and the strategic choice approach (SCA). The field of problem structuring is clearly more diverse than these three methods, but they are used here as exemplars of Soft OR practice to illustrate common themes.

Chapter 4 goes back to the origins of Soft OR and PSMs and presents a detailed history. This re-examination of the timeline starts with a detailed look at the critique of mainstream OR introduced in Section I and traces the emergence of a new paradigm, as Soft OR, over the period of covering roughly 1967–1981. The second, methodological, phase of the history covers the mid-1970s to 1989 and concerns the consolidation of the Soft OR paradigm around a new set of PSMs that provide the practical means of interventions in wicked/messy/swamp problems. The final part of the history covers the period of initial consolidation and spans the decade between RAfaPW and its second edition in 2001, and into the first decade of the new millennium.

Chapter 5 compares the core PSMs using a new analytical framework based on the combination of constitutive rules (CR), epistemologies (E), and affordances (A) of the modelling approach – the CR/E/A framework. This provides the means for comparing methods and discussing methodological choice in the context of wicked/messy/problems. The use of causal loop diagrams (CLDs) and issue-based information systems (IBIS) in problem structuring is introduced, and Chapter 5 concludes with setting out a generic description for the context/choice process of Soft OR.

Problem Structuring: Methodology in Practice, First Edition. Mike Yearworth.

Chapter 6 presents the practical matters that need to be considered in translating knowledge about PSMs into actionable interventions. This starts with the issues involved in building a relationship with clients based on trust. The value paradox is then introduced – the problem of valuing PSM engagements when, by nature, they are open-ended processes of enquiry. Stakeholder identification is then reviewed from a number of different perspectives. With all the elements in place, the process of facilitation is then discussed. Finally, questions of ethical practice are introduced and building relations to the wider PSM field is considered.

4

The Emergence of PSMs – A History

> *'What seems to emerge is not a moral reprimand of the management scientist, but rather a moral problem of the profession, a wicked moral problem. To what extent are we morally responsible to inform the manager in what respect our "solutions" have failed to tame his wicked problems? Does "inform" merely mean that we clear ourselves legally, or does it mean that we attempt to enter into a deep, mutual understanding of the untamed aspects of the problem?'*
>
> (Churchman, 1967, pp. B-142)

> *'Perhaps what Soft OR is – is what OR should be'*
>
> (Lesso, 2009)

Why go over old ground? A schism in Operational Research (OR), paradigm wars, incommensurability issues in systems approaches. Surely these are all behind us and we have moved on? Unfortunately, the situation is not clear-cut – it is both yes and no. There was not a clean paradigm shift from Hard OR to Soft OR in the sense of the previous paradigm being rejected (Kuhn, 1962), despite vigorous claims at the time (Ackoff, 1979a; Dando & Bennett, 1981). The situation was much closer to the fractal distinction over time described by Abbott (2001), where a small contingent of the OR community started to make opposite choices in matters of practice to mainstream OR along the six dimensions set out by Rosenhead (1989b, p. 12). As a consequence, their resulting position with respect to questions of ontology, epistemology and methodology meant a closer alignment with the disciplinary territory of the applied social scientists.[1] Ackoff (1979a, p. 93) notes the significance of the fact that this movement was mostly based in the United Kingdom, an observation that was underlined 30 years later by Mingers (2011), who comments on the relative strength of Soft OR in the United Kingdom

1 This is mapped out in more detail in Chapter 16.

Problem Structuring: Methodology in Practice, First Edition. Mike Yearworth.

compared to the United States, where it is practically non-existent. However, in the Kuhnian sense, as measured by success at attracting research grants, PhD student recruitment, access to publishing in top-tier journals and career progression, it has been very certainly constrained by paradigmatic laws. In this regard, the developments in Soft OR and hence problem structuring methods (PSMs) have largely relied on unfunded research or industry case studies, relatively few PhD students and new entrants to the field, and access to a few key journals to keep going.

In a commentary on a debate between Popper and Kuhn, Fuller notes that

> *'...as science has acquired more secular power, it has tended towards the self-perpetuation of existing regimes, as dominant research programmes are pursued by default, a situation that the sociologist Robert Merton has dignified as the "principle of cumulative advantage"... all scientists working in the same paradigm are equal, but some are more equal than others. ... While neither Kuhn nor Popper would care to deny that a specific paradigm may dominate the understanding of a particular slice of reality at a particular time, they differ over whether it should be treated as a source of stability (Kuhn) or a problem to be overcome (Popper). ...Kuhn and Popper represent two radically different ways of specifying the ends of inquiry: What drives our understanding of reality? Where is the truth to be found? Kuhn would have us look at dominant paradigms, the beliefs and actions of those who have come to be certified as knowers. It is ultimately a backward looking standard, one based on entitlement through survival. For his part, Popper proposed a more forward-looking perspective based on what enables us to think that our knowledge and actions are always subject to improvement.'*
>
> (Fuller, 2003, p. 30)

Reflecting this '*high table*' debate, the legitimacy of Soft OR within the wider OR field has remained a constant issue. Recent encyclopaedic reviews of the OR field have included Soft OR/PSMs with no question (see, for example, Petropoulos et al. (2024, p. 465–467), whilst others, such as Salhi and Boylan (2022, p. vii) have surfaced the fact that an issue exists '*[a]t a more fundamental level, should "soft OR" be in scope or out of scope?*' even if deciding that '*an inclusive stance is the most helpful, both for theory and practice*'.

The following history therefore (re-)presents some of the debate within OR, as it is instructive to set out the motivation for the emergence of this distinctive set of methodologies for the benefit of '*students of problems*' in STEM and the business and management fields who realise the need for a similar push-back against prevailing mindsets. For applied social scientists, it also explains how some interlopers have turned up without perhaps the same depth of grounding in the relevant literature that goes with the territory.[2]

In setting out this history, I identify two overlapping phases of emergence, followed by a period of initial consolidation characterised by some significant reflection on progress. The first phase, covering roughly 1967 to 1981, charts the emergence of a few, but highly influential, critical perspectives on OR practice up to the open declarations of the new paradigm of problem structuring. The second phase, roughly from the mid-1970s to 1989, marks the methodological period where the new approaches to problem structuring are developed and become known to the wider OR community. Although less clearly defined as a phase in terms of reflective pieces, it nonetheless culminates with Rosenhead (1989c) introducing and editing contributions for *Rational Analysis for a Problematic World: Problem Structuring Methods for Complexity, Uncertainty and Conflict* (RAfaPW), as a comprehensive multi-authored exposition of this new field. The period of initial consolidation spans the decade between RAfaPW and its second edition in 2001, and into the first decade of the new Millennium.

4.1 The Critique of the Mainstream OR Phase: Emergence of a New Paradigm

The emergence of Soft OR/PSMs as a new paradigm in OR can be charted chronologically, starting with Churchman's guest editorial in Management Science on dealing with wicked problems as a moral challenge to the profession. Continuing from the quote at the top of this chapter:

> *'To date, operations research and management science have been largely indifferent to the morality of the profession, perhaps because the profession has not yet taken itself seriously. That the profession has a moral problem, nonetheless, there can be no doubt. It might make us look more mature if we began to discuss it.'*
>
> (Churchman, 1967, pp. B-142)

Churchman explores *how* this challenge should be addressed by practitioners in a special article addressing educational issues in OR (Churchman, 1970). His argument is grounded in situating OR practice within a broad systemicity that recognises the correlative relationship between a system boundary and decision-makers, and that the consequences of interventions extend beyond just

2 What would be a suitable '*way in*' to prepare STEM students for reading in the social sciences? An attempt to answer this question is made in Chapter 8.

what was (narrowly) intended by addressing a client's needs. He also expresses the belief that the OR practitioner must develop their own accommodation to this inherent systemicity.

Ackoff returns to this ethical debate in a conference presentation on the social responsibility of OR (Ackoff, 1974c) and the personal choices facing the practitioner:

> *'A commitment to try to serve all the participants and stakeholders effectively distinguishes a professional from a servant. ... A professional seeks to serve his client as well as possible but in so doing seeks to serve all others affected by his service.'*
>
> (Ackoff, 1974c, p. 366)

The recognition by Ackoff of a broader stakeholding that exists beyond the ostensible purpose of an intervention on behalf of a client leads not only to an appreciation of the OR practitioner as an advocate for those excluded from the decision-making process but also to a much-enlarged conceptualisation of professional practice that encompasses the wider community, and hence to the idea of Community OR.[3] A corollary to this argument is that, as OR practitioners, we cannot simply assume that our professional clients will be on the lookout for these wider stakeholder interests. In fact, to do so would be a form of '*opt-out*' (Ackoff, 1977).

The roles of the OR practitioner as a dispassionate scientist *and* an engaged ethical practitioner are seemingly in conflict. This distinction and ensuing tension, in terms of achieving desirable aims in the social world, is pinpointed by Checkland[4]:

> *'It seems beyond the power of science, however, to cope with the unstructured problems of the real-world, as opposed to the explicitly defined problems of the laboratory. In the "unrestricted" sciences progress is slow and methodological problems abound. Other ways of thinking need to be explored.'*
>
> (Checkland, 1976, p. 127)

In the definition of this emerging new paradigm, another facet of the hard/soft split is opening up. Checkland begins to lay out a methodological approach that

3 I do not devote much space to the subject of Community OR as it is not inherently bound to the use of PSMs – its existence is not a consequence of the emergence of Soft OR. However, the relevance of PSMs to Community OR, not least because of the cognitive accessibility of the methods, can be appreciated in edited collections such as Midgley and Ochoa-Arias (2004) and the Special Issue on Community OR in the European Journal of Operational Research edited by Johnson et al. (2018).

4 Although not part of the historical narrative in OR, the failure of laboratory-based science to address the unstructured problems of the '*real-world*' directly mirrors the observations of Callon, Lascoumes and Barthe (2009d), who lament the scientist's '*obsession with seclusion*'.

sees a re-direction of analysis away from '*the problem*' but towards '*the situation within which a problem is perceived*', a switch in thinking away from problems as ontology towards problems as epistemology.

The report by Dando and Eden (1977) on the Second European Congress on Operations Research (EURO II) introduces a new perspective on the current state of OR practice. Their observations, reflections and analysis of the invited speakers programme conclude that there was gap between the '*introspection*' of the invited speakers' accounts of their OR practice and the '*technically bounded perspective*' of the main conference, characterised by a mathematical orientation towards technical problems and a disregard of human behaviour and power relations in organisations and society. The seeds of the UK orientation to the split[5] are signalled here.

> '*The problems our society faces will, we believe, require as much relevant technical expertise as science-based analysts can provide. Yet it is beginning to be widely appreciated that technical expertise alone is insufficient – or even counter-productive. We have noted previously the interest that this matter has aroused in the UK. ... if we wish to assist effectively with complex social problems (and we have expressed that desire often enough) then we must develop a much greater ability to reflect on our own studies. If EURO II is any guide this means a fundamental reorientation of much OR practice.*'
>
> (Dando & Eden, 1977, pp. 268–269)

In a trenchant critique of the '*violence*' of mainstream OR towards the system under study, Rosenhead (1978) pulls apart the focus on singular goals to be optimised within a framework of hierarchical decision making, wholesale quantification of the social world and the '*scientisation*' of political debate and suggests that OR should in fact be doing the opposite and adopting a social planning viewpoint. Ackoff (1979a) picks up on Rosenhead's work, and others, such as Friend at the Institute of Operational Research (IOR), and draws attention to a '*paradigmatic dilemma*' facing OR and the need to move away from a '*predict and prepare*' mindset to one that involves '*designing a desirable future and inventing ways of bringing it about*'. Of necessity, this requires a new perspective on objectivity as '*not the absence of value judgments in purposeful behaviour. It is the social product of an open interaction of a wide variety of subjective value judgments*'.

Eden and Sims (1979), in a work presented at an event organised by the Operational Research Society[6] articulate a position that is recognisably part of

5 I use split here, rather than shift, to indicate a *division* in the profession as a consequence of conscious choice over matters of practice as set out by Rosenhead (1989b, p. 12) and consequent position with respect to questions of ontology, epistemology and methodology.
6 '*The King is Dead, Long Live the King*' at the University of Aston, December 1977.

this new paradigm. Aligned to Ackoff's diagnosis, they recognise that OR has become proficient at offering solutions to a range of problems, but that these solutions have become decoupled from the processes of problem formulation, data gathering and implementation. They also observe that the mainstream process of OR enquiry ignores the co-creational aspect of problems…

> '*…by implicitly ignoring the complicated qualitative experience and wisdom of those who have endured the problem environment for much of their occupational lives but who unfortunately do not have the skills to construct "respectable" science-rational models of their complex environment. … It seems almost trivial to note that it is not possible to talk meaningfully about a problem without ascribing an owner or owners to it, because different persons will experience different problems in what might be thought of as the same situation.*'
>
> (Eden & Sims, 1979, pp. 119, 120)

They identify the different roles of the consultant (coercive, empathetic and negotiative) in the process of problem formulation with the client and set out a new set of bases on which to develop problem formulation techniques, including construct grids and cognitive mapping.

Ackoff (1979b) returns to his theme by setting out his view of an alternative paradigm for OR. Like Rosenhead, he adopts a planning viewpoint – Interactive Planning – grounded in a participative principle of reaching decisions collaboratively with stakeholders. The approach is directed towards achieving feasible idealised designs. On this basis, he then sets out an agenda for the transformation of OR practice, its organisation as a profession, and changes to OR education at the university. Setting out an agenda on such a broad front is clearly signposting the new paradigm.

The review by Woolley and Pidd (1981) presents the first empirical investigation into the new '*problem structuring*' in OR.

> '*The term problem structuring is used here to describe the process by which the initially presented set of conditions is translated into a set of problems, issues and questions sufficiently well defined to allow specific research action.*'
>
> (Woolley & Pidd, 1981, p. 197)

They emphasise the essential nature of problem structuring if OR practitioners are to address messes and wicked problems. They present the emergence of this new way of thinking as a narrative interpretation of the (published) evidence in much the same way as I have approached the narrative in this chapter. However, at this point in time, the new paradigm of problem structuring exists only in outline form, without any clear indication of new approaches to the way in which OR is to be conducted – the methodology – or whether this paradigm has emerged to sweep away the old.

By declaring a '*Kuhnian crisis*', Dando and Bennett (1981) clearly mark what they consider a necessary turning point for the profession. They realise that OR is not the same kind of science as the natural sciences, and there is not going to be a straightforward replacement for the old paradigm. They go on to consider whether this new paradigm is to be reformist or revolutionary, concluding their argument with an expectation that the profession will choose the former but that there may be '*heavy penalties for an incorrect choice*'.[7] They align their notions of reformist and revolutionary with the interpretive and critical paradigms of sociology (see Section 2.6) and clearly mirror the sentiments of Jackson (2019a) that I discuss in Section 16.1, i.e. that the OR profession will not be able to tackle wider stakeholder interests *unless* it follows a critical stance. However, there is not a sense of the profession '*as a whole*' making a choice at this juncture. It is clear that some part of the OR profession is starting to adopt a reformist (interpretivist or managerialist, as Jackson calls it) position; but it is also clear that whilst the situation shows

> '*...some signs of the decline in importance and status outlined by Ackoff, traditional groups are not falling apart all around us. It may well be that most people in O.R. are still happily doing "Normal Science" within the old positivist paradigm. From this, our conclusion is that, as yet, the crisis is still in a fairly early stage.*'
>
> (Dando & Bennett, 1981, pp. 101, 102)

The crisis may well have been at an early stage in 1981, but the situation of '*normal science*', i.e. the normal functionalism and positivism of OR, has persisted through to today without showing any signs of having been through any sort of crisis. With the benefit of hindsight, supported by Abbott's notion of a fractal disciplinary landscape, we now see the Soft OR '*project*' as a clear split from mainstream OR, but it is not the revolutionary (critical) split that perhaps some were looking for.

4.2 The Methodological Phase: Emergence of Core Methodologies

To realise this new paradigm, OR needed new methods that would address key requirements, especially with regard to simplicity of use, enabling wider stakeholder participation and dealing with social judgements and uncertainty.

7 Dando and Bennet are not clear about what is meant by heavy penalties. I return to a discussion of what these might have been in Section 16.1.

However, none of these properties were collectively known at the beginning of this phase, only with hindsight. Instead, the core methodologies were a product of, largely, action research and developing along congruent but independent paths[8] that contributed the ideas that went into shaping this new paradigm.[9] By viewing a structured problem as a theory of a problematic situation that will be put to the test through an intervention, a methodology must have a language to express that theory, one that is cognitively accessible to a broad range of stakeholders. The practice of mainstream OR is grounded in mathematical models, but in this soft paradigm, the models are diagrams or maps with a range of formalisms. The new paradigm was coalescing around emerging mapping techniques that would enable stakeholders to share a map as a representational form of the structured problem – a *conceptual* model.

The beginnings of soft systems methodology (SSM) are clearly traceable in Checkland (1972). An epistemic shift is signalled by the switch in analytical focus towards problematic situations (Checkland, 1976), grounded in ideas of Singerian (pragmatic) enquiry through the influence of Churchman (1971). Checkland (1981) provides the first full account of the development of SSM as a complete methodology and positions it clearly as aligning with an interpretivist sociology (pp. 277–285). The structured problem interpretation of SSM is a collection of human activity system models. These grew from the need to represent *any* system in diagrammatic form, noting on the way the lack of any well-developed techniques in the systems field to do this and thus inventing a way of building conceptual models of a human activity system (Checkland, 1979). These are, simply, sets of activities, described using verbs that are linked together in patterns of dependencies to concisely describe the system. A guide to building such conceptual models of human activity systems is included as an appendix in Checkland (1981, pp. 286–293). From this well-defined position, Checkland (1983) draws attention to the contribution that SSM can make in resolving the emerging '*crisis*' in OR.

The origins of the strategic options development and analysis (SODA) project[10] can be traced to a focus on eliciting clients' subjective models of problematic situations grounded in both sociological and psychological perspectives, with a strong orientation towards dealing with messy issues of policy (Eden et al., 1979; Eden, Sims, & Jones, 1979). The first conceptual models were based on eliciting

8 In their detailed construction of a bibliographic atlas of the PSM literature up to RAfaPW, Georgiou and Heck (2021, p. 783) confirm that there was a lack of common papers in their development. Whether this is true independence or a reluctance to cross-cite other work is not clear.

9 Before any wider recognition of the coherence between these different paths, there was not a unifying paradigm to ensure consistency; the new paradigm was an emergent phenomenon from its parts.

10 As a label for the developments of SODA *methodologies* over the years, summarised recently by Eden et al. (2020).

construct maps (Kelly, 1955; Eden & Jones, 1984) and then developing derived belief maps based on techniques grounded in the theory of directed graphs. Apparent similarities to cognitive mapping were noted, as were difficulties in dealing with large maps and the potential use of computers to support the mapping process (Eden, 1982). The notion of using cognitive mapping as an interactive modelling technique to be used with clients is fully formed in Eden, Jones and Sims (1983, pp. 39–59). The identification of feedback loops in the models is used as an essential technique for understanding patterns of stability and '*vicious cycles*'. The map, or conceptual model, is used with the client (a single person) to jointly construct the problem to be worked on. To work with teams, individual cognitive maps need to be merged into a single map (pp. 69–89). Computer assistance was made available with the original development of the Cognitive Policy Evaluation (COPE) software package (pp. 113–122). The use of cognitive mapping as an established computer-supported practical method to enable teams to deal with messy problems was recognised in the OR community by the Invited Review for the *European Journal of Operational Research* (EJOR) (Eden, 1988).

The perspective of strategic choice as a distinctive approach to decision making in policy systems emerged at the IOR (Friend, Norris, & Stringer, 1988) within the organisational framework of the Tavistock Institute of Human Relations (Friend, Power, & Yewlett, 1974). As an endeavour firmly rooted in OR from the start, the original focus was a development of ongoing approaches to multi-organisational planning and empirically constructed through case studies in conjunction with the agencies involved in planning decisions in England – local authorities, government departments, community interests, private enterprises and appointed agencies. The policy system was seen as consisting of decision problems, actors, patterns of relationships and rules, policies, and objectives (p. 26). Uncertainties were considered to exist in awareness of the operating environment, policies and choices (p. 32). The map of problem structure was an informal diagram of relationships between decision areas connected by decision links and existing within the boundaries of the problem area or action space (p. 48). Within each decision area exist a number of available options, and these can be linked to options existing in other decision areas, where it is considered that the combination of options would be incompatible. These '*option bars*' are an essential element as they act to help manage the large set of possible option combinations. This option graph (p. 50), consisting of options linked by option bars, maps routes to feasible solutions to the planning problem (p. 52). The map of the structured problem thus comprises a set of interrelated diagrams with different types of entity – decision areas, uncertainty maps, action space, options, option bars and decision routes. The elements of the strategic choice approach (SCA) are complete and recognisably the same as in the definitive accounts 30 years later (Friend & Stradspan Limited, 2002; Friend & Hickling, 2005).

Whilst the main activity in this phase can be seen in the rising awareness of methodological support for this new paradigm, there are continued returns to the debate about its necessity. Checkland (1983) provides the critical observation that the '*success*' of mainstream OR arises from its recognition of, and acting on, situational logics (introduced in Chapter 1) and that, as a consequence, the profession was in danger of becoming practitioner-free. Rosenhead (1986) takes the opportunity of his Presidential Address to the UK OR Society to again remind the profession that it should not be excluding itself from addressing major social issues, noting that the UK founders of OR associated themselves with the radical science movement of the 1930s.

The first summative accounts of the core methods are presented together as PSMs in RAfaPW as theory/methodology plus practice chapter pairs[11] for (i) SODA by Eden (1989b) and Eden and Simpson (1989), (ii) SSM by Checkland (1989a, c), and (iii) SCA by Friend (1989), and Hickling (1989). The dichotomous framing of problems that Rosenhead uses to set up the motivation for PSM development has been discussed at length in Section I. In his Introduction, Rosenhead (1989b) rehashes Checkland's practitioner-free argument into the metaphor of the profession becoming a '*sausage machine*'. If the only job of the analyst is means-definition for well-structured problems, then this is just a case of implementing best practices[12] and operating within the constraints of dealing with simple and complicated problems, *not* complex (Kurtz & Snowden, 2003). The introduction also positions PSMs as the '*new paradigm*' of analysis in contrast to the old. In discussing the principles and prospects for PSMs, it is interesting to note the following:

> *'Certain of the methodologies have a concentration on particular types of decision or problem situations, while others make fewer assumptions about the internal structure of situations for which they can be of use. In the latter category it is reasonable to place strategic options development and analysis (SODA), soft systems methodology (SSM), and strategic choice. Yet in another sense each of the methods does provide a general problem structuring facility. Each deploys a repertoire of methods, any one of which may (or may not) be applied at some point in the course of a study. Many of them are quite eclectic in their willingness to take other approaches under their wings.'*
>
> Rosenhead (1989a, p. 342)

11 As admitted in the Preface, I have decided to exclude coverage of Robustness Analysis, Metagame Analysis and Hypergame Analysis, which follow the same structure over the following six chapters in RAfaPW.

12 I return to this point in Chapter 11. If practice, for any expert, is reduced to (just) the implementation of best-practice, then it is unlikely that the professional is really dealing with the complexity of a client's problem situation. Rosenhead's sausage machine metaphor is highly relevant to today's concern about automation by AI.

The commonalities of this alternative paradigm are summarised through the six characteristics already presented in Section 1.6. Whilst these were assembled empirically from the evidence of early application, it requires a more refined analysis of the properties of the different methodologies to understand how they differ. It is possible to intuit some of this already through the different types of element from which maps of problem structure are constructed. Chapter 5 provides this through the introduction of the constitutive rules, epistemologies and affordances of the modelling approach (the CR/E/A framework). The survey of PSM applications in Chapter 13 then provides a further empirical update on these (early) observations by Rosenhead.

4.3 First Reflective Period

This first reflective period is characterised by filling out the details of these new methodologies, in terms of both the refinement of their description and their capabilities. Rosenhead (1996) presents an introduction to PSMs for a wider OR audience. Building on the foundations of SODA, and relying heavily on software support for cause mapping,[13] Eden, Ackermann and Cropper (1992) focus on the possibilities for different types of analysis afforded by the formal graph structure of the map. Unlike the problem structure maps of SSM and SCA, SODA methodologies benefit from the potential of formal graph analyses of complexity, such as the automatic analysis of loop structures such as virtuous and vicious cycles, identifying clusters and potent nodes, and various means of manipulating the display to manage navigating highly detailed maps through the use of views. Checkland and Scholes (1990) set out a refinement of the ideas of SSM, especially in areas such as the constitutive rules and models of the methodology in the language of purposeful activity systems models, which are relevant to the development in Chapter 5. Of considerable use to SSM practitioners and scholars alike, Checkland's reflections on 30 years of development and use of SSM appear as a '*retrospective*' in the re-published original texts (Checkland, 1999; Checkland & Scholes, 1999), as well as being separately published (Checkland, 2000b). The mid-1990s were a period of consolidation for SCA given its already long history in application to messy problems in planning. Software support for SCA – Strategic Advisor (STRAD) is introduced (Friend, 1992). The unique operating context of working with multi-organisations, right from the start, sets SCA apart from the other core methods that were developed primarily with single organisations

13 Dealing with merged individual's cognitive maps or with maps elicited directly from teams. The COPE software has been further refined and is capable of supporting graph-based analyses of a map (Ackermann, Eden, & Cropper, 1990).

in mind. The lack of appropriate theoretical underpinning for such methods is reviewed by Friend (1993) and a reflection on the development of SCA over 30 years is made by Friend (1995).

The principles and application of the core methodologies are again set out in the second edition of RAfaPW (Rosenhead & Mingers, 2001b). The introduction further reinforces the dichotomous framing of problems and presents PSMs as the answering new paradigm for addressing wicked/messy/swampy problems (Rosenhead & Mingers, 2001a). The Invited Review '*Problem structuring methods in action*' in EJOR (Mingers & Rosenhead, 2004) and the first Special Issue on PSMs in the Journal of the Operational Research Society (JORS) (Shaw, Franco, & Westcombe, 2006) show a certain accommodation of this new paradigm in the mainstream. Here, it has acquired additional supporters from within OR, and the actors who were the prime movers have had some time to reflect on progress. It is also possible to start setting agendas for ways forward (Franco, Shaw, & Westcombe, 2007). At this point, the core methodologies are now sufficiently established to be presented in textbooks without the necessity of lengthy discussions about their origins, as follows:

i. Checkland and Poulter (2006) provide a useful and short, practical guide to using SSM in *Learning for Action*,
ii. Ackermann and Eden (1998) release the first edition of the *The Journey*[14] *of Strategy Making* as a practical manual for facilitators[15] leading strategy workshops with the SODA approach, and
iii. Friend and Hickling (2005) publish the third edition of *Planning under Pressure*.

With the update to the second edition by Ackermann and Eden (2011a), we now have the complete, canonical references for the key methodologies of the new paradigm. However, the end of this third phase is also marked by the fact that the acceptance of the new paradigm was not universal; the US journals were showing solid resistance to publishing any Soft OR/PSM work (Mingers, 2011), despite authoritative entreaties as to the value (Ackermann et al., 2009).[16] At this point, the coexistence of two clearly defined OR streams is made manifestly clear.

4.4 Summary

The historical account presented in this section is clearly a selective reading from my own Soft OR perspective (from within the paradigm) and thus does not set

14 JOintly Understanding, Reflecting, and NEgotiating strategY.
15 With excellent accessibility through the Publisher's online version.
16 See the response to the Ackermann et al. (2009) letter by the (then) Editor of Operations Research stating that only work '*based on rigorous mathematical models*' is publishable.

out the other side of the argument. Whilst some of the leading actors in the split describe push-back on their positions there was no need for the old paradigm to defend its position, the new paradigm emerged in coexistence with the old, it did not replace it. Jackson (2019a, p. 188) observes that '*the paradigm shift achieved in Soft-OR has proved a step too far for those in thrall to the traditional positivist/quantitative version of the discipline*'. This is completely consistent with the fractal landscape thesis of Abbott (2001).

I have chosen to present a narrative reconstruction; an alternative presentation of the data in the period up to the publication of RAfaPW is made by Georgiou and Heck (2021) as an atlas of the journal literature constructed from a citation analysis of the key articles. However, what is abundantly clear from this history is that the paradigm '*split*' is located in debates around the *practice* of OR, in its professionalism and ethics, and through a concern for addressing broader stakeholder interests. This is not a split rooted in esoteric matters relating to philosophical underpinnings. However, at some point, we do need to make explicit the philosophical assumptions about the nature of social science and social order/change in order to explain the differences in thinking between practice in these (now) different paradigms (Burrell & Morgan, 1979, pp. 1–20). In this regard, their framework is useful – as an explanatory device – but it is not the philosophical assumptions that were at the centre of the debate. Another point of (early) divergence concerned the nature of publication practices between US journals that were even then completely technically focussed than in the UK OR Society journal,[17] which was far more open to case studies of practical application (Lawrence, 1966a; Rand, 2024).

17 Originally the *Operational Research Quarterly* (ORQ), which then became the *Journal of the Operational Research Society* (JORS).

5

Comparison of Core Methods

'Asking a client to signify how the elicited constructs would be used with respect to each element provides data which can be statistically analysed to identify the system of constructs and elements used by the client to describe the problem situation. For example, when provided with the following three elements: Soft Systems Methodology, Strategic Choice methodology, and Cognitive Mapping; I might differentiate them by arguing that cognitive mapping and SSM are similar because they both "involve some backroom work analysing data away from the client group" whereas Strategic Choice is always "done completely on the hoof in collaboration with the client group." I would then score each element, including the three that were used for construct elicitation, against the construct "involves backroom work ... all on the hoof." After several random selections of three elements from a repertoire of different "soft" OR approaches (or alternatively, a repertoire of OR consultants) I would have generated a list of constructs that should say something about the way in which I think about soft methodologies.'

(Eden, 1988, p. 3)

5.1 Analysis of the Three Core Techniques

The review in this chapter is grounded in an analytical framework with which to compare and contrast the emergence and development of the core problem structuring methods (PSMs), rather than a tutorial on the methods themselves. It is important to note that none of the methodologies analysed in this chapter have remained static since they were first developed and as they have been subsequently used in practice. One of the key aspects of all PSMs is the notion of methodological learning. Churchman (1970, pp. B-40) advocated the use of the label '*securing*' to indicate the process of delivering an OR intervention *and* learning from the

Problem Structuring: Methodology in Practice, First Edition. Mike Yearworth.

experience – '*[s]ecuring has a much deeper meaning than selling managers on OR findings. It also means observing the implementation process and controlling the change in the face of unforeseen circumstances. In general, it means "learning how to do OR better the next time"*'. Given that PSMs are deployed in unique contexts, then there is likely to be little from the outcomes from an engagement in terms of specific actions that are likely to be relevant to other applications of the same methodology. It is the positioning of problem structuring as a response to dealing with wicked/messy/swampy problems that complicates evaluation and means that learning from engagement-to-engagement focusses attention on *methodological* learning. This is discussed in more detail in Chapter 12 through the lens of practice theory. For now, the analytical perspective adopted to present the core PSMs is similar to that set out by Yearworth and White (2014) and grounded in the notion of causal realism. This is the strong assertion that the use of a PSM in a problematic situation *is* the causal means, or mechanism, by which actions are taken to improve the situation i.e. '*causal outcomes follow from mechanisms acting in contexts*' (Pawson & Tilley, 1997, p. 58). From this it is possible to state that the outcomes from a PSM intervention are a set of purposeful activities arising from the application of the PSM as the mechanism, in an appropriate context. Methodological choice is the intentional act of agents that are '*socially situated in embodied social relations*' (Little, 2011). Thus, problem context and methodological choice go hand in hand (Yearworth & White, 2014, p. 936). Although PSM use is associated with wicked/messy/swampy problem contexts as defined in Section I (Rittel & Webber, 1973; Ackoff, 1981; Schön, 1987), this should not default to another situational logic in operation. The establishment of such problem contexts is a co-creation with the act of intervening. However, regardless of this circularity, there is still a need to make a decision about context/choice as we have various methodologies to choose from. These are not just the core methods discussed in this chapter, but also causal loop diagrams (CLDs)/qualitative system dynamics (QSDs) and issue based information system (IBIS) introduced later, as well as hierarchical process modelling (HPM) as an entirely new PSM approach presented in Chapters 9 and 10. Also, as discussed in Chapter 13, different PSMs can be used in similar application contexts, e.g. strategic options development and analysis (SODA), soft systems methodology (SSM) and strategic choice approach (SCA), all used in urban planning, public administration and the sustainable built environment. Therefore, what guides methodological choice at this next level of detail? The key to unlocking the answer to this question lies in the analysis framework presented here. By analysing PSMs along the three dimensions of

1. Constitutive rules (CRs),
2. Epistemology (E) – the language through which the constitutive rules make sense (Checkland & Scholes, 1990, pp. 288, 289), and

3. Affordance of the modelling language (A) – following Checkland (1981), who defines a modelling language as '*a class, or set of classes, of elements used to construct models*'. For example, in the case of SSM, '*the modelling language suitable for making models of human activity systems is all the verbs in the language; an indication of logical dependency; indicators of flows, concrete or abstract*'.

These three elements, used in combination as an analytical device, are henceforth referred to as the CR/E/A framework. It is through this framework that it is possible to come to a better understanding of how exactly context and choice go hand in hand. Clearly, there will be an entirely reasonable choice dictated by what is actually known about or practiced regularly. A practitioner with long experience of using SSM might reasonably apply SSM in *any* context appropriate for a PSM and thus expect to be a meaningful causal agent in the resulting intervention. Practitioners skilled in many methodologies or a new practitioner or student would perhaps be less constrained in having a '*go to*' methodology and could look at the match of problem context to the constitutive rules, epistemology and affordance of the modelling approach with a more open mind. As argued below, for all the PSMs discussed, it is possible to construct a generic set of constitutive rules, therefore the distinction really lies in the epistemologies and affordances of the modelling approach.

The level of detail captured by the CR/E/A framework is not intended to provide a definitive or operationalisable account of the core methodologies; the reference texts should be consulted for this purpose.[1] However, the detail is sufficient to provide useful common ground to support the goal of comparing and contrasting methodologies with a view to resolving the context/choice question (Ormerod, 1997). It is very clear that there is a *simple* alignment between strategising and SODA, planning and SCA, and general organisational transformation and SSM. And this may be enough for methodological choice. However, it also very clear that the nuances of context would lead to engagements where, for example, SSM could be used for strategising or planning, mainly because the epistemology for SSM, compared to SODA and SCA, is relatively under specified (or under-constrained). The primary purpose of all PSMs could be summarised by achieving transformation of the problematic situation in context, and because SSM places the notion of transformation of entities (any and multiple) as a central concept, then it is easy and natural to use it in a very broad range of problem contexts.

The reason for stressing these points is that the ambition here is to foster methodological thinking rather than the slavish implementation of specific methodology. The CR/E/A framework for any of the methodologies provides a *generative* framework in which context/choice becomes more than choosing a

1 I Suggest Checkland and Poulter (2006) for SSM, Bryson et al. (2004) for SODA, and Friend and Hickling (2005) for SCA.

product (a specific technology), but rather a three-way *co-production* exercise between the originators and developers of the methodologies, the practitioner, as well as the problem context (the problematic situation with its stakeholders). For example, in Chapter 9, it is shown how HPM can be used to replace purposeful activity modelling in SSM, but HPM also shows similarities to the '*Eliciting Purpose*' forum in SODA or just general strategising. Methodological learning therefore begins right at the start of the context/choice process and is not just confined to post-engagement evaluation and reflection.

5.2 Constitutive Rules

The origins of describing what it is that practitioners were actually doing when using Soft Systems Methodology can be traced back to a critique of the methodology by Naughton (1977) as discussed by Checkland (1981, pp. 252–254). Naughton described soft systems research as a kind of '*social technology*' with a '*whiff of functionalism*'. Since then, a number of contributors have picked up on related themes, such as Talbot's functionalism in OR (Ormerod, 2020a), the performative idiom in PSM practice (White, Yearworth, & Burger, 2015; Yearworth, White, & Ormerod, 2015), OR as technology (Keys, 1998) and PSMs as mechanisms (Yearworth, 2015b; Yearworth, White, & Ormerod, 2015).

However, the approach adopted here is to view PSMs through an interpretivist framing. The modelling is essentially subjective, a phenomenological enquiry from the outset to structure a problematic situation. Interventions, loosely categorised as transforming, strategising, planning and so on, are assumed to take place within a social world of human affairs that is '*cohesive, ordered and integrated*' (Burrell & Morgan, 1979, pp. 28–32).[2] As originally proposed by Checkland (1981, p. 253), constitutive and strategic rules were defined separately to make a distinction between methodological rules that defined the conduct of an enquiry (that could be identified as SSM) and the personal interpretation of those rules. However, the key point about constitutive rules is that they are *generative* of method rather than prescriptive.

> *'Searle's original work on constitutive rules (Searle, 1995) and the example he provides of the emergence of constitutive (as opposed to regulative) rules for the game of chess (ibid, pp. 27–28), highlights the need for a more sophisticated appreciation of their value to analysis. Superficially, it would seem that a game is either a game of chess or it is not – i.e. a binary categorisation – and it is the constitutive rules that actually define the game "chess". However, many*

2 Else our gerunds of choice would be revolutionising, other-throwing, breaking, smashing, etc.

variants of the game of chess exist, and we see that a fundamental property of constitutive rules is to foster their continuous and ongoing interpretation i.e. the most important characteristic of constitutive rules is that they have generative properties.'

(Lowe, Espinosa, & Yearworth, 2020)

In subsequent work, the distinction between personal and strategic rules disappeared into a single set, but their role in defining SSM became more central.

'It is the Constitutive Rules which are of greater interest since they answer the stark question: What is SSM? If there are no such rules then in what sense can SSM be said to exist?'

(Checkland & Scholes, 1999, p. 285)

Given that SSM, SODA and SCA emerged over roughly the same period of time in the 1970s–1980s as a reaction to the functionalism of prevailing OR and systems approaches (Rosenhead, 1989b), it is reasonable to posit that the constitutive rules for a *generic* interpretive systems methodology developed by Jackson (2003, pp. 307–311) would have been a good starting point for analysis. However, Yearworth and White (2014) provide a more general set of rules for PSMs – the Generic Constitutive Definition (GCD). The need for the GCD was initially adduced from the Rolls-Royce case (Yearworth et al., 2012), and where a suitable mechanism was found in the notion of constitutive rules (Checkland, 1981, pp. 250–254).[3] The GCD was an attempt to put forward what that regularity might be for the general case when PSMs are used. Their work also opened up the possibility of non-codified problem structuring – i.e. problem structuring behaviours that are not explicitly labelled or identified as an existing method, yet appear to be following identifiable rules. The intention behind a researcher codifying the non-codified behaviours of expert groups was to bring the phenomenon of practitioners dealing with messy problems into a state of empirical investigation for those cases where the use of a specific PSM was not indicated – as in the cases analysed by Yearworth and White (2014).[4] When dealing with messy problems, any practitioner will most probably be doing something that follows the GCD. This assertion was put to the test in a rigorous ethnography of everyday problem structuring conducted by Reynolds (2021, 2023b). In a later development to encompass the use of Beer's Viable Systems Model (VSM) as a PSM, the

3 Checkland attributes Naughton (1977) for the original articulation of constitutive rules for Soft Systems Methodology and then goes on to provide a much clear articulation of a constitutive definition and associated epistemology for SSM in Checkland and Scholes (1990, pp. 284–290).
4 This is discussed again in Chapter 12.

Table 5.1 Constitutive rules for problem structuring covering SSM, SODA and SCA.

	Constitutive rules				
	SSM	CSP	GCD	VSM as a PSM	Core PSMs
i	1	1	1	1	Aiming to bring about improvements in a problematic situation
ii	2, 3b, 3c, 3d	—	7	2	Using models as learning devices
iii	3a	8	5, 6, 8	3	Making no assumption that an observer has an objective view or that different observers share the same worldview (interpretivism)
iv	4	7	9	4	Tailoring application to practical situations
v		2, 5, 6	3	5	Using methods and techniques creatively to examine the problematical situation
vi	5	9	4	6	Extracting methodological lessons through reflection
vii	2	1, 3, 4	2	7	Studying organisational function using distinctions and principles of explicit epistemologies and model affordances to identify necessary interventions

Source: Adapted from Lowe, Espinosa and Yearworth (2020).

constitutive rules set out by Lowe, Espinosa and Yearworth (2020) were further synthesised from the constitutive rules for SSM (Checkland & Scholes, 1990, pp. 286, 287), the generic interpretive systems methodology constitutive rules for critical systems practice (CSP) (Jackson (2003, p. 309), and the GCD for PSMs (Yearworth & White, 2014). They have been minimally re-interpreted in Table 5.1 to accommodate the problem structuring language of SODA and SCA.

5.3 Epistemologies

Inspired by, and following, the original derivation (or statement) of an epistemology for SSM (Checkland & Scholes, 1990, pp. 288, 289), similar derivations have been constructed for the SODA methodologies and SCA and are summarised in Table 5.2.

Table 5.2 Epistemologies for SSM, SODA and SCA.

SSM	SODA-1, SODA-2[a)]	SODA-3, SODA-4[b)]	SCA
See Checkland and Scholes (1990, pp. 288, 289)	See Eden (1989b, pp. 23–26) and Bryson et al. (2004)	See Ackermann and Eden (2011a) and Eden et al. (2020)	See Friend and Hickling (2005, pp. 67–69)
SSM	*Facilitative device*	Facilitating the social process in a forum	Shaping mode
Real world	The individual	Participation	Decision areas
Systems thinking world	Personal construct theory	Voice and choice	Decision links
Problem Situation	Language as common currency of organisational life	Consistency	Decision graph
Analyses one, two, three	Subjectivism	Clear promises	Problem focus
Rich pictures	Richness and complexity	Communication	Designing mode
Root definitions	System of concepts	Comprehension	Decision options
CATWOE	Nature of organisations	Personalisation	Option bars
The 3Es	Small teams and coalitions	Channel choice	Option graph
Conceptual model	Politics of problem solving in organisations	Leadership behaviour	Decision scheme
Activity	Organisations as a negotiated enterprise	Social support	Comparing mode
Measure of performance	Consulting practice	Relationship building	Comparison area
Control activity	Problem solving through facilitating negotiation	Identity renewal	Relative assessment
Monitor	Managing consensus and commitment	Using scripts	Advantage comparisons
Comparison	Facilitator/consultant	Facilitating groups[d)]	Working shortlist
Desirable and feasible change	Technology and technique	Building a facilitator/manager–client relationship	Choosing mode
Action		Deciding participation	Uncertainty areas
Use of SSM		Getting the environment right	Exploratory options
System to use SSM			Action schemes
			Commitment package

(Continued)

Table 5.2 (Continued)

SSM	SODA-1, SODA-2[a)]	SODA-3, SODA-4[b)]	SCA
	COPE Software[c)]	Cyclical goals	
	Quantitative analysis	Prioritising goals	
	Managing complexity	Introducing the forum	
	Cognitive mapping	Managing the participants	
		Keeping the group on track	
		Providing group support using software (Strategyfinder)	
		Providing closure	
		Managing key issues	
		Gathering (surfacing) issues	
		Facilitating divergence	
		Labelling clusters (of issues)	
		Chunking clusters into themes	
		Relating issues into hierarchical means-ends	
		Identifying feedback cycles	
		Finding hot-buttons	
		Creating a statement of strategic intent (SSI)	
		Eliciting purpose	
		Goal	
		Strategy	
		Action	
		Preferencing	
		Voting	

Understanding goals
Creating a goals system
Generic goals
Business goals
Negative goals
‘*Not our*’ goals
Reverse engineering goals
Attending to competitive advantage
Gathering and surfacing distinctive competences
Core competences
Distinctive assets (DA)
Distinctive Competence (DC)
Hierarchy of competences
Distinctive competence outcomes (DCO)
Distinctive/differentiated strategy model
Feedback loops
Categorised competence picture
Distinctive competence network
Managing stakeholders
Developing a power interest-grid
Creating an actor-influence network
Stakeholder management web

a) SODA-1 and SODA-2 in the sense meant by Eden et al. (2020); SODA-I and SODA-II by Pidd (2009, pp. 109–134).
b) SODA-3 and SODA-4 in the sense meant by Eden et al. (2020).
c) A precursor to Decision Explorer and most recently Strategyfinder.
d) Having a blank canvas to reconstruct the epistemology for SODA, it was natural to adopt a process-oriented language for the issue management forum.

5.4 Comparative Affordances of Modelling Approaches

The use of qualitative models as a representation of the structure of a problem is not the same as anything goes. Dewey's formal notation for semiotic analysis has recently been brought to the attention of the OR community by Ormerod (2023). However, the use of a diagramming convention in each of the core PSMs is covered by the same use-case in Dewey's classification (i.e. category (v) rhematic, iconic legisign). Therefore, to produce a means of discriminating affordance for the context/choice decision that links back to the related epistemology, a simple taxonomy was introduced based on the following elements: reference sources for the notation, underpinning philosophy, associated diagrams, expression of primary activity in the model – covering primary and secondary objects and their constituent elements, the graph morphology, feedback control mechanisms, how the problem boundary is expressed, whether any graph formalism exists, means of indicating action, useful heuristics in modelling and actors and referents. The details for each of these elements collectively describe the affordance of each modelling approach and are summarised in Table 5.3.

Note the simultaneous use of the term mapping (Eden et al., 2020) and diagramming. The map/diagrams in SODA and CLDs (below) and HPM (in Chapter 9) are actually graphs and have properties that can be analysed mathematically.[5] For example, cognitive maps in SODA can be analysed using graph theory for the presence of expected structural properties (Georgiou, 2009). The map/diagram is still a formal representation of the structure of the problem.[6] What we can express about the structure of the problem is therefore both enabled and constrained by the '*visual language*' that is used. We can think of this as – borrowing a term from anthropology (Gibson, 1977) – the *affordance* offered by the assemblage of elements that go to make up the map/diagram. Thus, while the different PSMs discussed here are broadly similar in terms of constitutive rules – as set out in Table 5.1 – it is their epistemologies and affordances offered by their mapping/diagramming that set them apart and make them more, or less, suitable for certain problem contexts.

5 Furthermore, they are potentially transmutable, in that it is possible to convert from one modelling representation to another. This property is made use of extensively in Chapter 9 where hierarchical process models (HPMs) stored in XML files from a no-longer maintained software package have been converted into a JSON format that can be imported into the Strategyfinder platform.

6 Intriguingly, pure morphological analysis of multidimensional, usually non-quantifiable, problem complexes by computer has been proposed as a means of problem structuring by Ritchey (2006). This is based on earlier work in Field Anomaly Relaxation (FAR) by Rhyne (1981, 1994, 1995), which also informed related work by Coyle, Crawshay and Sutton (1994) and Powell and Coyle (1997).

Table 5.3 Comparative affordances of the different modelling approaches for the core PSMs.

Modelling approach	SSM Purposeful, human activity system modelling		SODA Cognitive and causal maps		SCA Decision graphs – Analysis of interconnected decision areas	
Reference	(Checkland & Scholes, 1990, p. 294) (Checkland & Poulter, 2006, pp. 8189–8194)		Cognitive maps (Eden, Jones, & Sims, 1983, pp. 39–59; Eden, 1988; Ackermann, Eden, & Cropper, 1990) Causal maps (Ackermann & Eden, 2011a, pp. 315–330)		(Friend & Hickling, 2005)	
Philosophy	Social reality is continually constructed (Checkland, 1981, pp. 264–277) Vicker's appreciative systems (Checkland, 1981, pp. 261–264; Checkland & Casar, 1986) Churchman's enquiring systems (Checkland, 1981, pp. 258–261)		Cognitive mapping from the *perspective* of personal construct theory (Kelly, 1955) Negotiation (Fisher & Ury, 1981)		Management of uncertainty in decision-making	
Associated diagrams	Rich pictures					
Expression of activity (Primary objects) (Constituent elements)	Nodes[a)] Drawn as text on a canvas with an oval boundary Nodes are usually numbered	Activities described using verbs	Nodes Drawn as text on a canvas with or without decoration Nodes are usually numbered	Concepts (constructs, statements) e.g. ideas, beliefs, values and attitudes (poles – as expressions of psychological opposites) expressed in natural language	Nodes Drawn on canvas with a round boundary	Decision areas (groups of decision options) Usually decorated with a question mark → opportunity for choice Highlighting/colour coded as key/significant decision areas

(Continued)

Table 5.3 (Continued)

Modelling approach	SSM Purposeful, human activity system modelling		SODA Cognitive and causal maps		SCA Decision graphs – Analysis of interconnected decision areas	
					Sub-nodes Bullet points	Decision options Courses of action
Expression of activity (Secondary objects)	Links Drawn as arrows	Indicate logical dependencies	Links Drawn as arrows	Indicate causes (e.g. '*may lead to*', support, explain…). Can be negative (−ve) or positive (+ve). Undecorated arrows assumed +ve	Links Drawn as lines	Decision links Mutual relationship between decision areas (not directional) Problem inputs Problem outputs
					Sub-links Drawn as lines	Option bars indicate incompatible Decision options between decision areas
Graph morphology	Arbitrary sequences of activities, including loops Hierarchical decomposition into sub-systems to elicit further detail		Hierarchical structures with heads and tails (heads) Goals → Strategic directions → Potential options (tails)		Decision areas joined by decision links form a decision graph, the primary representation of problem structure Problem focus is a region of the decision graph selected for closer examination Decision areas can be grouped according to organisational structure	

Feedback control	Monitoring and controlling action at the top level of the model to ensure the system stays on purpose (also see Kotiadis et al., 2013)	Feedback loops can be identified, but not in the sense of monitoring and control of the system expressed in the model – can be '*vicious cycles*' or stabilising	
Boundary	Explicit system boundary Hierarchical by recursion (expansion of the first-level model and so on), i.e. nodes can contain whole models, but no links are shown between the levels of this hierarchy, they are disconnected.[b)] Boundary exploration upwards	No explicit system boundary	No explicit system boundary
Formalism[c)]	Informal Loops are possible	Formal – directed graphs Decision explorer – Analysis for Explanations, Consequences, Centrality, Clusters, Collapse, Composite tails, Domains, Issue focus, Hieset, Loop, Potency Strategyfinder – Count ingoing/outgoing links, Detect loops/loop potency, Find heads/tails, Hierarchical potency, Closeness centrality, Connection strength, Betweenness centrality	Informal STRAD – Highlight triads, pairs, singles
Acting	Action through comparing model with real world	Preferencing scores Voting scores[d)]	A Decision Schema is a combination of feasible Decision Options (i.e. those not excluded by an Option Bar)
Heuristics	7 ± 2 heuristic, arises from '*The magical number* 7 ± 2' (Miller, 1956)		

(*Continued*)

Table 5.3 (Continued)

Modelling approach	SSM Purposeful, human activity system modelling	SODA Cognitive and causal maps	SCA Decision graphs – Analysis of interconnected decision areas
Actors and referents	Collectives that are transforming an entity (very broad)	<u>Cognitive maps</u> Client's (singular) map of issues (mostly organisational) <u>Causal maps</u> Collectives that are strategising	Collectives that are planning
Group model building	Possible – examples exist when SSM is used in multimethodologies e.g. Simulation (Tako & Kotiadis, 2015), System dynamics (Lane & Oliva, 1998), SODA (Georgiou, 2012)	<u>Cognitive maps</u> No, by definition <u>Causal maps</u> Yes, expected. Enabled by Group Explorer and strategyfinder	Yes, expected
Tools[e)]	Possible, but tend to be hand-drawn	<u>Cognitive maps</u> COPE → Decision Explorer → Strategyfinder <u>Causal maps</u> Decision Explorer + Group Explorer → Strategyfinder	STRAD (Friend, 1992; Friend & Stradspan Limited, 2002)[f)]

a) For consistency throughout, graphs are referred to as consisting of nodes and links. Equivalent labels are nodes ≡ vertices or points, and links ≡ edges or lines. In any graph, the basic question is '*what do the nodes mean and what do the links mean*?' (Sowa, 1991).
b) A link to a node that contains a sub-level model can be thought of as a dependency on the whole sub-level model. Models can be nested.
c) In the sense of whether the model can be formally represented as a graph and analysed using graph theory tools.
d) Preferencing and voting were originally implemented in Group Explorer and the results overlaid onto the Decision Explorer map. In Strategyfinder, they are integrated into the overall platform.
e) It should be noted that none of the modelling approaches depend on or require software tool support. All models can be constructed using Post-it® notes, hexagons®, oval mapping®, whiteboards, flipcharts and Sharpie® pens – to list the usual panoply of the facilitator's non-IT tools.
f) Friend (2001) provides an overview of SCA and a useful demonstration of how the hand-drawn diagrams also appear in STRAD. The associated case study also provides a good representation of the diagramming used in SCA (Hickling, 2001).

It is possible to go further with this analysis through recourse to grounding in ideas from pragmatism.[7] Checkland and Scholes (1990, pp. 284–290) used a combination of rules and epistemology for their constitutive definition of SSM. Here, epistemology not just avoids ontological commitment but also provides something akin to a dictionary – in their words, '*the language through which its processes make sense*' (pp. 288, 289). An activity system model may come into existence by following constitutive rules that animate the methodology, but the '*blobs*' in the model, whilst associated with the concept of '*activity*', still require additional words to communicate what is meant by '*activity*'. This seems to me to be similar in notion to Peirce proposing the contingent reification[8] of concepts; that is, the names for concepts are not standing-in for immutable, universal truths (Menand, 2001, pp. 222, 223). When modelling using a PSM, the elements that form the model are symbols – what the symbols mean is what the participants engaged in the process agree the symbols to mean. After all, a lot of work in PSM workshops has been done with nothing more than '*Post-it® notes*', '*Sharpie® pens*', flipcharts and whiteboards, yet with these basic tools, it is possible to construct models that have different types of meaning depending on which methodology is being used.[9]

Peircean semiotics, as presented by Ormerod (2023), offers a detailed explanatory framework for why a model can be read one way in one context, and another way in a different context. It can be used to account for the way in which a screen full of symbols in Strategyfinder can be interpreted as either a causal map or a hierarchical process model.[10] Given the recency of this work, it is not possible to

---

7 I need to thank Richard Ormerod for pressing me on the point of addressing Peirce's semiotics (Ormerod, 2023).

8 I make even more use of this idea, as delayed reification, in Chapter 9.

9 When modelling using a PSM, the elements that form the model are symbols – what they mean is what we agree the things to mean. If I am running a group model building workshop using hierarchical process modelling (HPM) (see Chapter 9 for a complete description), then we have an agreement in the room that a post-it note with a verb on it in gerund form is standing-in for the idea of a process. When we place a Post-it® note on a whiteboard or flipchart (representing in effect the '*system boundary*') with one above another and draw a line between the two then the agreed meaning is that the '*superior*' '*process*' '*contains*' the '*inferior*' '*process*'. We all agree in the room that this is the case and we have a shared understanding on the basis that this is what the pen marks, bits of paper, and positioning on the board mean. In another context, the same sharpie® pens, post-it® notes, flipcharts and whiteboards could be used to develop a Causal Loop Diagram, where the post-it® notes are standing in for variables and the lines are causal links. I have done much the same thing with Strategyfinder online group support system (GSS). It was built for SODA/JourneyMaking, but I have been using it for HPM. It is just a question of deciding a convention about what the blobs mean and what the lines mean. The fact that the blobs and lines are also vertices and edges in a formal graph also makes them suitable for graph analysis. This is an extra layer of meaning that is not immediately apparent to the participants in the group model building exercise, but can be '*read*' by the adepts in the room (e.g. by the facilitator(s) using the built-in graph analysis features of the GSS) and the derived insights presented back to participants. I return to a discussion of these matters in Chapter 14.

10 As a process of semiosis, the generation of meaning from reading the symbols in the diagram.

comment on its impact, but the CR/E/A framework that I have used to analyse and compare methodologies might be enhanced or even supplanted by a more coherent semiotic analysis in the future. For now, mindful of the dictionary of meaning for the terms sense that Checkland and Scholes have used epistemology, I think it is sufficient to move a bit closer to Peirce's work by thinking of epistemology as the following – the meaning of terms is explicated in the process of taking action. Tables 5.3, 5.5 and 9.2 which describe the '*affordance*' of the different modelling approaches (in SSM, SODA, SCA, CLD IBIS and HPM), is basically listing all the ways that the '*blobs*' and '*links*' can be interpreted in each methodology and therefore the way in which meaning gets translated into action.

5.5 A Processual Analysis of Modelling in the Core Methodologies

Whilst the CR/E/A framework provides a descriptive basis for the comparison of the core methodologies, this picture is rather static. Problem structuring as a means of intervening in a problematic situation and acting in the world requires the constitutive rules to animate a *process* over a specific epistemology and affordance provided by the diagramming/mapping approach. The following provides a processual interpretation for each of the core methodologies.

5.5.1 SSM

In SSM, the problem situation is expressed as a root definition of a human activity system according to some Weltanschauung. Rather than a stateful description of system elements, human activity system modelling in SSM views the system as *transformational process*, the conceptual model of the system is a description of what the system must do to achieve the transformation. Checkland (1981, p. 170) emphasises the need to keep working at a conceptual level to avoid ending up with models of what already exists and thus excluding novel possibilities, observing that '*...the step from root definition to conceptual model is the most rigorous in the whole methodology, the nearest to being "technique"*'. Because the conceptual model is of an activity system, the only elements are verbs (activities), and the model should consist of the minimum list of verbs (activities) that are necessary to achieve the system (the transformational process) defined by the root definition. The sequencing of verbs (activities) arises from the logic of the transformational process to be achieved. The generic form of any activity system model is shown in Figure 5.1.

Core system ideas are expressed through (i) the notion of communication and control processes (i.e. feedback control), (ii) a layered structure that can

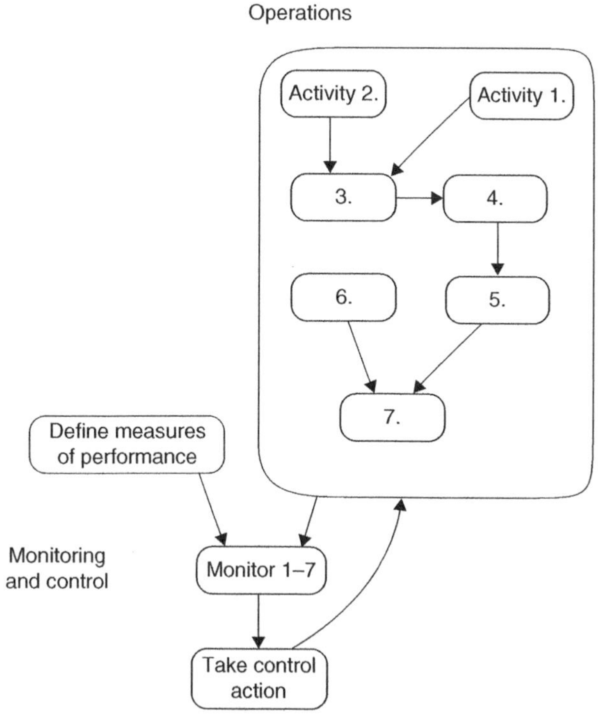

Figure 5.1 The generic form of an activity system model. Source: Checkland and Poulter. (2006, p. 8)/with permission of John Wiley & Sons.

include sub-processes, or the activity system being modelled can be viewed as part of a larger system, and (iii) the activities in combination should give rise to emergent properties, i.e. achieving a purpose beyond that attainable by the individual activities separately. An interesting property of the combination of a root definition and conceptual modelling of a human activity system is that it is possible to express SSM *itself* as a transformational process in its own modelling language. The generic model of the process of SSM is shown in Figure 5.2.

> *'Certainly the biggest difficulty in understanding SSM is to absorb its shift from assuming "systems" exist in the world (as in everyday language) to assuming that the process of enquiry into the world can be a consciously organized learning system.*
>
> *This is to say that* process thinking *is very unfamiliar to many people, and there is no doubt that teaching a way of thinking is harder than teaching substantive factual material'*
>
> (Checkland, 2000b, p. S44)

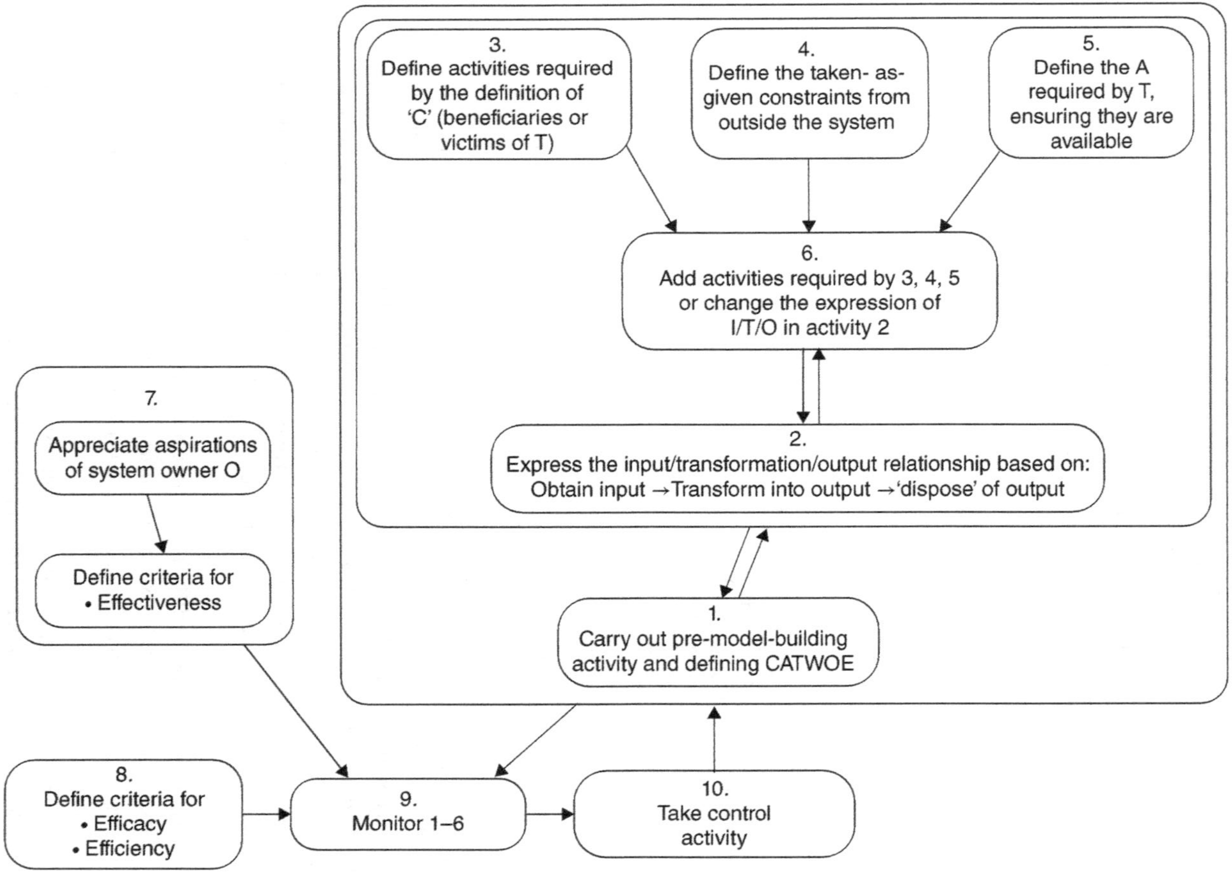

Figure 5.2 The generic model of the process of SSM. Source: Reproduced with permission of Checkland and Poulter. (2006, p. 194)/John Wiley & Sons.

Checkland's self-reflection provides far more detail on the process of using SSM (Checkland, 2000a,b).[11] The *Festschrift* Edition of Systemic Practice and Action Research in *Celebration of Peter Checkland's 70th Birthday* should also be consulted (Mingers, 2000). However, perhaps the best summary of SSM is to think of it as a methodology in use. From the generic model of the process of SSM shown in Figure 5.2, it is then possible to situate specific instances of using SSM within a much wider activity, that of the *system to use* SSM on an ongoing basis, i.e. viewing SSM practice as an ongoing action learning process. The activity system model that captures this broader practice-based view is shown in Figure 5.3.

5.5.2 SODA Project

The SODA methodology has developed over the years, hence the reference to the SODA project to mean any of the different instantiations of SODA (Eden et al., 2020). The origins of the modelling approach in SODA are grounded in the insight from consulting practice that managers use non-mathematical language around decision-making problems and that

> *'OR should be making a model building contribution to the way managers work with ideas. If OR models of any sort were going to have a real impact on the decisions they made then they must gently shift the way in which the manager sees his decision making world'*
>
> (Eden, 1988, p. 2)

The cognitive mapping technique in SODA originated in the theory of personal constructs (Kelly, 1955), and the method of eliciting repertory grids as a system of constructs used to make sense of a problematic situation. Whilst providing a solid theoretical basis, their practical use in consulting engagements was found to be limited due to the $O(N^2)$ scaling of time to elicit and analyse the constructs in a grid (Eden, 1988, p. 3).[12] Instead, cognitive mapping '*in the style of Kelly*' was developed as a '*negotiative device*' that could be used by a '*facilitator*' directly with clients. These initial maps were developed by capturing '*constructs*' represented by opposing or contrasting '*poles*'. Links between constructs are interpreted as being indicative of explanations and consequences, but not causality in a precise sense. The default link is positive if between two presenting poles and negative if between

---

11 Note that Checkland (2000b) also appears as the 30-year retrospective in Checkland (1999), which is a reprint (with corrections) of Checkland (1981). Likewise, it appears in Checkland and Scholes (1999) with the reprint of Checkland and Scholes (1990).

12 There are some passing resemblances here to the technique of Interpretive Structural Modelling (ISM) in dealing with the $O(N^2)$ scaling of time to examine all the possible element pairings (see Sage (1977, pp. 91–164) for an overview; for details about the properties of binary matrices and their use in systems modelling, see Warfield (1973, 1974a,b)).

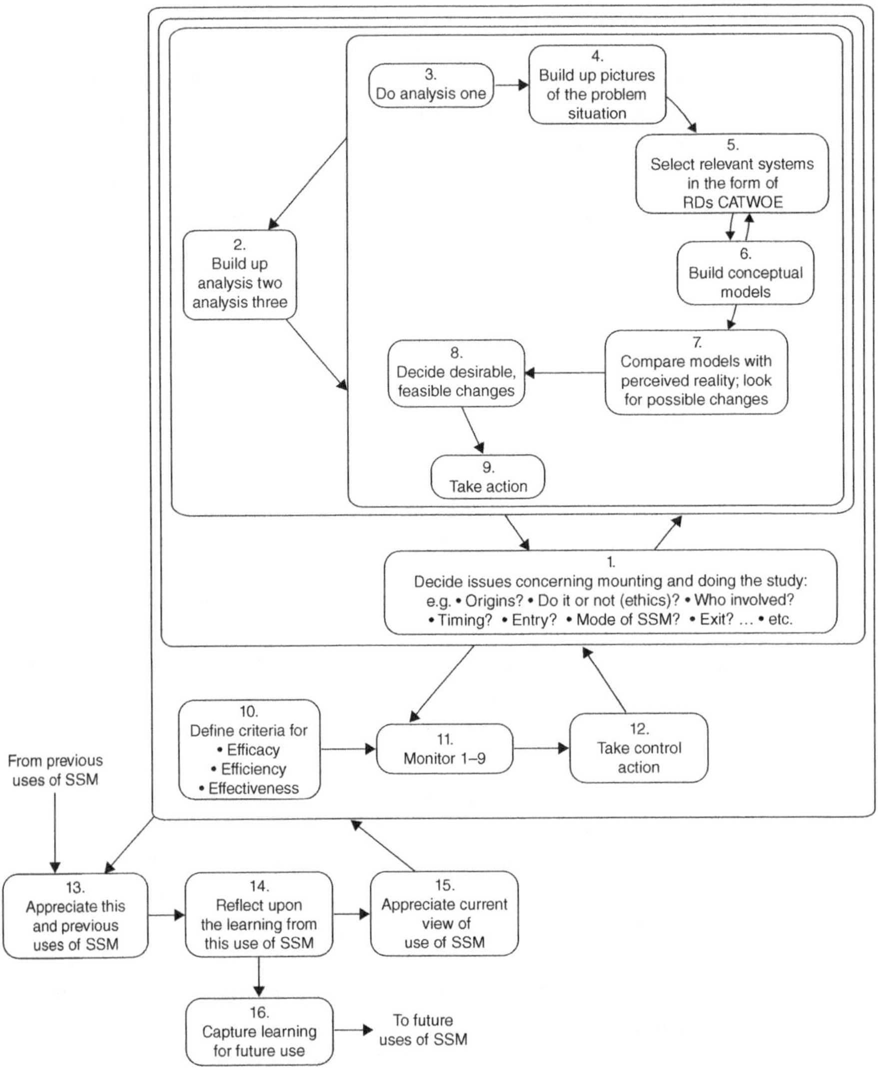

Figure 5.3 The system to use SSM, i.e. an expression of the ongoing use of the methodology as afforded by the language of an activity system. Source: Reproduced with permission from Checkland and Scholes (1990, p. 294) © 1999 John Wiley & Sons Ltd.

a presenting pole and an opposing pole (Eden, Jones, & Sims, 1983, pp. 39–51). The overall map is hierarchical, with goals appearing towards the top of the hierarchy, then strategic directions and issues, and at the bottom of the hierarchy options and actions as shown in Figure 5.4 (Ackermann & Eden, 2011a). Constructs can capture ideas, beliefs, values and attitudes. Once linked, the cognitive map can be analysed for clusters and feedback loops – both vicious cycles and stabilising (Eden, Jones, & Sims, 1983, pp. 51–56). Whilst the elicitation of cognitive maps is useful for structuring the problem from the perspective of an individual client, SODA really developed as an approach to structuring problems owned by teams in organisations. The next stage in SODA-1[13] is the merging of individual cognitive maps into causal or strategic maps.[14] A crucial part of the process is the analysis for loops, potent nodes, and clusters – this is where formal graph analysis enters the process. This not only identifies features of the structured problem where attention needs to be placed – e.g. dealing with '*vicious cycles*' or understanding the role of potent nodes – but also helps with managing the complexity of a map that may contain, as a result of merging, many hundreds of nodes. This analysis needs to be carried out before working again with the merged map with the whole team. The sequence of eliciting individual cognitive maps, the merging into a single causal map by the facilitator, and then the hosting of a facilitated workshop with the team constitutes the original SODA-1 methodology.

> *'The methodology uses causal mapping (or cognitive mapping when starting the process by working with individuals) to structure the material surfaced and provide the means for exploration and negotiation. Each contribution (regardless of the mode/technique adopted) is captured along with the relationships that link it to other contributions building up a network (directed graph) where items that are goals or aspirations appear at the top of the structure with those supporting them appearing below.'*
>
> (Ackermann & Eden, 2001)

SODA-2 skips the individual cognitive mapping step of SODA-1 and starts with a facilitated workshop to elicit a causal map directly with a team (Eden et al., 2020). Whilst these workshops can be achieved through the use of '*conventional*' means (i.e. using oval mapping on flipchart paper), the formal graph analysis of the causal maps can only be achieved through the use of software, hence the early adoption of the mapping software Cognitive Policy Evaluation (COPE) and then Decision Explorer.

Whilst the theoretical emphasis of SODA's origins was in achieving a cognitive buy-in to the structuring of a problem, there was also a need to establish

13 Eden et al. (2020) refer to SODA-1 to SODA-4 (see also the detailed presentation of SODA-I and SODA-II (sic) by Pidd (2009, pp. 109–134).

14 Note that a collective cannot share cognition (Eden et al., 2020) and therefore the merged maps are no longer cognitive maps.

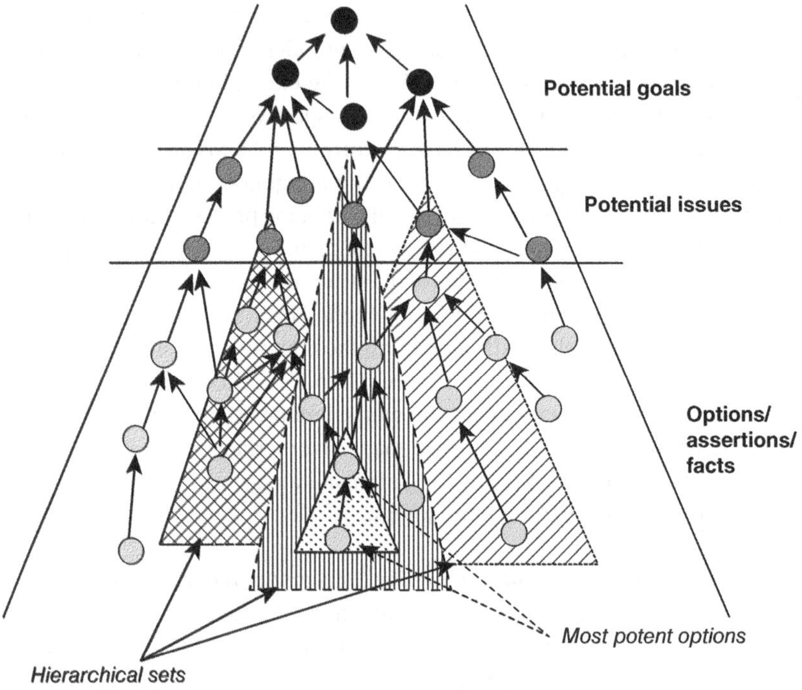

Figure 5.4 Goals, potential issues, options/assertions/facts and hierarchical sets in SODA causal mapping. Source: Reproduced with permission from Ackermann and Eden (2011a)/SAGE.

an emotional commitment from the team to taking action and therefore also included concern about procedural rationality and justice (Bryson et al., 2004; Eden et al., 2020). JourneyMaking (SODA-3) was a further development arising from extensive consultancy with top management teams leading to outcomes that were setting strategic direction, i.e. the focus was shifting away from problem structuring per se and towards strategy-making and the detecting of emerging strategy (Eden et al., 2020). The limited time available with senior leadership teams led to the adoption of a group support system (GSS), known as Group Explorer, to connect workshop participants' inputs directly, using a laptop or tablet, with the Decision Explorer mapping and analysis software. The tight integration of Group Explorer and Decision Explorer was a core enabler of SODA-3. The combination also offered additional benefits. With participants interacting with the model through a computer interface and observing the model displayed using a data projector, it was possible to enable a degree of anonymity in model contributions, thus facilitating potentially conflicting views to be surfaced that may have been suppressed in a workshop environment where statements were being gathered in plain view (e.g. by openly placing a Post-it® note on a wall). Further, the logs of participant interactions through Group Explorer also provide data suitable for post-workshop analysis about the performance of the methodology (e.g. see Yearworth and White, 2017; Franco and Greiffenhagen, 2018), and discussed further in Chapter 6. In 2019, the Group Explorer/Decision Explorer software combination was replaced by a new Web-based platform – Strategyfinder[15] (Eden et al., 2020). The switch to Web-based delivery thus opened up the possibility of delivering same-time/different-places workshops, as explored by Yearworth and White (2017, 2020).[16] JourneyMaking-2 (SODA-4) continued the strategy-making purpose of SODA with an extension to workshops addressing specific aspects of strategy, i.e. strategy as issue management, as purpose, as a competitive advantage, and as stakeholder management (Eden et al., 2020). This is reflected in the CR/E/A framework analysis by a the extended epistemology for SODA as shown in Table 5.2.

5.5.3 SCA

Modelling in the strategic choice approach (SCA) is of a different form than both SSM and SODA. SODA models are formal graphs, whereas SSM activity system models are not, but both have regular and singular structures, i.e. there is one *type* of model that represents the structured problem. SCA uses a collection of models to represent different aspects of the structured problem. An overall schema for all the different models that constitute the complete methodology is shown in Figure 5.5.

The primary philosophy behind SCA is decision-making under uncertainty, and therefore, a primary undertaking in the methodology is the identification of '*decision areas*' and their association with three types of uncertainty – uncertainties about the working environment (UE), uncertainties about related decisions (UR) and uncertainties about guiding values (UV). The methodology has four processual '*modes*' or aspects that define the learning process for SCA – shaping, designing, comparing and choosing.[17] Whilst Figure 5.5 seems to suggest sequential movement between modes, Friend and Hickling emphasise that

> '*...in describing the process emphasis of the strategic choice approach, the general orientation towards a learning process will be interpreted more specifically in terms of an operational guideline of cyclic continuity. In this, the*

15 See strategyfinder.com.
16 Fortuitously, as it happened, with the switch to working from home during the lockdowns arising from the COVID-19 pandemic (see Chapter 14).
17 See the SCOOPs model in Section 7.5 for a similar idea emerging in engineering practice.

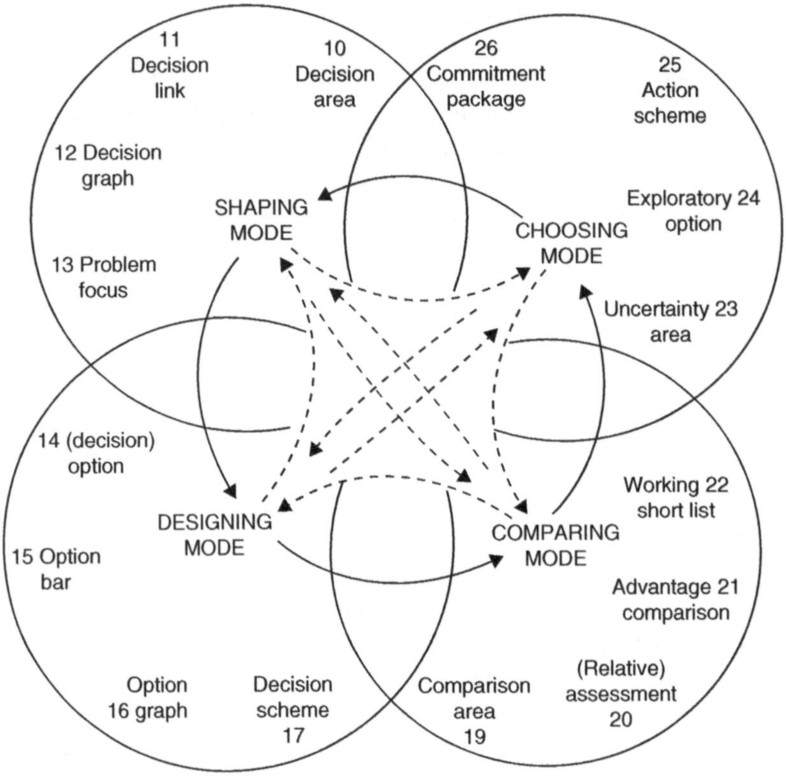

Figure 5.5 The vocabulary of Strategic Choice (Note: In effect the epistemology for SCA as shown in Table 5.2). Source: Reproduced with permission from Friend and Hickling (2005, p. 68) © Elsevier.

> *conventional representation of a process in terms of a linear sequence of stages is replaced by a more flexible set of possibilities for movement between one mode of decision making and another.*'
>
> (Friend & Hickling, 2005, p. 81)

In the '*shaping mode*' (Friend & Hickling, 2005, pp. 89–117), the map of decision areas connected by decision links *is* a graph. Links indicate a connection between decision areas, and more doubtful links are indicated by a dotted line; neither are directed, i.e. they have no arrow heads. An example is shown using the original Strategic Adviser (STRAD) software in Figure 5.6, where the normal links are shown in blue and doubtful links in red[18]. The decision graph is an essential

18 Normal links appear black and doubtful links appear grey as reproduced in Figure 5.6.

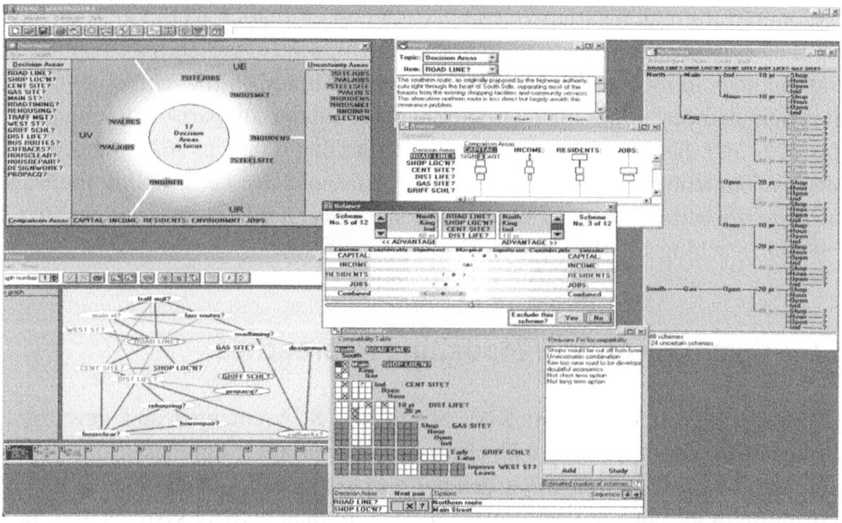

Figure 5.6 An example of the STRAD desktop for SCA (Friend & Stradspan Limited, 2002). From top to bottom, left to right: (i) uncertainty areas – overview, (ii) analysis of interconnected decision areas – focus graph, (iii) decision areas – descriptive notes, (iv) assessment, (v) balance, (vi) compatibility matrix and (vii) schemes. The example shown here corresponds to the '*South Side*' case study presented in Friend and Hickling (2005). Source: Reproduced with permission from Dave Friend.

element in the shaping mode and is intended to be used to bring out the structure of the relationship between decision areas. The emphasis is on rearranging the spatial relationship between decision areas and grouping decision areas into clusters. Extra information can be brought into the decision graph, such as highlighting more important decision areas, adding organisational hierarchies and identifying different problem focus areas (pp. 97–107). This is intended as a manual process and is not supported by any formal graph analysis (unlike SODA).

In the *'designing mode'* (pp. 119–145), decision areas are analysed for sets of mutually exclusive '*options*'. Ideally, this list should be exhaustive, but may also include options that are uncertain or potentially null, i.e. explicitly considering and recording the case of there being no option. Decision areas might be decomposed into two or more decision areas to deal with there being too many options or composite options. Once options are identified in decision areas, then it is possible to consider option bars – combinations of options that are considered to be incompatible or undesirable. The barring of options is important for reducing the number of options to take forward for comparison. STRAD does not show option graphs but does show the equivalent compatibility matrix (item vi in Figure 5.6). The resultant tree of decision areas and compatible options produces a list of potential '*schemes*'. In the comparing mode (pp. 147–177), a working shortlist of schemes is produced through a variety of comparison and evaluation techniques, including choices in the approach to relative assessment, ways of presenting advantages, and methods to incorporate numerical scoring. Finally, in the '*choosing mode*', the original uncertainties must also be factored into the comparison of schemes[19] (pp. 179–209). Similar to SSM, Friend and Hickling (2005, p. 362) mapped out the developmental challenge for the ongoing use of SCA, i.e. situating the SCA *methodology* within the wider context of its practice, not least the relationship to research, teaching and the impact of new technologies.

5.6 Summary of Core Modelling Approaches

The modelling approaches in SSM, SODA and SCA emerged and developed through a process of working with clients. It is important to emphasise this point. Checkland was searching for a general-purpose systems modelling approach to capture descriptions of transformations by human activity systems and, not finding anything suitable, was compelled to develop a modelling approach that followed a set of systems rules and was easy to draw and comprehend. Eden started with Kelly's repertory grids, but through the pressure of limited time availability with managers he had to adapt and simplify his approach to developing cognitive maps. SCA was always firmly grounded in the practical means of decision-making under uncertainty in the context of planning and working across stakeholder groups that included multiple organisations.

The epistemology and affordance of the modelling approach for each of the core methods show how the constitutive rules for a generic problem structuring approach are interpreted in specific instances of the methodology. Whilst SSM and SCA have remained relatively stable in terms of the CR/E/A framework, it is clear that SODA has moved towards application in strategy-making and detecting emerging strategy, as reflected in the considerably extended epistemology for SODA in Table 5.2 compared to SSM and SCA. Whilst the epistemology and affordance of the modelling approach in each of SSM, SODA and SCA are sufficiently broad to enable their use across a wide range of applications (see Chapter 13), there is nonetheless a strong association between methodology and its broad processual characterisation of context for problem structuring as follows:

- SSM → Designing transformations, Transforming (generally);
- SODA → Strategy-making, Detecting emerging strategy, Strategising (generally);
- SCA → Planning (generally).

19 More recent developments in SCA have combined the shaping and designing modes with different quantitative means for comparison, such as the use of MCDA in the new MuVAM platform (e.g. see Lami and Todella (2022) and the case study in Lami and White (2022)).

5.7 The Influence of Computer-supported Modelling

The introduction of computer-aided support for mapping – especially in terms of analysing directed graphs and identifying hierarchies and clusters – was both an essential development to support the modelling process in SODA and also opened-up creative possibilities. For example, Georgiou (2012) exploits the structural complementarity between SODA and SSM to integrate human activity system models with SODA. In Chapter 10, I make extensive use of the latest SODA project GSS platform, Strategyfinder, to support HPM incorporating IBIS-style argumentation. SCA also benefitted from the development of software (STRAD) to support the methodology. Here, the emphasis was on managing the complexity of keeping the different models consistent, which would have been difficult using traditional workshop techniques. The comparing and choosing modes in SCA have suggested new developments in multimethodology and new platform developments in MuVAM (Lami & Todella, 2022; Lami & White, 2022). In Chapters 14 and 15 I discuss developments in the use of GSS and the possibilities for a *problem structuring platform*, augmented by new developments in machine learning.

5.8 Related Approaches

Two further modelling approaches are introduced that have found application to support the structuring of problems – IBIS, used to support argumentation maps, and CLDs. Argumentation maps are introduced here as they find application in HPM discussed in Chapter 9, as well as being used as a PSM in their own right. CLDs are included because of their obvious application in problem structuring (see Chapter 13), their widespread association with systems thinking (Senge, 1990), and the contribution of group model building to the PSM field (Vennix, 1996; Andersen et al., 2007). Although neither is covered in RAfaPW (both editions), the main reason for inclusion here is to present the epistemologies and affordances of modelling approach for the purposes of comparison with the core methods and to support the survey of the range of PSM applications presented in Chapter 13.

5.8.1 IBIS

IBIS emerged from the work of Kunz and Rittel (1970) and set out an argumentation-based approach to dealing with wicked problems. Problems are structured through the arrangement of elements such as problem areas, issues, and facts into an argumentation map that supports collective decision-making.

In effect, it is a technique for mapping an argument or debate of issues to arrive at decisions as actions. Conklin (2006) presents a form of IBIS where the key elements are Questions, Ideas and Arguments organised into a Dialogue Map. Software to implement IBIS-style argumentation as problem structuring technique, Cohere and Compendium, is described by Buckingham Shum et al. (2006) and Buckingham Shum (2008). Options also appear in the epistemology of the SODA methodologies as statement categories in Strategyfinder and concept-style properties in Decision Explorer models. Both support user-defined category and concept styles, respectively, and therefore could be extended to fully support IBIS concepts (see Section 10.2).[20] The epistemology and affordance of the modelling approach for Conklin-style IBIS are summarised in Tables 5.4 and 5.5, respectively.

5.8.2 Causal Loop Diagrams

My purpose here is to explain how CLDs are used in the process of problem structuring, although the same process could well be a part of the first step towards developing a full system dynamics model. However, it is useful reflecting on the aims of full system dynamics, mainly the *endogenous* explanation of dynamic behaviour. Similarly, users of CLDs should also be aiming for the same explanatory goal, but for the purpose of structuring a problem not reproducing a reference behaviour from a simulation study. Originally, in system dynamics modelling, CLDs were used *after* the simulation studies had been completed as a means of representing and communicating the structure of the dominant loops (Lane, 2000). However, over time, CLDs started to be used from the beginning of a study for *conceptual* modelling of a system (p. 242). However, from the perspective of the aims of system dynamics modelling – i.e. gaining an understanding of system dynamics – it is possible to critique CLDs as possibly offering a misleading perspective (Forrester, 1994; Lane, 2008; Schaffernicht, 2010). To overcome these limitations, the partnering of a reference mode – i.e. a behaviour over time graph – with the CLD is considered to be essential (Schaffernicht, 2010, p. 665). Recent work has attempted to assess the scale of the problem (Jalali & Beaulieu, 2023) and to propose remedies (Turner & Goodman, 2023). However, despite the criticisms, the overlap between use of CLDs for describing system structure and their use as a modelling approach to support problem structuring is clearly growing (see Chapter 13). Forrester (1994) also comments on the value that

20 Also, see Rooksby, Sommerville and Pidd (2006), who integrate IBIS properties into cognitive mapping.

Table 5.4 Epistemologies for IBIS and CLDs.

Issue-based information system (IBIS)	Causal loop diagram (CLD)
See Conklin (2006)	See Lane (2000, 2008) and Schaffernicht (2010). Also see Sterman (2000, pp. 137–190) and Morecroft (2007, pp. 31–57)
Building a shared understanding	Describing system structure
Dialogue mapping	Feedback
Shared space	Causal loop
Collaborative display	Variables
Dynamic boundary object	Link, causal link, arrow denoting causal influence
Facilitation	Link polarity
IBIS notation	Loop identifier
Question	Delay
Deontic	Goals
Instrumental	Reference mode (behaviour over time graph)
Criterion	Archetype (Senge, 1990)
Meaning/conceptual	*Ceteris paribus* explanation[b)]
Factual	Endogenous explanation
Background	
Stakeholder	
Idea (can be a decision)	
Argument (pros and cons)	
Synchronous meetings/distributed group[a)]	
Transcriptive capture	
Interpretive capture	
Meta-map	

a) Through the Web using Compendium (available open source at https://github.com/CompendiumNG). The same concept is referred to as same-time/different-places in Chapter 14 (Yearworth & White, 2020).

b) In the case of variable A linked to variable B, the *ceteris paribus* explanation proceeds as follows. With all other things (variables) remaining the same, if variable A increases (decreases), then variable B increases (decreases) for positive link polarity (i.e. linked variables move in the same direction) and are indicated in the diagram by a + sign. Otherwise, with all other things (variables) remaining the same, if variable A increases (decreases), then variable B decreases (increases) for negative link polarity (i.e. linked variables move in the opposite direction) and are indicated in the diagram by a – sign. Link polarities are sometimes indicated with **S** and **O** to indicate same and opposite.

Table 5.5 Affordance of modelling approach for issue based information systems (IBIS) dialogue maps and CLDs.

Modelling approach	**Issue based information system (IBIS) – Dialogue map (Conklin, 2006)**		**Causal loop diagram (CLD)**	
Reference	(Conklin, 2006)		(Lane, 2000; Sterman, 2000, pp. 137–190; Morecroft, 2007, pp. 31–57; Lane, 2008; Schaffernicht, 2010)	
Philosophy	Creating a shared understanding and shared commitment to intervention (Conklin, 2006, pp. 29, 30) existence of wicked problems (Rittel & Webber, 1973)		Understanding that the dynamic behaviour of complex systems arises endogenously from system structure that includes feedback loops, accumulations, and delays (Forrester, 1961)	
Expression of activity (primary objects)	Nodes Represented with icons	Questions (question mark), ideas (light bulb) and arguments (pros as + and cons as –)	Nodes Drawn on canvas as text	Variables
Expression of activity (secondary objects)	Links Drawn as arrows	(pointing towards the main (initial) question)	Links Drawn as arrows	Causal links + (positive) (sometimes s for same) – (negative) (sometimes o for opposite) Delay markers (box labelled delay or double slash mark through link)
Additional activities	None		Loop identifier R (reinforcing)/+ (positive) (sometimes downhill snowball icon) B (balancing)/– (negative) (sometimes see-saw icon) Name Number	
Feedback control	None		Primary purpose of model is endogenous explanation – the model should be a representation of a feedback control system that can generate the reference mode (behaviour over time graph)	

Table 5.5 (Continued)

Modelling approach	Issue based information system (IBIS) – Dialogue map (Conklin, 2006)	Causal loop diagram (CLD)
Boundary	No explicit system boundary	No explicit boundary; the system is the model
Formalism	None	Directed graphs with cycles[a] Partial derivatives with respect to time Explanation proceeds via '*ceteris paribus*'
Acting	Decisions can be shown as a consequence of arguments	Leverage points (Senge, 1990; Meadows, 2008)
Heuristics	None	System archetypes (Senge, 1990) Variables named with a noun or noun phrase with a clear sense of direction and corresponding to real-world quantities[b] Dimensional consistency Goals of balancing loops should be explicit (unless there is a decay process) Indicate important delays
Application (referent)	Any conversation, dialogue, discussion… can be mapped	Systems exhibiting problematic dynamic behaviour (reference modes)
Group model building	Yes – see Buckingham Shum (2008) and De Liddo & Buckingham Shum (2010)	Yes – see Vennix (1996) and Andersen et al. (2007)
Tools	Compendium Cohere PERIMETA and Strategyfinder (see Section 10.2)	Vensim Stella AnyLogic

a) At least one cycle is required to make it a CLD.

b) The behavioural definition of a causal link in note (b) of Table 5.4 must also include consideration that a variable may add or subtract from the value of another variable, i.e. that there is a difference between conserved flows and purely informational links, the so-called '*extensive definition*' (Lane, 2000, p. 243). Even if there is no intention of developing a full SD model, a CLD should be consistent with a model that *could* generate the dynamics of a reference behaviour and therefore at least one variable per loop should correspond to an accumulation (i.e. a conserved flow, a '*Stock*').

Soft OR places in participant engagement and how that could contribute useful insights, through the use of CLDs, into system dynamics practice.

In his reflections on 60 years of system dynamics modelling since its introduction as '*Industrial Dynamics*' (Forrester, 1958), Lane (2022),[21] comments on the strong connections between the development of system dynamics and PSMs:

> *'Concerning the role that models play, the importance attached to engagement with mental models in Industrial Dynamics is significant. This is the idea … that managers "playing" with a model together can discuss, clarify, learn, improve their intuition and create new mental models which become the shared basis for policy making. … Here then is a first strong connection between Forrester's ideas and those of PSMs: the role of models.'*
>
> (Lane, 2022, pp. 308, 309)

Further links with the problem structuring community can be seen in, e.g. approaches to synthesising the modelling approaches of system dynamics and SSM (Lane & Oliva, 1998). There is an increasing intersection of interests between group model building (Vennix, 1996; Andersen et al., 2007) and the PSM field. For example, in comparing group model building and SODA, Rouwette, Bastings and Blokker (2011) conclude that the theoretical assumptions behind both methodologies are more similar than they expected. The differences really just reflect how the epistemologies and affordances of the two approaches influence methodological choice, i.e. it is entirely possible to combine the epistemology and affordance of CLDs with the constitutive rules set out in Table 5.1. The epistemology and affordance of the modelling approach for CLDs are summarised in Tables 5.4 and 5.5, respectively. Similarly, recent work by Naugle, Langarudi and Clancy (2023)[22] attempts to codify '*what is system dynamics*' and, in the process, covers similar ground to the epistemology and affordance for CLDs covered here.

21 Lane actually marks the 60th anniversary from the publication of the book on Industrial Dynamics (Forrester, 1961). However, the *Harvard Business Review* article is easier to find online, and the printed article is remarkable in that it includes fold-out pages of graphs of simulation results. Perhaps one of the earliest examples of a computer-generated graph for a business audience in print?

22 There is also a useful bibliography focussed on methodology.

5.9 Conclusion

Most descriptions of PSMs refer to both the rigour of the process and the cognitive accessibility of the modelling approach, the latter being the property that makes the process usable for most participants. Following Checkland's reflections on '*why methodology?*' (Checkland, 1981, pp. 161, 162; 2001, p. 87) has led to a focus on constitutive rules, epistemology and affordance of the modelling approach – the CR/E/A framework – to provide a means of explaining the inner-logic of PSMs, i.e. more about how they work, rather than a step-by-step guide on how to use them in practice (although see Chapter 6 for some practical considerations). This chapter has also put some considerable emphasis on rigour, grounded in pragmatic philosophy, to describe a general *process* of enquiry into a problematic situation. Rigour is interpreted here as the recoverability of the process of enquiry by anyone '*interested in critically scrutinising the research*' (Checkland, 2010, p. S42). This reminder of the qualitative, subjective nature of an interpretivist paradigm must also be accompanied with the appropriate criteria for judging knowledge claims, not an 'anything goes' mindset, i.e.

> '*[t]his means declaring explicitly, at the start of the research, the intellectual frameworks and the process of using them which will be used to define what counts as knowledge in this piece of research. By declaring the epistemology of their research process in this way, the researchers make it possible for outsiders to follow the research and see whether they agree or disagree with the findings. If they disagree, well-informed discussion and debate can follow.*'
>
> (Checkland, 2010, p. S42)

Whilst this originated from Checkland's reflections about SSM, the thinking is generalisable to other PSMs. In this regard, the epistemologies in Tables 5.2 and 5.4 provide complete coverage.

The general applicability of the CR/E/A framework to the context/choice question brings us to the level where it is now possible to restate a generic description for the context/choice process of OR[23] in Table 5.6.

---

23 Whilst it is tempting to declare this a metamethodology (following Jackson) or a logic (following Ormerod (2020b) and Peirce), perhaps these eight elements are nothing more than the means to capture a description of a generic process of enquiry for OR practice in terms that Dewey might have recognised? While other pragmatism inspired constructions are possible (see, e.g. Metcalfe (2008)) I would like to think that these eight elements guide a process of enquiry that has been '*lurking*' within OR all along, just not clearly visible.

Table 5.6 Generic description for the context/choice process of Soft OR.

Element	Description
1.	A generic process of enquiry in to a problematic situation animated by a set of constitutive rules (CR) conducted over a set of symbols relevant to the problem context that have been agreed, or are at least understandable, by the participants in the process.
2.	Constitutive rules are generative of behaviours, not prescriptive, allowing for specific processes of enquiry to be constructed to suit the context.
3.	The symbols used in the process are related or grouped into one or more models. The set of symbols and the rules for their combination can be described by the affordance (A) they provide to the modelling process and an epistemology (E), which is the language through which the process of modelling with the symbols used makes sense to the participants. The model may be more formal than a diagram and consist of a directed graph, either cyclic or acyclic.[a)]
4.	The meanings of concepts expressed using the symbols are contingent to the specific context. They may be interpreted differently at some later time by the same or different participants.
5.	The process of enquiry is not directed towards any particular higher goal. The meaning of comparators, such as '*better*', depends on context and interpreted/disputed by the participants in the specific process of enquiry. The meaning of comparators will almost certainly change over time.
6.	The selection of a symbol set and epistemology is matched to problem context, some combinations may be more or less suited to the problem at hand, but the choice is a matter for the participants in the process to decide.
7.	It is the specific combination of elements in the CR/E/A framework that defines a codified methodology and positions it within a particular community of knowledge.
8.	The intended outcome from the process of enquiry is a set of actions to improve the problematic situation, not the creation of knowledge per se. However, within the wider context of curating a body of knowledge that can be applied from one problem context to the next, i.e. within a community of practice, the constitutive rules, epistemologies and affordances (i.e. the CR/E/A framework) of the codified methodologies will over time be adapted and revised through the process of methodological learning.

a) It only requires the addition of a quantitative modelling approach here for the description to address Checkland's observation that a phenomenological process of enquiry can encompass quantitative modelling (1989b, p. 40). Also see the discussion about the capabilities of hierarchical process modelling (HPM) in Section 10.8.

6

The Practice of Problem Structuring

'In understanding how methods and people can be organized to achieve a desired transformation, it is not so much a question of what objectively can be said about competency and method; it is more a question of what one feels can be achieved – it is a judgement. The judgement reached depends on one's personal preferences in general and one's attitude to risk in particular. A project is after all a bet (pragmatically speaking). The project may not work out for any number of reasons and both clients and consultants will be risk averse. Clients will be worried that the project will fail in some sense and that their judgement in using consultants, and choosing these consultants in particular (you), will be proved wrong. In order to reduce the risk they will place considerable emphasis on the track record of the particular consultants who will do the job, their CVs, the recommendations of previous clients and so on .... However, they still find it difficult to assess whether consultants will perform as expected ...'.

(Ormerod, 2008, p. 1441)

6.1 Recap

The engagement context for a problem structuring method (PSM) is always unique; this naturally suggests that some flexibility in approach is required to deal with specific circumstances. This is the purpose of the constitutive rules described in Chapter 5, to be generative of behaviours suitable for addressing the problem context. We also need to learn from engagement to engagement, and that requires a certain amount of regularity in the process, not least so that as practitioners we are able to approach a problem with some degree of knowledge about how to proceed. Also, so that we can compare what we did on previous occasions from the point of view of reflecting on what worked, and what did not. Through this,

Problem Structuring: Methodology in Practice, First Edition. Mike Yearworth.

we build a coherent view of our practice, a form of action learning. Communities of practice can emerge and grow more effective through the sharing of knowledge. The core PSMs, as reviewed in Chapter 5, define specific methodologies – as codified around specific sets of constitutive rules, epistemologies and affordances of modelling approach – and collectively the field of Soft Operational Research (OR).

Rather pointedly, the generic constitutive rules presented in Section 5.2 exclude the notion of relying on situational logics. This puts Soft OR practice at variance with the greater successes of mainstream OR, but not so much as to be incompatible. It is almost certainly going to be the case that the use of any of the main quantitative OR methods (as listed in Section 1.4, or refer to Petropoulos et al. (2024)) will be clearly indicated in the process of problem structuring. The argument is not to exclude them, or reinforce the hard/soft divide, but to put problem structuring, the process of enquiry into a problematic situation, *first*. This steps outside some of the older debates about paradigm incommensurability that occurred when considering multimethodologies (Mingers & Brocklesby, 1997; Mingers & Gill, 1997) or simple sequential approaches (Pollack, 2009)[1] and also avoids the category mistakes and pathologies discussed in Section I.

By putting problem structuring front and central in any OR engagement and acknowledging that until we have a clearer idea about the nature of the client's problem context, the actual point of contact with the client and the '*owners*' of the problem is the first encounter with stakeholders' needs and the first opportunity to explore the co-creational nature of the problem, especially if wicked/messy. OR is, by its very nature, '*client oriented not solution oriented*' (Eden, 1989a, p. 43).

In terms of practical considerations, this chapter focuses on four key aspects (i) establishing a trust relationship with the client, (ii) dealing with the value paradox, (iii) identifying relevant stakeholders and dealing with multi-organisational contexts, and (iv) being ready to facilitate a group workshop. From the point of view of being sufficiently prepared to use any PSMs for a specific engagement, these four considerations roughly equate to the necessary processes that are required to bind the practitioner to leading one or more workshops with a client for the purpose of developing actionable interventions. Whilst phrased commercially, by using the language of consultant, client, engagement, contract and so on, these processes are relevant to any sort of context where PSMs are used, i.e. within collaborative projects, internal task forces and project teams, with communities, pro bono, etc. These four aspects also need to be considered as an adjunct to the CR/E/A framework presented in Chapter 5; they provide a processual '*wrapper*' for the methodological considerations to turn them into a practical way of working with clients. I should point out that these four aspects and the CR/E/A

1 I come back to this point in Chapter 7 in reviewing how an Engineering Systems programme dealt with the hard/soft question through a modelling approach described in Chapter 9.

framework are just my way of approaching the problem. There are other ways of looking at these practical considerations. For example, Ormerod (2008) presents the transformational competence perspective (TCP) as a model of OR practice that links the competence of the OR practitioner with the problem context, the methods they use and the transformation (in the sense of soft systems methodology (SSM)) that the client wants to achieve. Ormerod's perspective is a viewpoint borne out of extensive consultancy experience and reflection on what works in practice. On the other hand, Keys (2006) constructs a model of what is required to become an expert in the use of PSMs derived theoretically from Actor-Network Theory (ANT). Key's declarative model describes what is needed from the PSM consultant by way of intervention to achieve the requisite *translation* for the client. White (2009) takes this further by delving into a theoretical understanding of the phases of translation to be achieved during a PSM engagement. As originally defined by Callon (1984), these are – problematisation, interessement, enrolment, and mobilisation of allies. In effect, the co-identification of actors (stakeholders) and problematic situations (problematisation), negotiating actors' involvement (interessement), getting commitment from the actors to address the problematisation (enrolment), and getting buy-in from a broader stakeholding to cement the enrolment and achieve the translation.

6.2 Establishing a Trust Relationship

Since soft OR/PSMs moved away from the predict/prepare paradigm and the quantitative measure of an improvement or optimisation, the question about trust between a consultant and their client is not focussed on the verification and validation of the predictions produced by a mathematical model or simulation, but instead concerns the qualities of the practitioner to deliver a desirable and feasible intervention in a problematic situation. This requires a critical view on consultancy that is based more on the idea of a companion on a journey rather than a purveyor of solutions, and suggests that an appropriate perspective on trust is grounded in a social learning model (Nikolova & Devinney, 2012, pp. 396–398) and notions of demystifying expertise (Schön, 1983). Certainly, the cultural and cognitive theories of knowledge transfer that are traditionally attached to the management consultant and consultancies, where trust relationships are based on reputation and leadership in the prevailing techniques of the neoliberal reform agenda,[2] really do not seem to fit the PSM consultant at all (Faust, 2012). Except

2 Over time these have come in '*waves*' – Total Quality Management (TQM), Business Process Re-Engineering (BPR), balanced scorecard, lean... (Faust, 2012, p. 141) – and delivered as prefabricated, packaged consultancy solutions (p. 143), not unlike the situational logics discussed earlier.

that the reputational brands of consultancies are clearly suitable as a *vehicle* for delivering PSM techniques, should they choose to use them. Clients buy into relationships (rather than solutions) that give the consultancies a certain embedded access to '*early project development activities*' (p. 150) – just the sort of strategising, planning and designing of transformations that PSMs are well suited to address.

Blomqvist (1997) makes a broad review of trust across economics, law and contracts, and marketing – which all seem relevant to the realm of the OR consultant – and builds a conceptual framework that reduces to a two-dimensional working definition for business contexts as '*an actor's expectation of the other party's competence and goodwill*' (p. 282). Here competence is understood as ranging across technical expertise, skills and know-how, and goodwill captures notions of moral responsibility and positive intent. For an established relationship between a consultant and client, or an established reputation, expectations have at least been set or can be judged; but how are these expectations to be evaluated in the case of new relationships, especially one that is to be based on an open-ended exploration of a problem context through the use of PSMs? For the PSM consultant, technical expertise and know – how could be interpreted methodologically and skills might be viewed as the practicalities discussed in this chapter, not least in the process of facilitation considered below.

Castaldo, Premazzi and Zerbini (2010) provide a more operational view of trust with a model constructed from a network analysis of the co-occurrence of terms associated with trust issues published in a broad range of management literature – marketing, organisation, strategic management, psychology, and sociology. Their model has a processual interpretation with trust mediating the performative elements of acting and delivering outcomes. It is not too difficult to see their model as having a bearing on a number of dimensions that will impact the trust relationship between a consultant and client, not least the industry sector and domain knowledge, relative power, incumbency or being a newcomer, and past relationship history. Their model also introduces the notion of risky situations as another mediating factor, suggesting that the messiness of the problem context has a bearing on trust. For the PSM consultant this raises the question of openness about risk. As discussed in Chapter 1, it is a lot easier to quantify the value of problem solving than problem structuring. Does a PSM consultant build more trust by being honest with a client about the messiness of their problem context than a consultant that promises a solution?

Once the use of a PSM has been decided – i.e. the consultant–client relationship has already been formed – then there is evidence to suggest that the methodology itself feeds into the maintenance and further building of trust. For example, in the case of quantitative simulation models (Tako & Kotiadis, 2015; Harper, Mustafee, & Yearworth, 2021), addressing science and technology conflicts (Cronin, Midgley, & Jackson, 2014), community participation in reforming local government (Laouris & Michaelides, 2018), and enabling inter-agency working (Sydelko, Midgley, & Espinosa, 2021).

6.3 Dealing with the Value Paradox

Ormerod (2022) presents a general review of the economic logic of OR practice. Through a examination of transaction cost theory, agency theory, signalling theory and the more sociological embeddedness theory, Ormerod outlines a set of economic foundations that underpin OR practice, observing that

> *'[c]onsultants need to develop and maintain expertise in the commercial aspects of interventions as well as their expertise in the conduct of the analysis process. Thus, the examination of the commercial aspects can be seen as elucidating one aspect of OR practical expertise.'*
>
> (Ormerod, 2022, p. 697)

These foundations are the characteristics that matter in the establishment of a commercial relationship between a consultant and their client. Ormerod identifies eight specific characteristics as follows: (i) choosing a consultant is bounded by incomplete knowledge about their capabilities, (ii) choosing is always a subjective process, (iii) the relationship becomes grounded in a contract for service, (iv) innovating will lead to substitute offerings (e.g. data analytics) and/or new entrants, (v) clients will be choosing between embedding or outsourcing, (vi) seeking to take advantage of divisions of labour, (vii) signalling, and (viii) emerging new technologies[3] will provide opportunities on both sides of the consultant/client relationship with a corresponding repositioning of value.

Whilst Ormerod's foundations are applicable to OR practice generally, the rather mundane question of how much a contract is *worth* deserves some attention. Clearly, if a consultant is contracted on an hourly or daily basis, then that *is* the value of the consultant as judged by the paying client. Paying by the hour or day also provides a convenient flexible boundary for the client to manage the duration of the engagement. A more elaborate contractual basis would be a fixed price, where the consultant bears the downside risk that the amount of work specified in the contract is inadequate to deal with the problem to the satisfaction of the client, and thus the consultant must provide extra work for free. For an engagement grounded in the situational logic of a well-known problem and the predicted value of a delivered optimisation then a fixed rate contract might well be

3 The recent emergence of Large Language Models (LLMs – e.g. ChatGPT (OpenAI, 2023)) is a potential example of what Ormerod means by AI in this context. I discuss the possible impact of LLMs on PSM practice in Section 15.2.

suitable. However, what kinds of engagements involve clients with more complex problems? Problem contexts whose messiness only becomes revealed through the process of engaging with it and structuring? Clearly, the low-risk option for the consultant is hourly or daily billing, but is this fair?

The open-ended nature of problem-structuring engagement makes it difficult/impossible to value. I discuss evaluating PSM engagements further in Chapter 12, but here I want to limit the question to the '*value paradox*' first introduced in the observation by Checkland and Scholes (1990, p. 299) that trying to decide whether a PSM engagement *worked* is an undecidable question.[4] Tully, White and Yearworth (2019) dig into this question at length from the perspective of a management consultancy struggling to find the most profitable way to value a PSM delivery *capability*. Dealing with this paradox is at the heart of establishing a consultant/client relationship. Sufficient trust needs to be built so that an engagement can proceed. However, the process to establish that trust involves, to a certain degree, the consultant starting to (re-)problematise the client's requirements or brief. Consultants/consultancies that have an established relationship based on reputation and are already embedded with a client can trust that their problem structuring will lead to further work. Consultants trying to win a new client will have to do this initial problem structuring *at risk*.

The original version of the value paradox set out by Tully, White and Yearworth (2019) was that to sell a PSM engagement contract requires problem structuring *before* the contract can be established. The origins of the paradox were based on the work of Akerlof (1970). In the *market for lemons*, Akerlof argues that in the formation of any contract, there will be an information asymmetry about the quality of the offering and used the example of the used-car market and the problem of car dealers withholding knowledge about '*lemons*'[5] from their customers. For the PSM consultant trying to establish an engagement with a client, sufficient trust will have to be formed in an initial phase of relationship-building that will inevitably involve a degree of problem structuring, just for the very process of getting an understanding of the client's requirements. This informal process might involve short workshops and meetings with some of the stakeholders (Taylor, Tofts, & Yearworth, 2004). Some of the early ideas about the value paradox emerged from work trying to win business – but not solve the client's problem '*pre-sales*' – in the commercial context of a client presenting a '*simple*' problem that attracts a low price, but where the client's problem is in fact a lemon (i.e. a mess). For a large European defence aerospace company, this was the problem of not knowing what contracts were likely to turn out to be lemons. They wanted to work this

4 The question of '*worked*' from a counterfactual perspective. Was the use of a PSM better than any other technique that could have been used, or, indeed, just by muddling through? Given the unique context of messy problems such questions are not answerable.

5 Cars with known faults or defects.

out pre-sales to avoid bidding, rather than winning contracts and ending up with projects on their books that were going to lose money. For the case of a global IT company, the presence of a powerful intermediary trusted by clients forced the '*simplification*' of the clients' needs to bring about an apples-for-apples market for comparing suppliers' offerings; in effect simplifying clients' messes so that suppliers were steered into selling packaged solutions, but liable for dealing with messes later. Ideally, in each case, the use of pre-sales problem-structuring workshops would help avoid such messes.

Eden's account of his development of cognitive mapping, and hence, strategic options development and analysis (SODA) could be viewed as an account of how to build trust between a consultant and their client, especially in the early phases of a relationship (1988). However, how much knowledge will be gained – i.e. information asymmetries reduced – through these activities prior to any contract formation is difficult to judge and presumably down to the skills of the actors involved. In an experimental setting using group model building, not live consulting, there have been positive results showing that participation in the model building process itself can reduce information asymmetries (McCardle-Keurentjes et al., 2018).

The value paradox can be understood in the sequence of four scenarios shown in Figure 6.1 for the consultant trying to sell a PSM-based engagement to a client (Yearworth et al., 2018). In scenario 1, a consultant bids against an original problematisation from the client – as specified in the brief or requirements document – without any further attempt at engagement with the client to ascertain knowledge about the problem context. This is an unlikely scenario of interest to a PSM consultant, therefore scenario 2 shows a more realistic situation where the consultant engages in a degree of problem-structuring with the client *pre-sales* to obtain a better idea of the problem context and, perhaps in the process, helps the client better understand their problem. The consultant then bids a more informed proposal than imagined in scenario 1. However, the consultant runs the risk that the pre-sales re-problematisation for the client was sufficient for the client to either (i) not proceed and reject the bid (scenario 3) – perhaps because they (now) realise that the original requirements were no longer relevant or (ii) worse for the consultant,[6] that the re-problematisation has meant that another consultant is now relevant or preferred (scenario 4).

The analysis of the paradox by Tully, White and Yearworth (2019) not only flagged the risk that PSM consultants face in working with clients pre-contractually (pre-sales), but also found a way of addressing the undecidability problem – at least for consultancies able to conduct A/B-type testing across a broad range of their clients. Based on the idea of using pre-sales problem

6 It is one thing to lose business, it is another to actually make life easier for your competitor(s). Unless, of course, it is clear that the prospect is a '*lemon*'.

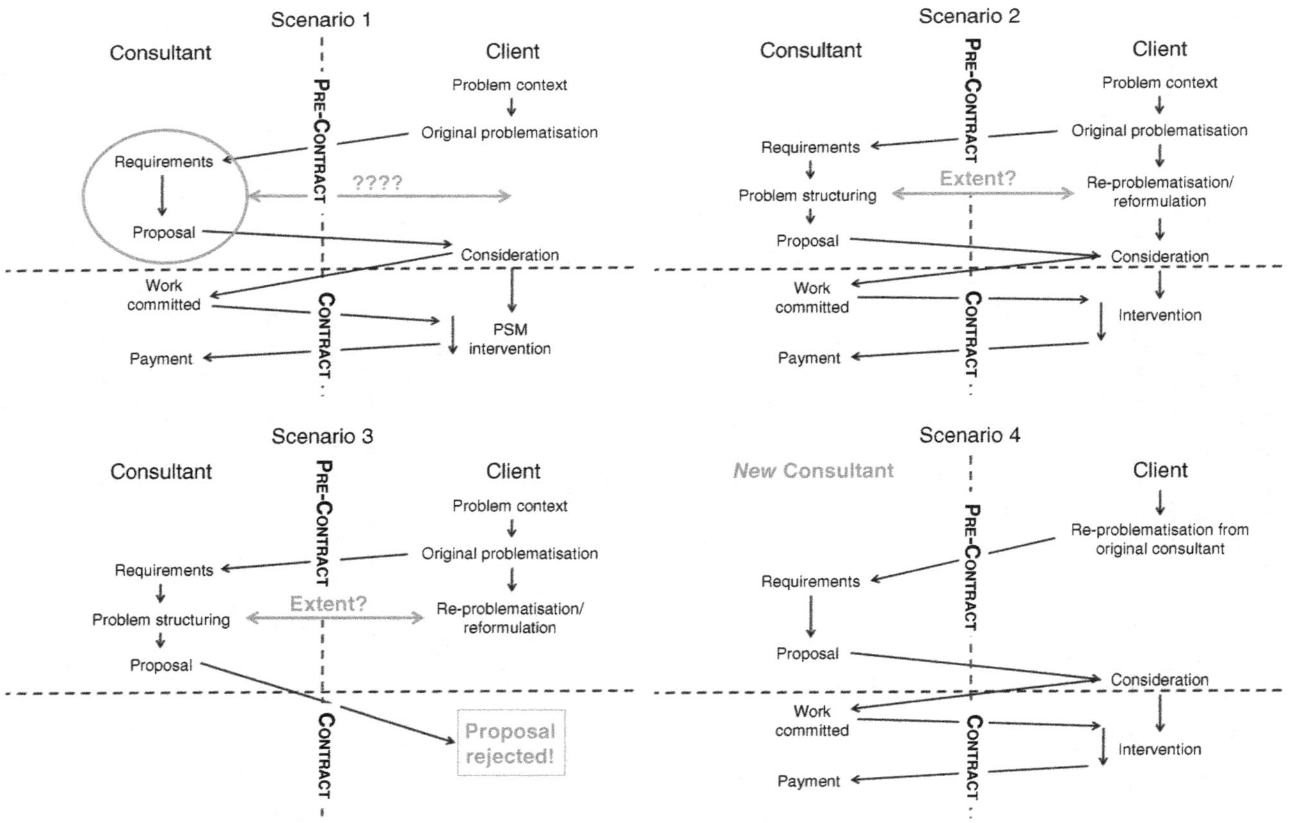

Figure 6.1 Engagement scenarios arising from a consultant's pre-contract problematisation proving to be of value to a potential customer, potentially leading to a proposal for work being rejected and a new consultant being selected – to work on the original consultant's (free) re-problematisation.

structuring workshops as a deliberate differentiator (e.g. see the Rapid Scenario Planning (RaSP) proposal by Taylor, Tofts and Yearworth (2004) and Taylor and Tofts (2008)) and assuming the ostensible purpose would be trying to avoid scenarios 3 and 4, there might be a return on investment argument that can be made for determining how much at-risk pre-sales problem structuring to do with clients. By measuring the cost of sales, success rates and project outcomes from pursuits with, and without, pre-sales problem structuring workshops with clients two testable null-hypotheses are presented[7]:

1. H_0 – Pre-sales PSM workshops have no impact on win rates (either positively in the case of scenario 2 or negatively in the case of scenario 4);
2. H_0 – Pre-sales PSM workshops have no impact on intervention delivery issues (i.e. they make no change to the rate of acquiring loss-making projects).

These propositions would be eminently testable by consultancies with sufficient volumes to risk the deliberate introduction of formalised pre-sales problem structuring workshops, for some customers, as part of their offering. However, until such work is undertaken, these remain untested and the value paradox remains a conundrum for the PSM practitioner.

6.4 Identifying Relevant Stakeholders and Dealing with Multi-organisational Contexts

Ackermann and Eden (2011b) present a detailed analysis of stakeholder engagement practices based on an action research (Eden & Huxham, 1996) with access to an organisation's data from client engagements spanning 15 years. Grounded in a stakeholder theory perspective (Freeman, 1984), their analysis was conducted of the three main techniques that were used in the organisation as the processes for identifying and managing stakeholders: (Technique A) the power-interest grid, (Technique B) the stakeholder-influence network and (Technique C) the stakeholder management web. The presentation of the key conceptual contributions arising from their analysis is made *processually* here and grouped accordingly as follows: 1. Working at a disaggregated level of stakeholder identification (A), 2. Separating power from influence and identifying '*players*' (A), 3. Identifying stakeholders at the nexus of the influence network (B), 4. Identifying formal and, significantly, informal relationships (B), 5. Maintaining confidentiality (B), and 6. Using the stakeholder management web to surface manifest forms of power and

7 Post sales, a third testable null hypothesis is suggested: H_0 – Post-sales PSM workshops have no impact on the levels of trust that exist between a consultant and client. If rejected, the Harper, Mustafee and Yearworth (2021) facets of trust model (for example) would be supported.

Table 6.1 Characterisation of modes of engaging stakeholders in OR.

OR tradition	Stakeholders' perceptions of the situation	Stakeholders' goals	Desired outcome from OR study
Optimising	Homogeneous, known	Homogeneous, known	Implementing an optimum solution
Balancing	Homogeneous, known	Homogeneous, partly known	Achieving compromise between conflicting goals
Structuring	Incomplete, idiosyncratic	Homogenous, partly known	Clarifying issues and actions, creating commitment to action
Involving	Incomplete, idiosyncratic	Conflicted	Managing conflicts (not necessarily consensus) and commitment to action

Source: Adapted from and extending de Gooyert et al. (2017).

bases of interest (C). Whilst oriented specifically towards strategising, the findings from their research are nonetheless applicable across a broad range of PSM-based consulting engagements.

de Gooyert et al. (2017) also develop a stakeholder theory perspective on the main modes (traditions) of engaging with stakeholders employed by OR practitioners based on a content analysis of the OR academic literature that specifically mentions dealing with stakeholders. They identify four modes that they associate with optimising, balancing, structuring and involving. Their framework is summarised in Table 6.1.

Roughly equating these four modes with (i) using standard OR optimisation techniques, (ii) multi-criteria techniques, (iii) problem structuring techniques and (iv) managing boundaries (critical techniques), we see reasonable agreement with the ontological position set out in the SOSM discussed in Chapter 2. The framing of the position of the PSM consultant in relation to stakeholders is clearly one where the first three cases in Table 6.1 are covered. Whilst the stakeholders (as classified) seem to be a given, the actual identification of stakeholders in the sense of searching them out as part of the process of engagement with a client is not addressed. How this might be approached is set out by Wang, Liu and Mingers (2015) who define a systematic process based on the use of SSM – in effect, through a process of developing activity system models to achieve strategic transformation and *recursively* identifying relevant stakeholders for key sub-activities (p. 567). They observe that

> *'[t]o some extent, an organisation tends to consider the stakeholders from an "instrumental view", to concern those stakeholders who have power or directly affect the business operations, such as shareholders. However, a single*

> *instrumental perspective may result in ignoring some stakeholder groups which have relatively less power or do not directly affect the operations, for example nearby residents of a chemical plant. Thus, the organisation not only needs to achieve its strategic objectives through a series of key activities which are including management and even some supporting activities, but also needs to consider their roles in moral and social responsibilities.'*
>
> (Wang, Liu, & Mingers, 2015, p. 563)

Despite elaborate schemes such as this to identify relevant stakeholders, there still remains the problem of group homogeneity. Cunico, Zimmermann and Videira (2024) propose a new facilitation technique, the *'New Devil's Advocate'* (NDA), where a facilitator deliberately drops a neutral position with respect to the problem context and attempts to fulfil the role of *'missing stakeholders'*. The problem of identifying wider stakeholder constituencies also needs to be addressed. Gregory et al. (2020) present a framework for stakeholder identification and reporting based on the use of critical-systemic perspective and social identity theory. Developed in the context of community OR engagement, their framework nonetheless has applicability across OR generally.[8] Aubert and Lienert (2024) review the processes involved in working with citizen stakeholder groups. Müller, Groesser and Ulli-Beer (2012) describe an iterative process for assembling appropriate expert groups.

Dealing with multi-organisational groups also brings a unique set of challenges concerning stakeholder identification. Of course, the strategic choice approach (SCA) was designed with multi-organisational planning groups in mind right from its inception (Friend, Power, & Yewlett, 1974; Friend, 1993), whereas SSM and the SODA project were developed more with single organisations in mind as clients. However, multi-organisational collaborations have become the norm (Ackermann et al., 2005), and group support systems (GSSs) that support problem structuring, such as Decision Explorer/Group Explorer and now Strategyfinder, make multi-organisational collaboration a little easier when the burden of travelling to workshop locations[9] can be removed if desired. Despite the positive contribution that GSSs can make, the consultant's decision-making about stakeholder involvement is clearly more complicated. Is there a lead organisation in the collaboration that is defining membership, or is it a collaboration of equal peers? The powers of decision-making or authority to commit their organisations to action may be unequally represented by participants (Ackermann et al., 2005, p. 324), or completely misunderstood by the participants and facilitators alike (Freeman & Yearworth, 2017). The task of facilitation is also made more complicated by the need to additionally play the roles of *'arbitrator, referee, moderator, etc.'*

8 See, for example, the teaching example in Section 10.3 and its underpinning case study.

9 i.e. Same-time/different-places workshops.

(Ackermann et al., 2005, p. 324). Their enhanced multi-organisational collaboration teams (MCTs) framework (p. 325) provides a comprehensive map of considerations and could be interpreted processually as a purposeful activity model for achieving effective multi-organisational collaboration.

6.5 Being Ready to Facilitate

The process of facilitation translates the methodological knowledge in a specific embodiment of the CR/E/A framework into practical action with a stakeholder group in a workshop. The nature of the process can be understood from a number of perspectives that are interrelated and interdependent. I treat facilitation primarily as an observable process that can be analysed for the purpose of learning and as a skill that can be acquired and developed into expertise. Theory then provides additional insights into the micro-processes of facilitation for the purpose of improving; for example, processes of deconstruction and co-counselling (Taket, 2002), conversational processes (Franco, 2006), facilitators' questioning behaviours (McCardle-Keurentjes & Rouwette, 2018), and on the conversational object '*formulation*' (Franco & Nielsen, 2018). The range of theories relevant to facilitation and the ongoing interest in the analysis of micro-processes are likely to lead to the introduction and use of further theoretical perspectives on facilitation performance.

Observing facilitation in practice requires ethnographic techniques applied to the collection of data concerning the interactions between the facilitator and participants in a workshop in order to reveal the micro-processes in operation. Such fine-grained data enables temporal analyses to begin to disentangle the interwoven nature of process and content. Workshops based on the use of SODA methodologies have benefited from GSS from almost the beginning.[10] Recent analyses of the micro-processes of SODA methodology workshops have made use of the data logging facilities in the GSS to match statements made by participants (through the '*Chauffeur*' interface of the GSS) with audio and/or video recordings of the workshop in progress. For example, Franco and Greiffenhagen (2018) conducted a '*moment-by-moment*' analysis of a group workshop using the ethnomethodology of Garfinkel (1996). Such a fine-grained analysis technique reveals the *doings* of an OR engagement that is able to specifically address questions about the process of facilitation – i.e. what actions the facilitator is making at every moment of the process. They contrast this ethnomethodologically informed way of learning from following scripts, but also comment on how resource intensive it is to conduct such analyses.

10 COPE and then Decision Explorer/Group Explorer.

Yearworth and White (2017) used a similar analytical technique for an online same-time/different-places workshop to examine how the online setting '*de-centres*' the workshop facilitator.[11] This repositioning was an attempt to '*demystify*' the role of the facilitator. They observed a facilitator spending time managing the overall workshop process and then stepping-back so that the participants could concentrate on using the platform to work on the content of the task. This clear demarcation of the facilitator role with respect to process and content stimulated ideas of enabling groups to work autonomously (of facilitation) in online settings if given sufficient '*scaffolding*' in the methodology. Some of the practical implications of this are discussed further in Chapter 14.

An alternative approach to researching the process of facilitation and learning is set out by Ackermann (1996). Building on extensive experience of engagements around the use of the SODA methodologies, Ackermann conducted a qualitative analysis of surveys administered to workshop participants and therefore provides a different perspective on facilitation from that obtained by researchers observing the process itself. The survey data covered facilitation during workshops as well as the preparation by the facilitator prior to the workshop and post workshop actions. The analysis elicited 26 learning points that can be used as a '*useful checklist, either for reinforcing existing practices, or adopting new ones*' (p. 109). Using the modelling approach developed by Lowe, Espinosa and Yearworth (2020), it is entirely possible to interpret these learning points *processually* as a model for ensuring the quality of facilitation[12] and is presented in Figure 6.2. The model can be evaluated for performance by participants and/or facilitators using the measurement schema set out in Section 9.3.

The use of predefined scripts in PSM workshops is a direct result of the influence of group model building in System Dynamics (see Section 5.8). The use of scripts by novice facilitators was analysed by Tavella and Papadopoulos (2015a) with a view to understanding how they enabled the consultant to switch between the acquisition of skills and competences, and managing the process of the workshop in order to achieve desirable outcomes for the client. Further work by Tavella and Papadopoulos (2015b) on comparing expert and novice facilitators found common facilitation patterns, refuting prior claims that they use knowledge differently and lending empirical support for the notion of being able to teach facilitation skills to clients – thus enabling them to '*self-facilitate*'.

11 The novelty of not having the facilitator appear '*in front*' of the workshop participants as in the usual setting for a same-time/same-place workshop provoked a new perspective on analysing the micro-processes of the group workshop.

12 This is made particularly straightforward for hierarchical process modelling (HPM) – as described in Chapter 9 – by the use of gerunds in the summary of learning points (Ackermann, 1996, p. 95).

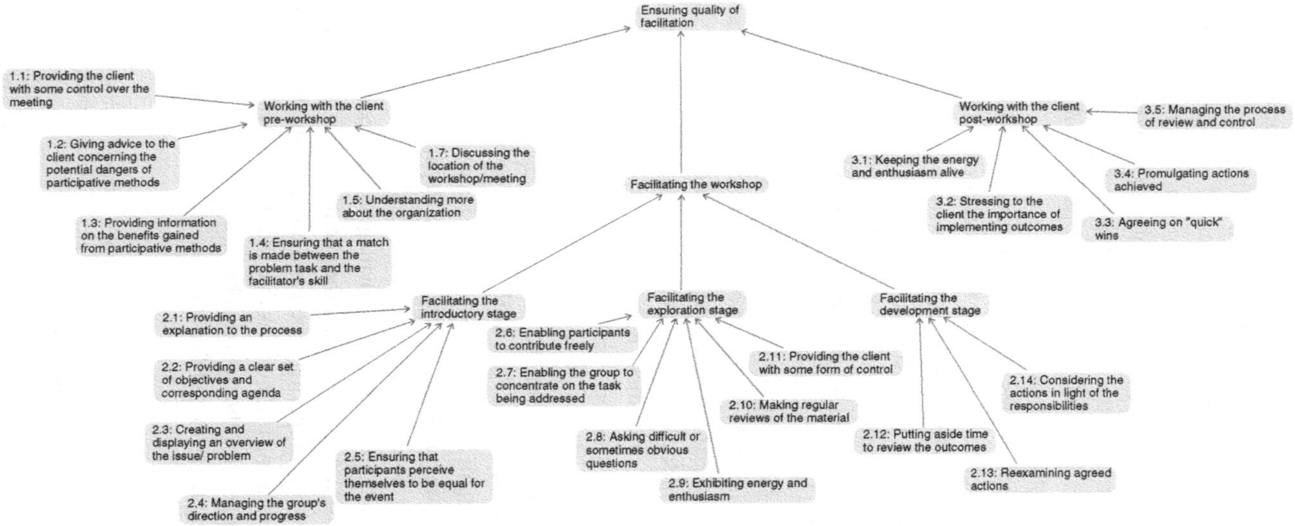

Figure 6.2 An interpretation of the learning points on facilitation (Adapted from Ackermann, 1996, p. 95) as a process to achieve the purpose – ensuring the quality of facilitation. The numbering of processes matches the original model.

There is an interesting tension emerging here between viewpoints that are advocating a de-centring of the facilitator role and alternative views where the facilitator takes a contributory position with regard to content, as in the NDA role described above. This is picked up again in Chapter 14 when considering the role of GSSs in problem structuring.

6.6 Ethical Practice

The review of ethics by Ormerod and Ulrich (2013) leads us to the conclusion that OR should be viewed as a discipline that must be thought of as a humanity. It is not disconnected from practice in the same way as a '*pure*' science. If we choose to define our profession as the '*science of the better*', then we should have a very good understanding of what better means. Better for whom? Who decides what is better? Who has a voice? It is therefore up to the practitioner when working with their client to decide what is an ethical outcome from the intervention, i.e. the meaning and likely impact, the human '*cost*', of any planning, organising and strategising. This cannot be left to the function of any algorithm. Questions of bias need to tackled head-on with a critique of the actions of the practitioner, not as an exercise in debiasing algorithms. If the optimisation algorithm is likely to be deployed in ways that have a negative impact on some stakeholders, then was the practitioner aware that was the intended purpose of their work? As argued by Ormerod, Yearworth and White (2023) the practitioner cannot, and must not '*opt out*'. There will always be a place for inventing, designing, improving, crafting etc. the techniques, models, algorithms, and tools for optimisation and this activity can take place away from the real world working on well-known or test data sets i.e. on solving puzzles. However, engineers, operational researchers, applied social scientists are all practitioners. Their practice has a direct impact on the world, on people's lives. As argued in Chapter 11, this is so much more than practice that is simply not-malpractice. Avoiding malpractice is obviously and clearly a minimum standard. Nor is not-malpractice the same as '*best practice*', although for complicated, known problems (see Section 2.5) this is perhaps enough, i.e. practice according to situational logics. Problem structuring, as a means of conducting OR engagements, is therefore the process by which ethical consideration becomes embedded in everyday practice.

6.7 Relation to the Wider PSM Field

The practical considerations covered in this chapter have mostly addressed the issues associated with turning knowledge of the CR/E/A framework into the delivery of a PSM-based engagement process with a client. This focus has come at the cost of not considering the implications on a number of related fields, such as Community OR (Jones & Eden, 1981; Parry & Mingers, 1991; Johnson et al., 2018), Behavioural OR (Franco & Hämäläinen, 2016; White, 2016); use of less common PSMs such as value-focused thinking (VFT) (Keeney, 1996; Françozo & Belderrain, 2022), drivers-pressures-state-impact-response (DPSIR) (Bell, 2012; Gregory et al., 2013), visioning choices (O'Brien & Meadows, 2007), and decision analysis (Rosenhead & Mingers, 2001b, pp. 267–288); systems approaches used in a PSM sense (e.g. viable systems model (VSM) (Lowe, Espinosa, & Yearworth, 2020)); the relationship to multi-criteria decision analysis (MCDA) (Marttunen, Lienert, & Belton, 2017); and, finally, multimethodology (Mingers & Brocklesby, 1997) and examples such as partisim (Kotiadis & Mingers, 2006; Tako & Kotiadis, 2015). Some of this deficit is rectified in the survey of PSM applications in Chapter 13. However, the process of constructing the epistemology and affordance of modelling approach within the CR/E/A framework for the methodologies covered so far (SSM, SODA project, SCA, issue based information system (IBIS) and causal loop diagram (CLDs)) is easily extended to other approaches, as is demonstrated in Sections 9.4 and 9.5 for hierarchical process modelling (HPM).

6.8 Summary

Developing a trust relationship between a client and their consultant, navigating the value paradox, identifying and engaging stakeholders, being proficient in the facilitated delivery of problem structuring workshops and practicing ethically constitute a *processual* view of Soft OR practice. On becoming an expert in the use of PSMs, Keys recognised that the idea that a successful problem structuring engagement would be reflected in a more favourable configuration of the actor-network.

> *'The degree of detail that experts use to distinguish between the meanings associated with aspects of situations is much greater than that used by novices. That is, experts are more aware and better able to appreciate and respond to subtle points and qualities. When using PSMs this is critical, for effective interpretation of meaning in sophisticated client behaviour and language will influence the design and content of an intervention. As interventions proceed, the model(s) constructed change in response to events and new learning, irrespective of whatever method is being used. That is, change in one part of the network constructed by an analyst leads to change in other parts as the user maintains its cohesion and internal consistency. Expert users are adept at recognizing how these changes occur and interpreting their impact upon the model and any analysis it informs. As novices become expert in designing and*

> *managing actor networks for interventions they learn to recognize slight shifts in the elements of that network that may lead to its destabilization and how to correct the network in response to them.'*
>
> (Keys, 2006, p. 826)

The demands of developing skills in the use of PSMs prompted Ormerod to revisit his Transformation Competence Perspective (Ormerod, 2008) with a revised schema for the core competences of OR practice (Ormerod, 2014). This comprehensive review produced a useful summary of the unique character of PSM practice within the broader OR field and that

> *'[o]ne recurring theme in the PSM literature reviewed here is the requirement to be a skilled facilitator (sub-competence: facilitation) and being able to recognize when it could be useful (sub-competence: mental models). More generally the competence, designing and managing the intervention process, is critical for all OR, traditional or soft. Facilitation is one distinguishing feature of PSMs. Modelling (usually quantitative) is one of the distinguishing features of traditional OR. We have seen here that PSMs require mastery of new (often qualitative) analytical methods (sub-competence: problem structuring), and an increased emphasis on contextual understanding (competence: appreciating context) and ethical concerns (sub-competences: personal and professional ethics; ethics in analysis, intervention ethics, ethics and society). These new emphases are already having an impact on the way that more conventional approaches of traditional OR are conceived and understood. On the one hand, distinctions between traditional and soft/PSM approaches are apparent, while on the other the boundaries between the two are blurring.'*
>
> (Ormerod, 2014, pp. 333, 334)

Despite Ormerod's optimism that boundaries are blurring, there is still a sense that in Soft OR/PSM consulting the idea of the consultant as a companion on a learning journey – skilled in process, of course – is still a radically different proposition to hiring a consultant who knows how to deliver a solution according to some situational logic. Conventional views of the expertise of management consultants can read almost like through-the-looking glass accounts: '*faking it so that they can make it*' as Bourgoin and Harvey (2018) summarised their study to a general management audience. The particular nature of PSM consulting is still not widely understood outside of operational research. The specific circularity of the process of deciding messiness as bound up with the process of addressing such messes complicates both analysis and an appreciation of its value.

6.9 State of the Art

The historical review of the emergence of the core PSMs in Chapter 4 finished at the end of what I characterised as the first reflective period, at roughly the end of the first decade of this century. At this point, the core methodologies were generally known about and in a relatively stable form, as published in the standard texts. Reviews in the OR literature had established their broad applicability. Chapter 5 provided a comparison of the core PSMs using a novel analysis based on the CR/E/A framework leading to a general statement of the process of problem structuring in OR. This chapter has focussed on the aspects of turning methodological knowledge into practical application with clients. These three chapters in Section II have provided an overview of the state of the art as I see it. The emergence of a fully *processual* view of OR practice, not just Soft OR, can be glimpsed through the reinterpretation of the learning points from Ackermann (1996) as a process model. The revised schema for the core competences of OR practice (Ormerod, 2014) could be re-interpreted processually without too much work in much the same way as the pragmatism viewpoint from Ormerod was interpreted as a process model in Chapter 3.

Of course, other researchers have set out different opinions about the methodological construction of PSMs and future developments in a number of recent reviews. For example, see Smith and Shaw (2019), Gomes Júnior and Schramm (2021), Franco and Rouwette (2022), and, specifically on Soft OR, White (2025). Ackermann et al. (2020) set out their view of the teaching of Soft OR to students. There have also been renewed questions about their direction of travel. The viewpoint piece by Harwood (2019) and the responses from Lowe and Yearworth (2019) and Ackermann (2019) were existential in nature, but it was an important exchange in that it flagged the problem of visibility of PSMs and PSM practice in the main OR literature as judged by publication volumes. However, whilst this may be an issue for PSM scholars, it was not a reflection on the value of PSMs or their prevalence as measured and judged by other means than the academic literature.

Section III

Problem Structuring in Engineering Practice

Having set out the Soft OR/PSM landscape in the first two sections, Section III concerns the development of a problem structuring sensibility in engineering practice. The motivation for inclusion is that the narrative shows similarities to the emergence of Soft OR/PSMs and thus suggests that disciplinary splits of this kind will be observed whenever a STEM-based subject begins to engage with wicked/messy/swampy problems of practice. In the case presented, the situation led to the development of a new modelling approach in hierarchical process modelling (HPM) that has the potential to avoid some of the paradigmatic wrangling that was attached to the emergence of Soft OR outlined in Chapter 4. Using HPM as a PSM is explored in full and leads towards the conclusion that this new processual turn in engineering has implications for OR practice.

Chapter 7 charts the emergence of a problem structuring practice in engineering originating from the initial development of HPM and then the award of grant funding for an Industrial Doctorate Centre (IDC) in systems and an action research programme of over 400 person years of doctoral work. A new contextual model for engineering practice is introduced. The narrative in Chapter 7 can be interpreted as an account of a similar process that led to the development of SSM.

Chapter 8 presents how this new problem structuring sensibility has implications for the theoretical foundations of Systems Engineering.

Chapter 9 introduces HPM as a new PSM. The modelling approach starts with a strong process epistemology and focusses on eliciting process models of entities enacting a purpose through a how/why questioning dialectic. A measurement process is introduced that explicitly accounts for uncertainty – in knowledge about process performance and enactment. Finally, argumentation based on IBIS ideas is included. Using the CR/E/A framework, similarities between this new PSM and

Problem Structuring: Methodology in Practice, First Edition. Mike Yearworth.

the core methods are discussed. One of the strengths is the potential for the wider application of HPM to problems of strategising, planning and transformation.

Chapter 10 provides examples of HPM used as a PSM through three case studies and three teaching examples.

Chapter 11 discusses the implications on practice of this new problem structuring sensibility. Complementary to Chapter 6, the approach here is grounded in asking questions about engineering '*know why*' and based on notions of anticipation, injunction and translations, widening participation and broadening scope, moving away from singular outcomes, working with multiple viewpoints, being aware of conceptual limitations and valuing process knowledge as much as content (engineering '*know how*').

7

Developing a Problem Structuring Sensibility in Engineering

'A methodology I take to be an explicit, ordered, non-random way of carrying out an activity. As such it is independent of the content of the activity and can be considered separately from content, although in practice this is not easy to achieve: most people, asked to describe how they tackled a problem will quickly slip into describing the content of the problem itself. The reason is that methodology, once adopted, tends to become invisible; it becomes, in both a logical and a behavioural sense, simply the way the activity is carried out, and is taken for granted. Most practising scientists and engineers show little interest in methodology because the pattern of their activity, both in its logic and its displayed behaviour has long been established.'

(Checkland, 1972, p. 93)

7.1 Introduction

I start with the same question I ask of OR practitioners at the start of Section I. What is it that engineers are doing when they are dealing with a complex problem? I take a methodological viewpoint, grounded in the *actions* of practitioners dealing with the complexities of the world, not as something that is taking place in a laboratory, and attempting to reveal what might be taken for granted and implicit. We should be able to observe the effects of engineers structuring problems empirically in much the same way as OR practitioners, through their actions as *interventions* in messy problem contexts.

This section is an account of what a specific group of engineers has been doing when they have been bringing a problem structuring sensibility to their practice. It is grounded in practice, not a theoretical treatise, and builds towards a methodological interpretation through the emergence and development of a new PSM. It presents a summative, reflective account of the results from the EngD in Systems

Problem Structuring: Methodology in Practice, First Edition. Mike Yearworth.

Programme at the University of Bristol from roughly 2009 to 2020. No doubt other accounts will present different interpretations reflecting different research perspectives and traditions.[1]

7.2 From Systems Engineering to Soft Systems Methodology – and Back Again

The journey that Peter Checkland and his colleagues undertook at the University of Lancaster in the 1970s in applying the methods and tools of Systems Engineering, as they were known at the time, to complex problems in business and management directly led to the development of soft systems methodology (SSM). This *action research* program (Checkland, 1972, p. 93) was an effort that spanned roughly 10 years and over 300 engagements (Checkland, 1981, 2010, 2012). The diffusion of SSM has played out over many years and has been applied in a range of disciplines, but the key *methodological* publications have largely been in Operational Research (OR) or systems journals. Its eventual *adoption* as a Problem Structuring Method by the Soft OR community has meant that this methodological knowledge has largely remained within an academic community centred within Management Schools, and largely in the United Kingdom. As a body of knowledge, and practice, it has on the whole not really found its way back into engineering. It has remained a project outside of engineering despite the obvious links to its origins.[2]

The emergence of a *soft systems* sensibility in the Department of Civil Engineering at the University of Bristol must be viewed as a parallel development sharing no common intellectual roots (Dias, 2013). The work of Blockley, Davis, and Hall in the 1990s provided the basis for the publication of *Doing it Differently* by Blockley and Godfrey (2000), motivated in part to address problems in the construction industry and the publication, by the UK Government Department for Trade and Industry, of the Egan Report – *Rethinking Construction* (Egan, 1998). The former set out a practically oriented guide to a systems approach for the student Civil Engineer at a Masters Degree level. Further work in the subsequent decade by (i) PhD students Rosenberg (2003), Marashi (2006), MacDonald (2010), and others,

---

1 See for example, the valedictory impact report (University of Bristol & EPSRC, 2020) and Blockley & Godfrey (2017), oriented towards developing a systems capability for rethinking infrastructure.

2 Many of the early publications were in the Journal of Systems Engineering, published by the Department of Systems at the University of Lancaster. See Jenkins (1969), who sets out the starting point for the systems approach at Lancaster. This is not the same journal as the current Systems Engineering. The Journal of Systems Engineering morphed into the Journal of Applied Systems Analysis (1969–1991).

(ii) several research projects and consultancy engagements, (iii) the development and teaching of the Civil Engineering Systems Modules on the MEng Civil Engineering undergraduate Programme, and (iv) securing an Engineering and Physical Science Research Council (EPSRC) grant to establish the Systems Centre as an Industrial Doctorate Centre (Godfrey, 2009) must be seen as the foundations for the work presented in this section. However, the contents cannot be viewed as a simple linear accumulation of what is now more than three decades of research. The core of the systems approach at Bristol, and in particular the use of purposeful systems models, created through a process of hierarchical process modelling (see Chapter 9) for complex problems in engineering, eventually led to the recognition that a coherent methodological approach had emerged that shared similar characteristics to SSM, as alluded to by Dias (*ibid*), but perhaps of more significance was its development as a problem structuring method (Davis, MacDonald, & White, 2010; Yearworth, 2021). The 2010 publication by Davis et al marked the first occasion where the methodological development in systems and problem structuring in the Faculty of Engineering at Bristol was brought to the attention of another, but related, discipline – OR. The scholarly link to the Soft OR/PSM tradition was made by White, who at the time was Professor of Management Science in the Department of Management at the University of Bristol.

The £4.9M EPSRC funding for the Systems Centre in 2009 led, in part, to my own recruitment (Godfrey, 2009). The interview panel, which included both the Dean of Engineering and the Dean of Law and Social Science was a formative experience; setting me on a mission to build academic bridges in the University between researchers in engineering and the social sciences. It was also the opportunity to further develop the modules on Advanced Systems and Research Methods in the Engineering Doctorate in Systems Programme, building on the work of Davis and Edwards amongst others, and drawing on material from the wider systems literature and research methods for business and management. This activity brought together several strands of thinking that previously had remained apart. I believe it is quite reasonable to think of 2009/2010 as the start of a return journey to the one started by Checkland at Lancaster, bringing together the knowledge that had developed independently and largely outside of engineering scholarship as mentioned, but also drawing on other centres of systems activity – such as the University of Hull (see for example Jackson (2019a) and Midgley (2000)) and MIT (see for example Senge (1990) and Sterman (2000)) – and creating a coherent picture of what is fundamentally an interpretivist viewpoint relevant to the practice of engineering when dealing with complex problems.

This assembly of knowledge followed no roadmap, there was no unified schema – except perhaps the system of systems methodology (see Chapter 2) – and is probably better described as a bricolage. There was also the ongoing revelatory process resulting from the accumulation of empirical evidence from the Engineering Doctorate in Systems projects, which was building at roughly 10–15 new industry engagements per year, and of course, the corresponding recruitment of the students, the Research Engineers, to deliver these projects and work on their Doctorates. It is only today that I can think of the summation of this work as forming the basis of a new research agenda (Yearworth, 2020).

The Systems Centre, whilst funded by EPSRC as one of the United Kingdom's Centres of Doctoral Training (CDTs), was constituted as an Industrial Doctorate Centre (IDC) (EPSRC, 2011) and existed as a partnership between the Faculty of Engineering at Bristol, and the School of Management at the University of Bath. Projects were solicited from prospective industrial partners and Research Engineers were recruited to the project under a unique three-way contract – between the academic supervisor at the respective University, the industrial partner, and the student (Godfrey, 2009; EPSRC, 2011; Godfrey, 2012). The funding and establishment of IDCs were largely in line with the recommendations of the Parnaby Report in 1990:

> *'Central to the debate has been the need to recognise the differences between pure or basic science research and engineering research and development with its application focus, as well as the contrasts between the science laboratory and the multidisciplinary engineering development department'*
> *'PhD graduates in Engineering are overly narrow and overly reliant on mathematical analysis in their approach. They, therefore, do not integrate as well with the engineering environment in industry with its mix of empiricism, custom and practice, analysis, and multidisciplinary team interaction as say, the Dr. Ing. graduate from a German technical University. Indeed, the very label Doctor of Philosophy is seen in industry to be at odds with the mission of the contemporary research and development process in engineering.'*
>
> (SERC, 1990)

The taught component of the Programme consisted of Modules that were taken by the students in intensive week-long blocks scheduled during the first two years of their four-year Programme. The remainder of the Research Engineers' time was devoted to their industrial project. The industrial partner was encouraged to treat the Research Engineers as if they were employees and the overall experience, for the first two years, was not unlike that of an Executive MBA Programme.[3]

The three-way contract arrangement generated considerable tensions. The projects, originating from the industrial partners, were driven by practical need and application. Right from day one, there was a complex problem that

3 We also recruited a few employees of organisations, where the analogy to an Executive MBA programme was even more apt.

needed addressing. Whereas the University had a duty to deliver a programme of doctoral training and to ensure that the students' research was conducted with the rigour of research methods and novelty of contribution appropriate for the award of a Doctorate. Furthermore, academic supervisors needed to generate academic publications to warrant their time investment.[4] These objectives were not necessarily aligned, and a considerable amount of time was spent debating amongst the delivery team how best to deal with these tensions (Yearworth, 2011; Yearworth, Edwards, & Rosenberg, 2011). It was only through the recognition that these could be addressed through (i) a guided process of problem exploration and translation to achieve an alignment of stakeholder needs, (ii) a finessing of research questions that supported this alignment, and (iii) agreement over ways forward/action plans that we had the final pieces of a coherent programme (Yearworth et al., 2013). There was also the (dawning) realisation that Problem Structuring Methods, generally, should be an essential element of practice for *any* engineer dealing with complex, messy problems.

The emergence in 2011 of an internal project within the System Centre on Systems Practice in Engineering was in part an attempt to mirror the action research programme undertaken by Checkland et al. at Lancaster University (Checkland, 1972) and is summarised in Figure 7.1.

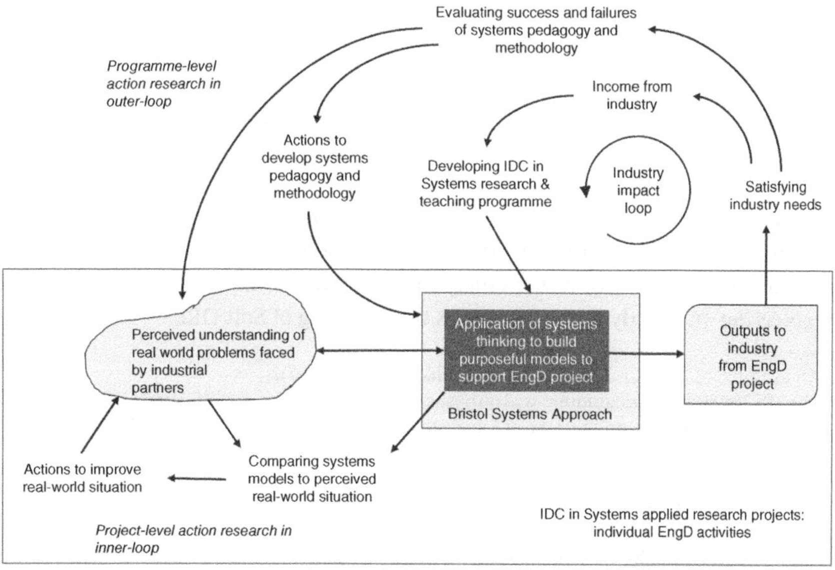

Figure 7.1 The EngD in Systems at Bristol conceived as a programme of action research. Source: Adapted from Yearworth, Edwards and Rosenberg (2011).

4 See the '*Dancing with the Devil*' commentary by Guide Jr and Van Wassenhove (2007) on the challenges faced by scholars partnering with industry to tackle actual problems, but still needing to fulfil normal academic publishing requirements.

However, this programme was perhaps premature, wrestling more with trying to construct an ontology and epistemology of systems that were meaningful for engineers (Buckle-Henning, Wilmshurst, & Yearworth, 2012), rather than realising that we were formalising our learning processes as a distinctive methodology. The arguments and debates that had taken place in OR concerning the development of systems and problem structuring approaches relevant to complex problems in business and management (see Chapter 4) were highly relevant to engineers too, and perhaps where our attention should have been placed sooner. If we had, it would not have taken long to discover that struggles of this type were not new. For example, Assad (2011, p. 176) provides an account of the failure of Churchman and Ackoff to establish an Institute for Experimental Method in the Department of Philosophy at Penn to apply what was essentially a pragmatic (Singerian) approach to a range of issues in management, education and city planning. Their later success at the Case Institute of Technology's Department for Engineering Administration describes a setup not too dissimilar to the IDC in Systems (*ibid*, pp. 177–178).[5]

The commercial basis for the projects supported by the IDC in Systems Programme was different from the action research programme at Lancaster[6] that led to the development of SSM. The Engineering Doctorate[7] students were funded through stipends drawn from the EPSRC grant. They were not fee-earning commercial consulting engagements; although the industrial problems were just as real and the research engineers were often working on customer-facing commercial projects in their sponsoring organisations. The volume of projects, whilst not approaching the same level achieved at Lancaster, was nonetheless still a respectable number and also meaningful from an action research perspective. By the end of the EPSRC grant funding, 100 Research Engineers had graduated with EngDs from the Programme. Of course, not all the projects were embodiments of systems thinking or problem structuring methods, but they all contributed to a greater or lesser degree to the emergence[8] of what could be thought of as '*Soft Engineering*' in exactly the same way as the meaning of Soft OR.[9]

5 An example of structural isomorphism (Frank & Meyer, 2007)?

6 The IDC in Systems programme was almost exclusively based on EPSRC stipends for Research Engineers, the programme at Lancaster appeared to be more diverse in its funding, including commercial consultancy engagements (Checkland, 1972).

7 Post-nominal/designatory letters – EngD.

8 Not least through the setting and assessment of reflective assignments in the core modules. Reading and providing feedback on these assignments provided valuable insight into the '*struggles*' to translate the classroom experience into practice.

9 This will be the only mention of Soft Engineering. Despite occasional use in the context of the EngD in Systems Programme. Here, the meaning is engineering in the sense of Soft OR (for example, see Yearworth and White (2015)). Unfortunately, for my purposes, the term is more generally used to refer to engineering of shoreline and coastal defences, the realm of engineering with *soft materials*.

7.3 Philosophy of Engineering

The first attempt to position PSMs within the curriculum was grounded in the characterisation of the philosophy of engineering along dimensions of (i) theory/practice, and (ii) an ostensible hard/soft axis; as shown in Figure 7.2. What we were trying to achieve in the programme was a shift in emphasis towards understanding the role of an engineer as a designer and humanist in this scheme. At the theory end, we saw this as a need to teach Research Methods from a social science position.[10] This was entirely novel to the students, whose first degrees in engineering had only equipped them for either laboratory-based research experiments or limited fieldwork. At the practice end, we saw the need to teach problem structuring methods as shown in Figure 7.3. In addition to dealing with competing tensions, especially for academic supervisors' attention given publication pressures, the EngD in Systems programme also needed to satisfy the practice-oriented professional registration requirements of the Research Engineers. To this end, the programme also secured accreditation from the UK Engineering Council through a specially convened joint panel.

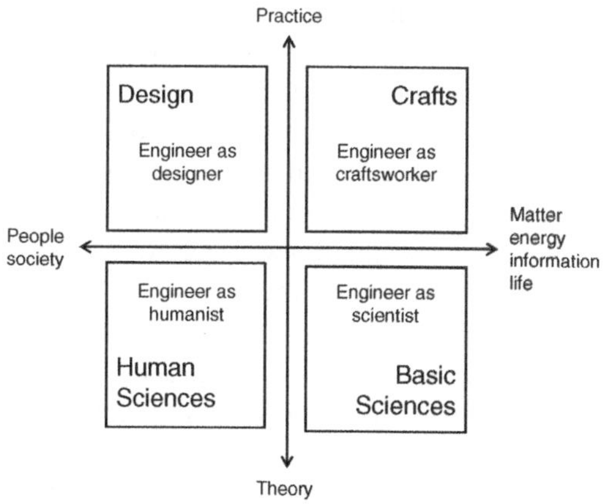

Figure 7.2 Philosophy of engineering summary. Source: Adapted from The Royal Academy of Engineering (2010). Note that the theory and practice axis has been flipped vertically from the original to illustrate consistency with Figure 16.1, i.e. following the layout in Biglan (1973a,b).

10 Informal classroom polling of engineering students joining the programme revealed that there was little to no understanding of the philosophy of science. The meaning of terms such as ontology, epistemology, phenomenology, interpretivism, etc. needed to be taught as part of the Research Methods course. Is it the case that since STEM students are mostly being taught from within a dominant positivist paradigm that their teachers see little need for conceptual language to describe and debate alternative paradigms?

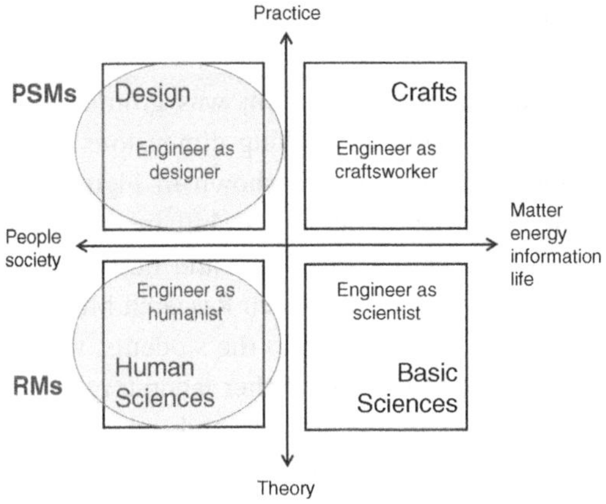

Figure 7.3 The teaching of PSMs and Research Methods addressed the needs on the left of this schema. Engineers joining the programme were already sufficiently trained as scientists, as the dominant mode of instruction, and as engineers, through accredited undergraduate programmes for UK Chartered Engineer status.

7.4 A New Definition of Soft Systems for Engineers

A frequent problem I came across when discussing hard and soft systems views with engineers was that the terms '*hard*' and '*soft*' were rarely defined clearly when they were used in engineering. A frequently stated position in the Systems Group at Bristol was characterised by the statement '*all hard systems are embedded in soft systems*' (Yearworth, Edwards, & Rosenberg, 2011). However, I considered the epistemic shift that Checkland and Holwell (2004) describe as a more pedagogically useful way of characterising hard and soft systems views. Instead of the rather vague association of soft with the social world, people, and human intentionality, the soft systems view moves away from an ontological commitment to the system and treats the definition as a question of epistemology, i.e. what can we know or find out about the world? The following quote spells out this epistemological position in a way that I always felt was a complete revelation to the engineering students on the Programme.

> *'Observer O[ntologist] – positivist, functionalist, working at the level of ontology – sees systems in the perceived world: some of which can be made to work better; or new systems can be designed and implemented; or the systems may be part of the regularities of the universe (frogs, foxgloves, chemical reactions); or they may have an existence deriving from the human logic of situations (manufacturing operations, fishing, refuse collection) ...*
> *Observer E[pistemologist] – phenomenologist, social constructivist, avoiding ontological commitment – sees the perceived (social) world as: culturally extremely complex; capable of being described in many different ways; and sees the "system" as one useful concept in ensuring good-quality debate about intentional action. The two observers both agree that the notion "system"*

> *can be useful, O seeing it simply as a name for (parts of) the real world, E seeing it as a useful intellectual device to help structure discussion, debate and argument about the real world'.*
>
> (Checkland & Holwell, 2004, p. 58)

This is all usefully summarised in the following chart that I used with my students (Table 7.1).

Checkland and Holwell's paper appears in an edited volume (Pidd, 2004), which brings together the ideas developed in the EPSRC funded Interdisciplinary Research Network on Complementarity in Systems Modelling (INCISM) network. This complementary framing can also be applied to systems modelling approaches such as System Dynamics

> '*... system dynamics modellers do not spy systems. Rather they spy dynamics in the real world and they organise modelling as a learning process, with the project team, to discover the feedback structure that lies behind the dynamics'*
>
> (Morecroft, 2007, p. 148)

Reflections on this hard/soft complementarity and the work of the INCISM network at the System Dynamics conference in 2004 are captured in the notes from the record of the plenary session *Working Ideas, Insights for Systems Modelling: The Broader Community of Systems Thinkers* (Checkland, Pidd, & Morecroft, 2004). All of this is neatly summarised by Checkland himself in a colloquium delivered at Lancaster University in 2012. In this short video, Checkland (2012) outlines

Table 7.1 Hard and soft systems thinking as an asymmetric complementarity.

Strand	Hard systems thinking	Soft systems thinking
Orientation	Goal seeking	Learning
The world...	contains '*systems that can be engineered*'	is problematical but can be explored using systems models of concepts of purposeful activity to define '*action to improve*'
Systems models are	... Models of (part of) the world (*ontologies*)	Devices: intellectual constructs to help debate (*epistemologies*)
Talks the language of	'*Problems*' and '*solutions*'	'*Issues*' and '*Accommodations*'
Philosophically	Positivistic	Phenomenological
Sociologically	Functionalist	Interpretivist
Systemicity lies in	the world	the process of inquiry into the world

Source: Adapted from Checkland and Holwell (2004, p. 56).

the development of SSM emphasising its role as methodology, not method, and the origin of this particular definition of soft systems as a systemic learning system designed to help us deal with the complexity of the world. Morecroft (2007, pp. 146–152) also offers a useful reflection on the work of the INCISM network especially on the point of whether a CLD is being used from either the epistemological or ontological position.[11]

7.5 Systems Engineering in the Context of Systemic Cooperation

Through simultaneous connection to networks in the International Society of Systems Sciences (ISSS) and the Systems Science Working Group (SS-WG) of the International Council on Systems Engineering (INCOSE), I was invited to take part in a workshop organised as part of the 17th Conversation of the International Federation for Systems Research (2014, pp. 4–13). This was brought about specifically in the context of a cooperation agreement between the ISSS and the INCOSE SS-WG.[12] The product was the OODA-SCOOPs model[13] (Yearworth et al., 2015). From a starting point of developing a model of engineering practice based on several existing views of systems and systems engineering practice, the main progress in this week-long workshop was made by resorting (yet again) to a set of dichotomies that characterise the nature of decision-making in a particular context.

The primary dichotomies used were between (i) the nature of the role in a transactional relationship and posed between Providing and Consuming, and (ii) the nature of the prevailing behaviour exhibited in the context and posed between Understanding and Intervening. The full reasoning behind these decisions is discussed by Yearworth et al. (2015). What was innovative, and what has led to the utility of the SCOOPs model here, was that the four quadrants defined by the two dichotomies could be associated with Roles and functions within a broader systems engineering Context[14]. These have been summarised in Table 7.2 and the resulting model is visualised in Figure 7.4.

To loop around the SCOOPs model is to move from (i) a framing problematic situation associated with the sponsoring role, to (ii) problem structuring associated

---

11 This in turn has contributed to the decision-making about what examples of CLDs to include in the survey of applications of PSMs in Chapter 13.

12 See Chapter 8 for more on this relationship.

13 The Orient-Observe-Decide-Act (OODA) labelling has been dropped here to emphasises the presentation of the main dichotomies that were used in the development of the Systems Engineering in a Context of Systemic Cooperation (SCOOPs) model.

14 There is some passing similarity with the 4 A's – Appreciation, Analysis, Assessment, and Action in Mingers (2001, pp. 291–293).

Table 7.2 The contexts and roles arising from the behaciour and transaction dichotomies in the SCOOPs model.

Behaviour	Transaction	Context	Role
Understanding	Consuming	Situating	Sponsoring
Understanding	Providing	Planning	Managing
Understanding	Providing	Planning	Managing
Intervening	Providing	Investing	Engineering
Intervening	Providing	Investing	Engineering
Intervening	Consuming	Effecting	Using
Intervening	Consuming	Effecting	Using
Understanding	Consuming	Situating	Sponsoring

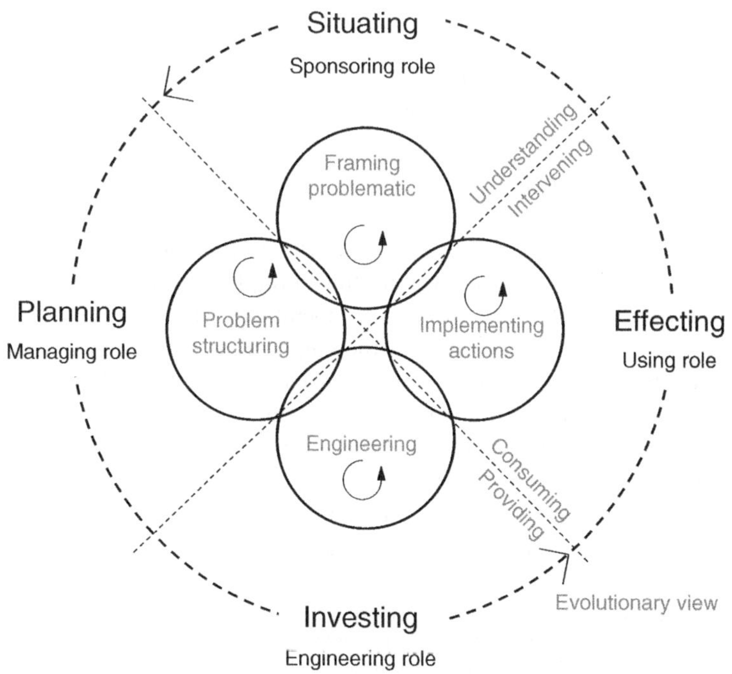

Figure 7.4 The SCOOPs model. Source: Adapted from Yearworth et al. (2015).

Table 7.3 Methodological issues confronting each of the four processes in the SCOOPs model are shown in Figure 7.4.

	Providing	**Consuming**
Understanding	*Planning* Conceptual modelling for understanding Dealing with worldviews and subjectivity Directed towards defining actions Usual methodological focus: plural, interpretivist, phenomenological	*Situating* Strategic analysis for understanding Engaging with a messy problem context Directed towards initiating action (securing budgets, political support…) Usual methodological focus: concerned with theories of power
Intervening	*Investing* Quantitative Modelling Dealing with specific requirements The scope of traditional Systems Engineering and '*Hard*' OR modelling Usual methodological focus: unitary, functionalist, realist, positivist	*Effecting* Implementing actions Dealing with customers Directed towards operational issues Usual methodological focus: Measuring

Source: Adapted from Yearworth et al. (2015).

with planning and managing the engineering activity, to (iii) the actual investment in delivering engineering artefacts, and to (iv) implementing actions, the deployed use of the products of engineering in operations. The final closing of the loop is to understand that the effect of the engineered objects may not be what was desired thus leading to a new framing problematic situation. This evolutionary loop was conceived as spanning multiple organisations and to be plural and recursive i.e. many of these loops operating simultaneously corresponding to programmes, which in turn are comprised of projects.[15] Furthermore, each of these four Contexts (Situating,[16] Planning, Investing, Effecting) would be associated with their operating paradigm as summarised in Table 7.3.

15 At the time the OODA-SCOOPs model was developed it was not interpreted fully as a process model, although the problem structuring input contributed to the workshop was grounded in thinking in terms of the process models introduced in Chapter 9.
16 It is important to note that the notion of Situating, as a context*ualising* process, is a long way from the problems of situational logic discussed in Section I. It is much closer, as an idea, to boundary critique and approaches such as CSH (see Chapter 16).

Table 7.4 Logics of transitions from one context to another and expected behaviour of the functional role.

Transition in context	Functional role	Expected behaviour	Description
Effecting → Situating	Sponsoring	Understanding consumption	How is deployed capability meeting needs? Is the Problem Suppression System (PSS) suppressing? Re-evaluating POSIWID (i.e. the Purpose Of the System Is What It Does). Strategizing behaviours.
Situating → Planning	Managing	Understanding provisioning	(Re-)Problematisation, action planning
Planning → Investing	Engineering	Providing intervention	Engineering (generally), i.e. building, making, constructing, …
Investing → Effecting	Using	Consuming intervention	Capability put to test through deployment

The expected transition from one Context to another in the SCOOPs model was a natural consequence of its development using dichotomies. How the transitions would be observed '*in the wild*' is not discussed by Yearworth et al. (2015) but it is not too difficult to surmise what they might be *logically*, and these transitions are summarised in Table 7.4.

Also not discussed by Yearworth et al. (2015) were the consequences should these transitions not lead to the expected behaviours. This led to the notion of *pathologies*, developed in conversation with Lowe (2020). Whilst the original OODA-SCOOPs model was developed to account for an ideal situation, no attempt was made to consider counter-examples of where this schema was not being followed. The model was idealised and a reflection of the logical combination of existing models and frameworks. The value of the workshop was in bringing together knowledge about these different conceptualisations and finding a way of assembling them into a single framework. Following the concept of pathologies introduced in Section 1.2 and discussed further in Section 3.2, it seemed reasonable to consider that failure to transition from one Context to another in the SCOOPs model would be reflected in the behaviours exhibited by actors fulfilling the roles indicated and that these behaviours could be observed empirically.

These notional pathologies are presented in Table 7.5 as hypotheses to be tested, they are currently without support from any specific evidence. They are presented as both strengths and weaknesses. For example, '*short-circuiting*' the

Table 7.5 Pathologies arise from missing processes in the SCOOPs model.

Shortcut transition	Absent/weak role	Symptoms – weaknesses	Symptoms – strengths
Effecting → Planning	Sponsoring	Lack of sponsorship Lack of scrutiny Situations not appraised *Cause of weaknesses below*	Autonomy Resilience Pace (e.g. Urgent Operational Requirements [UORs]) *Cause of 'saves' below*
Situating → Investing	Managing	Plans instead of planning Inadequate problematisations Wrong solutions and/or options Political '*pet projects*' ungated	Political expediency
Planning → Effecting	Engineering	No new provisioning Loss of capabilities (forgetting)	(Re-)servitisation Expediency
Investing → Situating	Using	Customers are ill-equipped Waste of money Not fit for purpose Obsolete before deployment	Avoid fiasco Avoid embarrassment Avoid defeat/loss

engineering process by *re*-servitising an existing engineering artefact might well be highly expedient; however, the longer-term impact might be a loss of capability to perform the engineering task due to a lack of ongoing investment to retain suitably qualified personnel.

7.6 Conclusions

The end of the Engineering Doctorate in Systems Programme[17] left a legacy at the University of Bristol with the ongoing teaching of systems approaches in

17 Although technically funded through to 2018, nonetheless had a slow ramp-down in activity with the final graduations in 2020.

Civil Engineering Systems and Sustainable Systems modules, with the latter being available to all engineering students across the Faculty. Another legacy was the involvement in the UK Collaboratorium for Research on Infrastructure and Cities (UKCRIC)[18] arising from previous work in the International Centre for Infrastructure Futures (ICIF) (Collins et al., 2013). Perhaps the closest to a research group with a similar intent was the creation of the Engineering Systems Division at MIT, formed as a partnership between the Faculty of Engineering and the Sloane School of Management between 1998 and 2015 (de Weck, Roos, & Magee, 2011; de Weck, 2016). A recent review of engineering education in the United States has started to look for evidence of the skills development necessary to address socially complex problems (Beagon & Bowe, 2023). It would seem reasonable in response to point to these two innovative programmes (at Bristol and MIT) as examples of what is needed to produce such skills.

18 https://www.ukcric.com/.

8

On the Theoretical Foundations of Systems Engineering[1]

In his recent address the President of the International Society for the Systems Sciences (ISSSs) raised some important observations about the state of Systems Engineering and need for theoretical foundations (Rousseau, 2019). The description of problems in this industry is frank and should be welcomed (Rousseau, 2019; section 2). Although his address also concerned the general (fragmented) state of Systems Science, its relationship to Complexity Theory, and the requirements for General Systems Theory to act as a foundation for Systems Engineering and Systems Practice generally, my response concentrates specifically on its relevance to the foundations of Systems Engineering. It is pertinent to position both in the context of the relationship that exists between the ISSS and the Systems Science Working Group (SS-WG) of the International Council on Systems Engineering (INCOSE), which was formalised by the Relationship Agreement that was signed between the two organisations in 2011 and as emerged from a joint meeting held in January 2011 (International Society for the Systems Sciences & International Council on Systems Engineering, 2011; Lawson, 2011; Martin, 2011). This agreement has been periodically reviewed as a memorandum of understanding (MoU), most recently in 2017 (International Society for the Systems Sciences & International Council on Systems Engineering, 2017).

This most recent statement of the relationship sets out that its principles are (i) to jointly further practices in Systems Sciences and Systems Engineering, (ii) to state that INCOSE members are interested in gaining foundational knowledge in systems science concepts that are relevant to systems engineering practice, (iii) to express that ISSS members are interested in seeing systems theories applied in practice, and (iv) to define how the relationship should be coordinated. Therefore, with regard to the MoU the president's address is well aligned. The relationship

1 The original text for this chapter appeared as an Open Access article in Systems Research and Behavioral Science and is reproduced according to the terms of the Creative Commons Attribution License, which permits use, distribution and reproduction in any medium, provided the original work is properly cited (Yearworth, 2020).

Problem Structuring: Methodology in Practice, First Edition. Mike Yearworth.

provides an excellent opportunity for good science to take place as it addresses the necessary praxis that occurs at the intersection between the empirical and the theoretical. A succinct diagrammatic summary of this praxis can be found in Singer et al. (2012).

The realm of Systems Engineering is ever expanding, as a systemic sensibility percolates across engineering practice, and Systems Engineering itself produces distinctive new branches of knowledge such as Model Based Systems Engineering (MBSE) and System of Systems Engineering (SoSE). However, is Systems Science and General Systems Theory[2] the only route to take to provide theoretical foundations for Systems Engineering? At face value, Systems Science would appear to be an obvious way forward. However, the key knowledge assets suggested by Rousseau (2019, table 3) seem disjoint from the phenomena of the introduction to his paper.

I suggest there is another place to look for Systems Engineering foundations. Almost 50 years ago Peter Checkland joined the Management School at Lancaster University, signing up to the agenda set by Gwilym Jenkins to apply Systems Engineering principles to complex problems of management (Checkland, 2000b, 2012). Systems Engineering principles came off rather badly in the encounter, but the ensuing action research programme over many years and many hundreds of engagements led to the emergence and development of soft systems methodology (SSM) (Checkland, 1981; Wilson, 1984).

SSM was taken up by systems groups, albeit outside of engineering, and eventually became part of a wider endeavour in Management Science more generally known as Soft Operational Research (Soft OR) and problem structuring methods (PSMs). However, in the 30 years since the publication of '*Rational analysis for a problematic world : problem structuring methods for complexity, uncertainty and conflict*' (Rosenhead, 1989c), which first brought together SSM, strategic choice approach (SCA) and strategic options development and analysis (SODA) under the PSM label, this knowledge has remained largely confined to the discipline of Management Science, the pages of a few journals,[3] and essentially disconnected from its engineering roots,[4] and even those of GST (Checkland, 2000b, p. S11). It is perhaps time that this rich – and growing (Lowe & Yearworth, 2019) – repository of theoretical, methodological, and practical knowledge found its way back to

2 As General Systems Theory * (Rousseau, 2019; section 4),

3 Notably Systems Research & Behavioral Science, European Journal of Operational Research, Journal of the Operational Research Society, Group Decision and Negotiation, and until about 2018, Omega – The International Journal of Management Science.

4 Although attempts were made in the Faculty of Engineering at the University of Bristol to reconnect them through teaching social science research methods to postgraduate engineers on the Engineering Doctorate in Systems Programme that ran from 2006 to 2017 (see Chapter 7 and Yearworth, 2011, 2016; Yearworth, Edwards, & Rosenberg, 2011; Yearworth et al., 2013, 2015; Yearworth & Edwards, 2014; Yearworth and White, 2015).

Systems Engineering. There is enough material here to establish a useful research agenda to test these contributions in a modern Systems Engineering context, thus taking the Lancaster Group's original agenda in the reverse direction.

However, whilst Soft OR/PSMs can be seen to be broadly systems-based in approach and therefore highly relevant to Systems Engineering it is also worthwhile to ask the question what else should Systems Engineers know about that is relevant to developing foundational knowledge? Systemic Intervention (Midgley, 2000) and Critical Systems Thinking (Jackson, 2001, 2019a) are highly relevant. Jackson (2019a) looks at the strengths and weaknesses of 10 systems methodologies (including Systems Engineering) and organises them according to whether they emphasise technical, process, structural, organisational, people/social, or coercive complexity. The argument is that '*wicked problems*' (Rittel & Webber, 1973) typically involve all these forms of complexity, and so an enhanced Systems Engineering needs access to a range of methodologies and the ability to use them in combination. However, I believe it is also incumbent on Systems Engineering to look to the wider social sciences for foundational knowledge.

In my teaching of postgraduate engineers at the University of Bristol, I used a critical reading exercise on a research methods course to introduce material from a broad range of social sciences. One of my favourite papers to set was Diane Vaughan's account of NASA during the Columbia accident investigation published in the American Journal of Sociology (Vaughan, 2006). Written in a lively first-person rhetorical style she presents an engaging ethnography of her experience and its relation to her earlier work during the Challenger accident investigation (Vaughan, 1996). The reaction of the engineering students was interesting. Predominately this was an initial claim that the work was '*unscientific*' because of the ethnographic method and writing in the first person. However, and this was truly interesting, Vaughan's notion of the normalisation of deviance resonated with the students. It was a concept that, no matter how it was derived, was believable and, crucially, they could see how it might apply in other engineering organisations, especially where safety was of paramount concern.

It is not too difficult to see that once the language of social science and a variety of essentially qualitative research methods is mastered there is a rich source of knowledge available and relevant to Systems Engineering. The opportunity is vast, and here I can only point to a few examples, in addition to Vaughan, that I have found useful and relevant to the phenomena that Rousseau (2019) touches on in his paper – large infrastructure project failures (Flyvbjerg, 1998; Flyvbjerg, Garbuio, & Lovallo, 2009), strategic misrepresentation (Kutsch et al., 2011), strategic ambiguity (Denis et al., 2011), and paradox (Poole & van de Ven, 1989; Schad & Bansal, 2018). This literature bears little relationship with Systems Science and GST but is likely to be highly relevant to Systems Engineering practice.

Perhaps Systems Science needs Systems Engineering more than Systems Engineering needs Systems Science and General Systems Theory. I suggest that Systems Science/General Systems Theory is just one of several foundations and that Systems Engineering needs a much broader research agenda for its foundations. In my capacity as Systems Field Editor/Systems Engineering for Systems Research and Behavioral Science, I would like to encourage submissions to this journal that make theoretical, methodological and practice-based contributions to this debate. I would also like to see the debate spread into the pages of Systems Engineering and other related systems-oriented journals in engineering. The specific case of reconnecting SSM to engineering practice (generally) via Soft OR/PSMs is my area of work. The reason for advocating this reconnection is the recognition of the *messiness* of the real world that engineers face and is frequently stated as the characteristics of '*wicked problems*' enumerated by Rittel & Webber (1973). This recognition then sets up the need for methods that enable Systems Engineering practitioners to deal with problem situations involving many interested parties with different perspectives (worldviews), problem situations that are not well defined, where there is difficulty agreeing on objectives and success requires creating agreement among parties involved, and that there may be many uncertainties and lack of reliable data (Mingers, 2011). However, in addition to this, I firmly believe that future developments of Systems Engineering foundations need to draw from the wider social sciences, not just Soft OR/PSMs.

9

Hierarchical Process Modelling[1]

'Here again we see the emergence of a generic principle, namely that the nature of the whole, whether state or individual, is determined by interactions between its constituents, whether groups or elements. These ideas have significance when we seek to model products and the organizations or processes that create them. The distinction between an organization and a process will be blurred purposely here since the broad functional definitions of a process and of an organization are similar; namely the achievement of a particular purpose.'

(Dias & Blockley, 1994, p. 418)

9.1 Motivation

The original motivation for the development of hierarchical process modelling (HPM) can be viewed as an attempt, in an engineering context,[2] to deal with the problem of integrating hard and soft systems thinking through a common modelling approach. Not in the paradigmatic, complementarity sense of Checkland and Holwell (2004), but in terms of thinking of hard systems (products) as being *entwined* with soft systems (conceptual organising) in that both are expressions of purpose. For example, a building and the conceptual organising that leads to the necessity of constructing a building are processually bound together in achieving a purpose. The material in this chapter follows the historical development of HPM from its functionalist roots through to its application as the modelling approach to

1 Material from this chapter is drawn from teaching Hierarchical Process Modelling on the EngD in Systems Programme discussed in Chapter 7 and as originally developed by Davis, from the training course in HPM delivered to the consortium members of the STEEP Project (STEEP, 2011), and workshops using the methodology held in Bristol and San Sebastian.
2 Specifically in Structural/Civil Engineering.

Problem Structuring: Methodology in Practice, First Edition. Mike Yearworth.

support a problem structuring method (PSM). Based on the CR/E/A framework presented in Chapter 5 the use of HPM-as-a-PSM can be thought of as using the generic interpretivist constitutive rules of a PSM but where the epistemology and affordance of the modelling is built around the concepts of HPM presented below.

9.2 Origins

HPM was developed at the University of Bristol during the 1990s and brought together the means of combining evidence of process performance, expressed using interval probability theory (Hall, Blockley, & Davis, 1998), with their representation as a system[3] comprising a hierarchical structure of processes (Dias & Blockley, 1994).[4] That is, bringing process structure and evidence of process performance together into a single model of purpose. The original motivation for the use of interval probability theory was driven by the need to model systems in Civil Engineering with a view to calculating their performance, with explicit accounting for the uncertainty of evidence or measurement. The core elements are

1. Model of engineering systems as a process with an overall purpose
2. Processes consist of sub-processes defined by a composition or containment relationship – hence a hierarchical representation when drawn on a canvas
3. Processes have an associated performance measure with explicit representation of uncertainty

The original thinking is illustrated in Figure 9.1 and starts with a decision problem, as in OR, where the process model and associated data model are integrated. From this basis, a hierarchical process model (HPM) can be established, as shown in Figure 9.2. This early example of an HPM illustrates modelling at a conceptual level. If we were to ask the question at the start of Chapter 7 '*What is it that an engineer is doing when dealing with a complex problem*?' then this process model provides the beginnings of an answer.

The original idea from this example model is easily extended to consider the broader scope of such a modelling approach. If we consider the following two processes

1. Constructing flood defences to a certain specification in accordance with predicted conditions at a specific location, and
2. Being resilient to floods in that location.

3 Consistent with the origins of HPM in a systems group in a Faculty of Engineering, HPM are frequently referred to as systems models. Similar to SSM, it is possible to think of an HPM as representing a system of processes to enact a purpose. However, as presented in Section 9.4 below, there is no explicit mention of a system in the epistemology of HPM and a more precise statement would be a '*process of processes*' to enact a purpose. The advantages of mono-category models is discussed later.

4 Blockley (2010) provides an historical account of the original development.

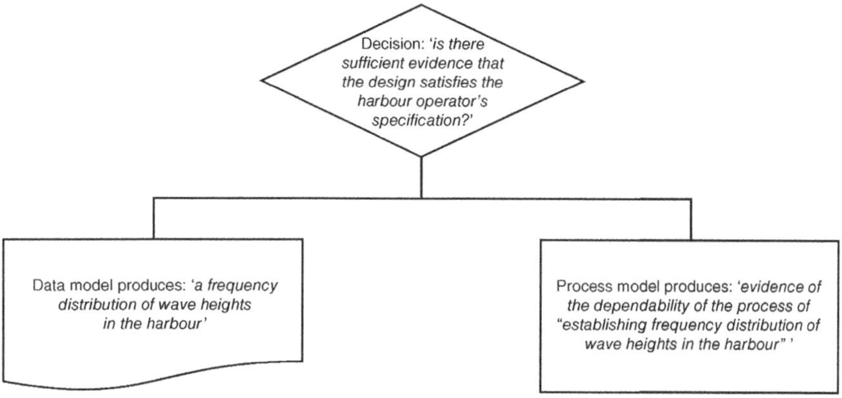

Figure 9.1 The origins of hierarchical process modelling (HPM). Source: Adapted from Hall, Blockley and Davis (1998).

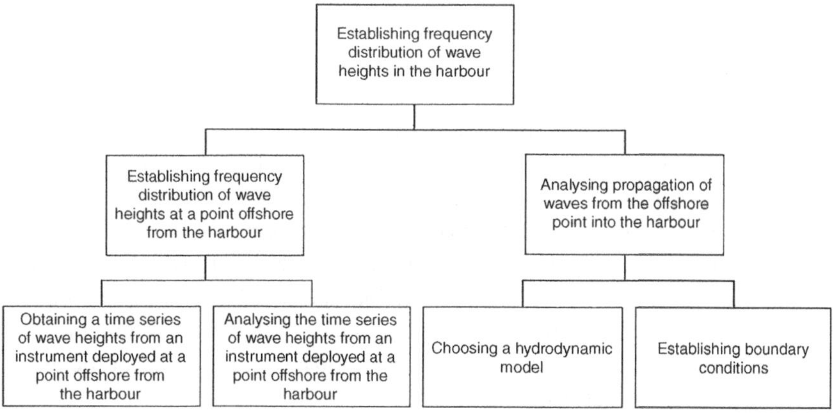

Figure 9.2 An example of a hierarchical process model obtained by using the basic elements described in Figure 9.1. Source: Hall, Blockley and Davis (1998)/with permission of ELSEVIER.

Process 1 would be considered well within the scope of Civil Engineering and would be a perfectly reasonable answer to the question of *why* we were trying to establish the frequency distribution of wave heights in the harbour. Whereas Process 2 is more strategic and could encompass Process 1, although it would also be reasonable to consider the possibility that resilience to floods could be achieved through other means – i.e. constructing flood defences is just one possible process answer to the question of *how* resilience could be realised.

Dias (2013), writing in Civil Engineering and Environmental Systems, suggested the simultaneous discovery of soft systems by Blockley – but only in terms of the similarity of hierarchy in system models and modelling conceptually, but I believe that Blockley (or Dias) did not engage fully with Checkland's meaning of soft systems as set out in Table 7.1 – i.e. as a profound epistemic shift. On the other hand, Dias argued that Checkland's modelling only focussed on the process of enquiry, whereas Blockley's approach addressed both hard and soft systems modelling.[5] However, this is a 10-year-old argument and I just want to highlight the fact that there is a useful starting point for the development of a systems modelling approach that supports *a form of* soft systems methodology.

Refinement of the approach was achieved through funding from two Engineering and Physical Science Research Council (EPSRC) projects '*Condition Monitoring and Asset Management*' (CMAM) (Hall, Davis, & Taylor, 1999), and '*A Generic Process for Assessing Climate Change Impacts on the Electricity Supply Industry and Utilities*' (GENESIS) (Watson, 2003). Much of the developmental thinking is captured in the published output (Fletcher & Davis, 2002; Marashi & Davis, 2006, 2007; Davis, MacDonald, & Marashi, 2007; Marashi, Davis, & Hall, 2008) and three PhD theses (Rosenberg, 2003; Marashi, 2006; MacDonald, 2010).

Around 2009–2010 the development of HPM at Bristol took a turn towards Soft OR (Davis, MacDonald, & White, 2010). This work, published in the Journal of the Operational Research Society, marks a distinctively interpretivist approach by demonstrating how the hierarchical process modelling approach could be used as a PSM. This was the first outcome of the joint research between the departments of Civil Engineering and Management at Bristol through the collaboration of Davis (Civil Engineering) and White (Management). Davis was starting to use hierarchical process modelling to support his consulting engagements with industry and had shown it could be used to model what is essentially a purposeful activity that was more strategic in purpose than the original focus in HPM towards products (as engineered systems) with purpose. It was this work that drew White into conversation with the Civil Engineers and I believe this was the genesis of the use of HPM as a problem-structuring method.

9.3 What Is HPM as a PSM?

There are four foundations to the use of hierarchical process modelling as a PSM; (i) the process epistemology, (ii) modelling oriented towards achieving purpose, (iii) the how/why modelling dialectic, and (iv) measuring process performance.

5 Incorporating both into a single diagram would seem to be a form of category error or a problem of paradigm incommensurability as discussed in Section I. However, a grounding in processual philosophy appears to finesse these objections and is discussed further in Section 16.3.

9.3.1 Process Epistemology

The origins of HPM were grounded in a strong process epistemology where everything and anything in a model is treated as a process. The process model is simply a network of '*nodes*' and '*links*', much the same as for the affordances of modelling approach for soft systems methodology (SSM), strategic options development and analysis (SODA), and strategic choice approach (SCA) described in Section 5.4. For HPM, the '*node*' is conceived as a purposeful entity, a process. The '*links*' describe *part-of* or *composition* relationships.

Identifying and labelling processes is helped by thinking of them as an entity that is enacting a purposeful process and constructing sentences such as <*This entity is enacting the process* establishing the frequency of wave heights>. This should then become so familiar and automatic that we do not need to keep thinking of describing processes in complete sentences. We can drop the first part in italics and use a capital letter (if preferred, it is not necessary), such that the process label reduces to <Establishing the frequency of wave heights>. Here, the verb phrase <Establishing the frequency…> is being used as the object of <*This entity is enacting…*> and is therefore a gerund.[6] This non-finite verb construction is used because it gives rise to a sense of purpose that is continuous and ongoing. This '*ongoing-ness*' is an essential element of HPM to create a *diachronic* model of purpose – i.e. literally a purpose that is occurring over time. Even physical entities can be considered as processes. For example, '*the bus is a process*' is a perfectly reasonable statement to make in this process epistemology because the bus is enacting the process <Transporting passengers> (literally <The bus entity is enacting the process Transporting passengers>). The most important advantage of modelling with processes is that avoids over-specification in workshops when the amount of available time with stakeholders, particularly senior leadership teams as discussed later, is very short. Perhaps when an enactment of the process is needed, we do not need a bus; e.g. it could be a car, or a horse and cart; what matters is that the purpose <Transporting passengers> is being achieved.

There is also an elegant symmetry between modelling processes using gerunds and ways of conducting qualitative research. Imagine that we are studying a change process in an organisation using a qualitative research method (e.g. semi-structured interviews and thematic coding) – we would expect to elicit the same enacted processes as those we '*designed*' in our problem structuring

6 In English the gerund form of the verb is constructed, give or take a few irregularities, by the addition of the suffix '*-ing*' to the infinitive. The full implications of the processual interpretation of modelling using gerunds, non-finite verbal nouns, in other languages is yet to be fully explored. Experience in facilitating group model building workshops using HPM with stakeholder groups modelling in Turkish and Basque, and reports from self-facilitated groups working in Italian, have demonstrated the use of the corresponding gerund form of verbs but not whether the meaning is the same as expressed here.

modelling. For example, an HPM starting with a top-level process of <This entity is enacting the process Transforming the business to achieve a new function>, suitably modelled hierarchically, and where the entity is a project team that is enacting the processes, then we should expect to recover the process descriptions through a qualitative analysis. Charmaz (2006), citing Glaser (1978), specifically suggests the use of coding with gerunds for the '*identification of enacted processes rather than static topics*'.

9.3.2 Modelling Purpose

The overall purpose, as enacted by the hierarchical composition of processes, can be thought of in many different ways. For example,

- <Projecting intent into the future>, where the entity is a project
- <Projecting many intents as a coherent assembly into the future>, where the entity is a programme
- <Changing an organisation>, where the entity is a change management team
- <Contracting for buying/purchasing>, where the entity is an organisation, or part of
- <Contracting for selling/supplying >, where the entity is an organisation, or part of
- <Developing a new product/service>, where the entity is a development team
- <Engineering/constructing/building…>, where the entity is an organisation, or part of

The generic purpose (purpose type) is expressed as the gerund – i.e. projecting, changing, modifying, contracting, developing, engineering, and so on. It is this gerund that would be expected to be seen as the single, top processes in the hierarchy. Immediate similarities to the modelling approach in the core PSMs can be made through the purpose type of the top process in the hierarchical model, i.e.

- *Planning* (SCA)
- *Strategising, making strategy, detecting strategy* (SODA methodologies)
- *Transforming* (SSM)

This purpose/process/gerund formulation of HPM sheds further light on the CR/E/A framework and opens up some creative possibilities. For example, since both HPM and the cognitive/causal maps of the SODA methodologies are graphs, it is possible to express an HPM that is modelling the purpose *contracting* using a platform such as Strategyfinder that was designed to support modelling purposes such as *strategising* (as is demonstrated later in this chapter). Conversely, it would be possible to use a root definition/CATWOE for a transformation from SSM but to model the activity system as a *transforming* HPM.

SSM stipulates that the subject of transformations is entities (Checkland & Scholes, 1990, p. 34), and that multiple entities can be affected by the ostensible transformation as designed.[7] HPM also allows for the same interpretation, but there is also an interesting asymmetry in that whilst processes are enacted by entities, the actual entity enacting the process is not a necessary consideration when modelling[8]. Like SSM, the purpose modelled by the HPM remains completely conceptual, although without paradigmatic wrangling. Again, like SSM, by ignoring which entities might enact the process it is possible to explore how purpose might be achieved in an unconstrained manner; so-called creativity in enactment.

9.3.3 How/Why Modelling Dialectic

Starting with a single top process, which embodies the purpose of the HPM, further detail is added to the model by the method of decomposing a process into '*contained*' sub-processes. Modelling proceeds through the how/why questioning dialectic of examining a process and asking *how* success for that process can be achieved. The answers to '*how*' questions reveal further processes. However, this is not a reductive exercise in modelling, these are not processes of finer detail, the answers may reveal processes that are necessary for the overall purpose to be achieved. All of the processes revealed by such '*how*' questioning, no matter where they appear in the hierarchy, are in effect contained within the top process and are therefore necessary and/or sufficient, to some degree, for achieving the overall purpose. Asking the question '*why*' a process exists can be answered by looking upwards in the model towards superior processes.

The same '*why*' question can be asked of the top process. *Why* are we transforming, projecting, strategising…? Asking '*why*' questions of the top process to examine possible wider purpose(s) and continuing the how/why dialectic of modelling opens up two further benefits from the approach

1. by exploring '*how*' questions that are associated with these '*higher*' purposes, the answers will reveal new process hierarchies parallel to the original model and thus potentially identify different stakeholder groups, and
2. relationships between these new process hierarchies and the original can be explored for potential synergies – i.e. can processes be shared between process hierarchies? Can processes exist to meet different purposes?

The first benefit bears some relationship to boundary critique in Critical Systems Thinking (for example, see Midgley, Munlo and Brown (1998)). This '*why, new*

7 For example, see the business transformation example given in Section 10.3 where ostensible purpose is contrasted with several other possible transformations that might occur.
8 It is the legacy of a substance ontology. The processual turn discussed later moves us away from the necessity of thinking about entities at all. We can just think about processes.

how' questioning explores the boundary of the existing problem and potentially finds different enactments that would achieve purposes related to, and potentially identical to, the original problematisation. Exploring synergies is more difficult but can be understood by finding the same process answers to '*how*' questions in two different process hierarchies. For example, <Building a new local railway station> might be the ostensible purpose for a programme entity, but exploration of the '*why*' question might reveal further purpose, such as <Reinvigorating the local economy> with different process answers to further '*how*' questioning. A new local railway station might be more than a transport interchange and could provide functional space for many different activities that might benefit a local community.

The insistence on using processes when modelling and the concomitant delayed reification – i.e. turning processes into things – also enables an important language game to be played when exploring models of purposeful activity. Delayed reification allows for the simultaneous exploration of '*why*' and '*how*' questions without getting lost in specific but different enactments – i.e. delayed reification can produce the same answer to '*how*' questions arising from different purposes. For example, consider the case of a United Kingdom police force using a (double-decker) bus – as the process <Being a mobile observing platform> – to catch drivers using their cell phones whilst driving; so-called '*operation top-deck*' by South Yorkshire Police Force.

9.3.4 Measuring Process Performance with Explicit Representation of Uncertainty

In the original functionalist design of HPM, understanding process performance was an important consideration when modelling. By attaching a measure of process performance to each process in the model it would then be theoretically possible to calculate the performance of the overall purpose. However, from a pragmatic engineering perspective, there is no such thing as certainty, therefore an important part of this measure was an explicit representation of uncertainty in process performance. The *affordance* offered by this explicit representation of uncertainty made HPM well-suited to modelling risk as shown in Figure 9.3.

The examples offered to the United Kingdom Government in the Blackett Review seem a little fanciful but illustrate the point. A more serious, consequential, case of modelling with uncertainty relates to some of the original examples of HPM, resilience to the sea.

> *'Once it is accepted that for low probability events in complex systems there will always be incomplete knowledge, then it should be good practice to consider what we do not know and develop strategies that are insensitive to the lack of knowledge. For example: in the meltdown at Fukushima Daiichi Nuclear*

Annex 4

"We need the wisdom to know what we do not know"[38]

With sufficient knowledge and informed judgement uncertainty can be characterised statistically. It follows that strategic surprises arise from lack of knowledge or the inability to perceive the consequences of what is known. An evidence based process to elicit what we do not know can be very helpful. The 'italian flag'[39 40] is a simple evidence based tool for that purpose.

The method invites consideration of the evidence for failure 'Red' and evidence for success 'Green' and what is not known 'White'. Its intention is to elicit what we do not know and by making it explicit invite strategic approaches to mitigation. It is a means of encouraging transparency, focusing on evidence and away from the fear of being wrong.

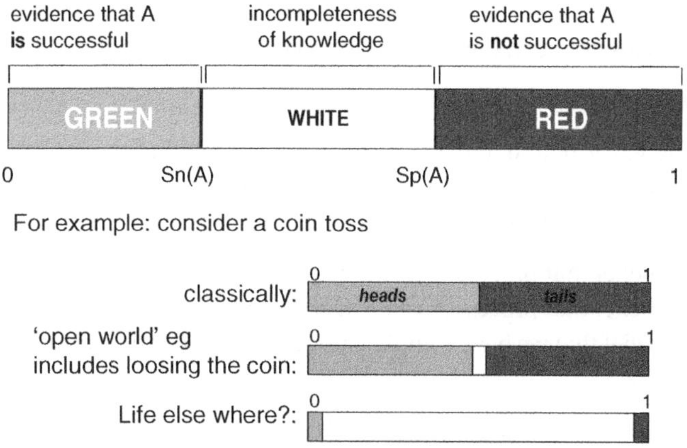

Figure 1 The Italian flag method of articulating in complete knowledge

[38] Theaetetus of Plato
[39] Blockley D and Godfrey P (2007) "Integrating soft and hard risks" Int. J. Risk Assessment and Management Vol. 7, Nos. 6/7, 798 - 802
[40] Hall J Blockley and Davis (1998) "Uncertain inference using interval probability theory" International Journal of Approximate Reasoning 19 (1998) 247-264

Figure 9.3 Application of the Italian Flag scheme for articulating uncertainty in assessing risk. Source: Government Office for Science (2012, p. 37)/UK Government/Public Domain/CC-BY-3.0.

Plant, the height and damage from a tsunami wave is uncertain. If the cooling systems for the fuel rods had failed to a passively safe mode of operation such as convective heat transfer to the adjacent Pacific then the system could have been much less vulnerable to what occurred. Alternatively if sufficient time exists between the evidence of existence of the hazard and ability to act, a rigorous observational risk management approach can be used as in geotechnical engineering'

(Government Office for Science, 2012, p. 38)

Note the seamless consideration of processes such as a cooling system, that are enacted by physical entities, and an observational risk management programme entity enacting a process geotechnical engineering.

Process performance is based on evidence that a process is succeeding, not succeeding, or there is uncertainty in our knowledge. These three values can be combined into a single closed interval number as follows

$$P(A) = [S_n(A), S_p(A)]$$

where $S_n(A)$ is the lower bound and $S_p(A)$ is the upper bound of the probability $P(A)$, and can be interpreted as a measure of belief in the performance of process A (Hall, Blockley, & Davis, 1998, p. 248). There are three bounding cases where

$P(A) = [1, 1]$ – the belief that process A is succeeding
$P(A) = [0, 0]$ – the belief that process A is not succeeding
$P(A) = [0, 1]$ – the belief that we have complete uncertainty about the performance of process A

When drawing process models, coloured bars have been used to represent these numbers. Green is used to colour the range corresponding to evidence that the process is succeeding or performing well, white to indicate lack of evidence and hence uncertainty, and red that it is not succeeding or performing badly. Using the colours in this order has led to the mapping being referred to colloquially as an '*Italian Flag*' and is illustrated in Figure 9.4 for several different values of $P(A)$.[9]

The interest in a quantitative account of uncertainty in engineering projects meant that much of the theoretical development in hierarchical process modelling during the 2000's was directed towards evaluating different ways of combining measures, or quantified evidential beliefs, about performance of processes into an overall performance measure for a system. Since all processes in HPM are linked to other processes, the links from sub-processes to their containing superior processes are *moderated* by necessity and sufficiency conditions, not causality. It is

9 As the reproduction of figures in the book is in monochrome, green will be the lighter shading to the left and red will be the darker shading to the right. Colour versions are available in the online version of the book.

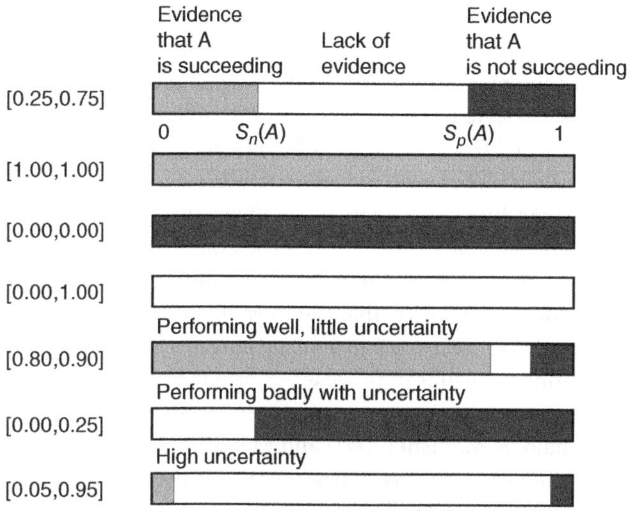

Figure 9.4 Illustrating the interpretation of evidence, expressed as an interval number $[S_n(A), S_p(A)]$, as a coloured flag – hence the notion of an Italian Flag.

possible to consider how necessary and/or sufficient are these sub-processes to the performance of the superior process. In practice, this means that the calculation of the performance of a superior process depends on the number of sub-processes, their measures of process performance, and the degree to which each sub-process is necessary and sufficient for the success of the superior process. The first approaches to the means for this calculation by interval probability theory were presented by Hall, Blockley and Davis (1998), and as subsequently updated by Davis and Hall (2003). During this development phase, many more schemes for the combination and propagation of evidence upwards in an HPM were incorporated into the approach (Marashi, Davis, & Hall, 2008). PhD work undertaken by Marashi (2006) led to the development of the Juniper algorithm, which was a compromise that *was* more suited to the messy consulting engagements of Civil Engineering. This was incorporated into the PERIMETA[10] software that was developed to support consulting engagements by capturing the process structure, the measurement model, and the calculation with its associated parameters.

The results from comparing several different propagation algorithms are shown in Figure 9.5. The effect, as can be seen, is subtle, but as model size increases in the number of sub-processes that need their evidence combined, then the potential for inconsistency between model structure and process evidence increases. There

10 A Java-based implementation under development at the University of Bristol and with industry partners roughly 2005–2008, based on a previous HPM tool called JUNIPER (versions 1–3, 1998–2001).

is a trade-off between the uncertainty propagated through the model from the combination of evidence of process performance and uncertainty about whether the hierarchical structure of the processes in the model is itself meaningful. The Juniper algorithm gave more wriggle room to avoid those conflicts. The implications are discussed by Yearworth, Lowe, et al. (2015), but boil down to the recognition that uncovering inconsistencies between model structure, evidence of process performance, and the understanding of necessity and sufficiency conditions generates additional insight into the structuring of the problem. It also suggests a straightforward remediation – the addition of more processes into the model with a degree of uncertainty; in effect recognising that the most likely reason is inadequate elicitation of all the processes necessary to achieve purpose.[11]

For physical entities, eliciting measures of process performance has been fairly straightforward and based on generating an interval number from data – see Hall et al. (2004), Davis, MacDonald, & Marashi (2007) and Marashi and Davis (2007) for a range of different techniques to do this. These were then used for calculating overall system performance via a suitable propagation algorithm. In practice, the necessity and sufficiency conditions on the links from sub-processes to superior processes are difficult to decide when modelling a complex transformation. Except in circumstances where it is possible to model necessity and sufficiency categorically (i.e. definitely equal to 1.0 or 0.0 – and the consequences of this are discussed by Yearworth, Lowe, et al. (2015)), the values shown in Figure 9.5 were a compromise worked out in conjunction with using the Juniper algorithm.

However, as HPM has been re-purposed to support problem structuring, the entities enacting processes such as strategising, planning, and transforming are more likely to be organisational – such as project teams, change teams, businesses, etc. – and the purposeful gerund is unlikely to be reducible to a single numerical measure of performance, even with uncertainty. As a consequence, the overall structure of the HPM is unlikely to be certain, processes may be non-existent at the time of modelling and therefore measures of process performance will be completely uncertain. The net result is that it becomes less meaningful to think in terms of calculating a measure of performance for the overall purpose or coming up with precise numerical measures of processes where they do exist at the time of modelling. Instead, the '*Italian Flag*' colouring can be used as a judgement or estimation of process performance elicited from participants during a workshop. This results in a simple action planning process from the HPM as the structure of a problem. Where the process performance is judged to be poor it is coloured red and an indicator that the performance of this process needs to

---

11 Modelling with uncertainty, by the use of interval numbers, reflects a process of dealing with known unknowns. An even more realistic model can be achieved by the addition of processes into a model with unknown purpose and either unknown performance or worse i.e. unknown '*confounders*' of purpose. These could be thought of as a recognition of '*unknown unknowns*'.

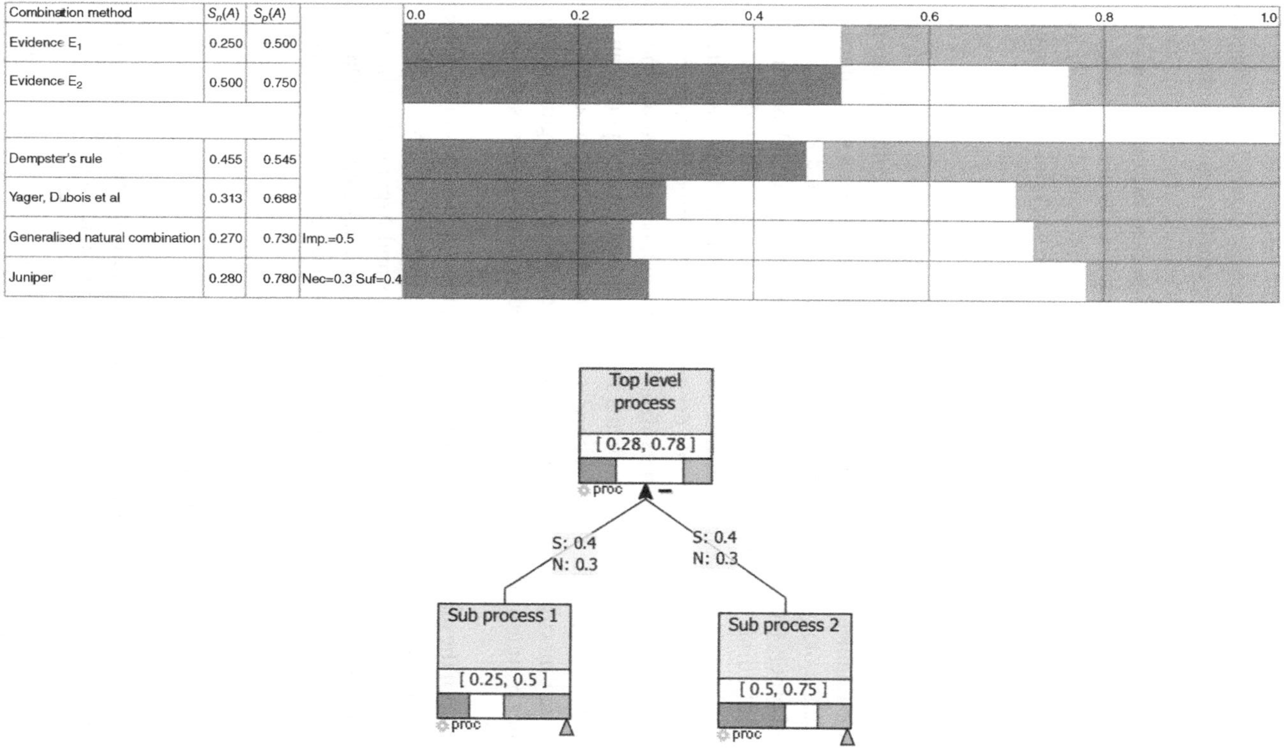

Figure 9.5 The effect of using different algorithms to propagate evidence *'up'* through the model. Source: Adapted from Marashi, Davis and Hall (2003, p. 364) and including the calculation and PERIMETA model for the case of the Juniper algorithm (Marashi, 2006).

be addressed (⇒ '*intervening*'). Processes where there is complete uncertainty are coloured white and an indication to carry out investigations into why the uncertainty exists (⇒ '*finding-out*' or '*enacting*'). Processes coloured green require no attention. In practice, it is still possible to use judgements similar to those as shown in Figure 9.4 – i.e. performing well, little uncertainty; performing badly with uncertainty – without getting into precise numerical measures with the workshop participants; although an example is presented below of combining two forms of judgment into a numerical process performance measure as a convenient way of generating an indication of process performance to workshop participants.

9.3.5 Argumentation

The final element of HPM is the addition of argumentation that is similar to the dialogue mapping of IBIS presented in Section 5.8. Originally called Evidential Discourse for Engineering (EDEN), three argumentation elements were added to the HPM epistemology – Issues, Options, and Arguments – rooted in the work of Conklin and others (see Conklin, 2003; Buckingham Shum et al., 2006; Conklin, 2006). By continuing with finding answers to '*how*' questions through eliciting sub-processes, a group is delving further into understanding how the purpose can be achieved. At some point in this decomposition the process answers will start to run into specific Issues of enactment[12]. Issues can be resolved by Options (i.e. specific enactments), and Arguments for and against each Option can be made. They are added to the HPM graph as a different type of (non-process) node element, which means that hierarchical process modelling with argumentation can use other platforms than PERIMETA. A worked example of this is shown in Section 10.2.

The essence of the dialogue mapping or argumentation approach is to recognise the Process to Issue boundary, in effect the edge of the Hierarchical Process Model, and then shift thinking to mapping Issues to Options and Arguments. The final step is to take the group through a process of voting or preferencing on the Options available, taking into account the Arguments that have been presented, and to start work on an action plan.

9.4 Epistemology for HPM

The epistemology for HPM is shown in Table 9.1, and is comparatively simple when compared to SSM, SODA, and SCA (as shown in Table 5.2).

12 There is a tension here between the benefits of delayed reification, keeping modelling at the level of processes, and moving into a consideration of enactments. The addition of argumentation capabilities to the original modelling approach reflects its engineering origins.

Table 9.1 Epistemology of Hierarchical Process Modelling.

Element	Description
Purpose	(Initial) Top-level process description
Process	Described using the gerund form of a verb
Entity	Enact processes. Implicit in modelling but not stated. Provide the first part of a process statement <This entity is enacting the process...> so that the verb phrase that follows, using the gerund, is the object
Hierarchical Process Model	Hierarchical composition of processes to achieve an overall purpose
Evidence of Process Performance	Formal As an interval number $[S_n(A), S_p(A)]$. It is possible that the numerical evidence could be derived from actual measurement of a process performance (see for example Hall et al. (2004), Davis, MacDonald and Marashi (2007), Marashi and Davis (2007)) Informal Performing well (green) Performing badly (red) Performing unknown (white)
Containment link	Informal Shows that the two processes are related. One process will be the superior process (shown at the arrow head of the link and, usually, higher up the page in the model), and the other will be the contained process (or sub-process) (shown at the tail of the link and usually positioned lower down the page in the model Formal As in the informal case but with numerical Necessity and Sufficiency values
Argumentation	Optional non-process addition to the modelling process to include Issues, Options and Arguments (Marashi & Davis, 2006)
Issue	Non-process Description of an issue that needs resolving. Usually arising from a process labelled with poor performance (red) or uncertain performance (white)
Option	Non-process Description of an Option that could potentially resolve an Issue, expressed as an Enactment
Argument	Non-process The argument that qualifies an Option. Usually positive or negative, but could be anything of use in making an argument.

Deprecated. As HPM moves away from a substance ontology towards a complete processual interpretation entities enacting processes do not need to be considered when modelling.

9.5 The Affordance of the HPM Approach

The affordance of the modelling approach for HPM is presented in Table 9.2. Again, this is comparatively simple.

Given that the affordance of the modelling approach in HPM, and associated epistemology, has had relatively little exposure in the Soft OR/PSM arena it is worthwhile comparing some of its features to a more well-known approach. Modelling with processes enacting a purpose in HPM shares some similarities with purposeful activity systems modelling in SSM.

The insistence on the use of the gerund form of verbs rather than verbs (generally) to express activity sets HPM apart from the activity system models in SSM. In examples of purposeful activity system modelling in SSM, verbs express activities that can complete, whereas the gerund form suggests an ongoing process. This reflects the strong process grounding in the development of HPM arising from its origins in structural engineering – one would expect a process like '*being a structural support*' to be an activity one would hope never completes. Blockley (2010) argues that '*things*', objects in the world, should be modelled as processes too. This seamless integration of hard/soft modelling into a single process view in HPM is well suited to the iterative and ongoing nature of the ideal of PSMs – problematising in a context never ends – by avoiding (or delaying) reification when modelling.[13]

The feedback control to ensure that a system is doing what was intended – i.e. the monitoring and controlling activity – is quite explicit in SSM modelling usually appearing at the outermost level of the model (see Activities 9 and 10 in Figure 5.2). In HPM, feedback control it is embedded within the model itself, both with explicit representation of the performance of every process but also with explicit modelling of monitoring and controlling processes at whatever level of detail is required.[14]

The idea of system boundary is both similar and different. It is similar in that more detail about activity can be found by delving into, or within a process – something called downward containment and achieved through continued '*how*' questioning until sufficient process description has been obtained. Boundary exploration can be achieved by going in the opposite direction, through '*why*' questioning. The only real difference is that in HPM we do not explicitly draw a

13 From these roots, it would not be too difficult to imagine a new conceptualisation of engineering through a processual grounding similar to the manifesto for biology set out by Nicholson and Dupré (2018). Davis, who first taught HPM on the EngD in Systems Programme at Bristol (see Chapter 7), would introduce his lectures with the provocation attributed to Heraclitus, *panta rhei* – everything flows.

14 Kotiadis et al. (2013) put forward a very interesting idea for monitoring and control in SSM that comes quite close to the combination of modelling process structure with process performance expressed in Hall, Blockley and Davis (1998).

Table 9.2 The affordance of the modelling approach for HPM.

Modelling Approach	Hierarchical Process Modelling	
References	Hall, Blockley and Davis (1998), Fletcher and Davis (2002), Marashi and Davis (2006, 2007), Davis, MacDonald, and Marashi (2007) and Marashi, Davis and Hall (2008)	
Philosophy	Everything can be modelled as a process Hierarchical model of processes enacts purpose Delayed reification Process performance is uncertain	
Expression of Activity (Primary objects) (Constituent elements)	Nodes (processes) Process description as text Optional items: Boundary Numbering Decoration with process performance (as interval number and/or '*Italian Flag*')	Processes expressed as gerunds Non-finite verb phrases Have a sense of not completing
Expression of Activity (Secondary objects)	Links Drawn as arrows Optional items: Decoration with necessity and sufficiency (as numbers)	Containment relationship between superior process(es) and inferior process(es)
Associated diagrams	Nodes (NOT processes) Issue description as text Option description as text Argument description as text Optional items: Boundary Numbering	IBIS/Argumentation Map additions
Feedback control	Each process can have a performance measure Monitoring and controlling modelled as additional processes within the overall process model	
Boundary	Hierarchical by downward containment ('*how*' questioning) Boundary exploration upwards ('*why*' questioning)	
Formalism	Directed Acyclic Graph (DAG) The original need for calculation forces use of DAG, hence loops are not allowed	

(Continued)

Table 9.2 (Continued)

Modelling Approach	**Hierarchical Process Modelling**
Acting	Action Intervene (processes with poor performance, '*red*' processes) Enact or find-out (processes with unknown performance, '*white*' processes)
Heuristics	7 ± 2 heuristic Necessity, sufficiency, and dependency between processes
Application (referent)	Extremely broad – any purposeful activity, including the notion that physical entities can enact processes. Examples – Conditioning monitoring complex infrastructure (e.g. Hall, Davis, & Taylor, 1999), Projecting (e.g. Davis, MacDonald, & White, 2010), Energy planning (e.g. Yearworth, Schien, et al., 2015)
Group Model Building	Yes. Not formally defined, but easy to implement. See Chapters 10 and 14.
Tools	PERIMETA and Strategyfinder (see Section 10.2)

boundary around a system. With HPM there is a more idealised sense of a model being a system and therefore does not require it.

Checkland has been clear that there was never any particular graph-based formalism invoked or required to develop activity systems models nor was computer support necessary. This persisted throughout his work by the continued use of hand-drawn models.[15] However, HPM did emerge from a very formal set-based approach to the combination of evidence from process performance measures, and therefore models must be restricted to directed acyclic graphs – i.e. there can be no loops. Whereas in SSM loops are possible. This is not a loss in HPM as the emergence of a process epistemology, that treats all activities as ongoing processes, does not require them[16].

Action in SSM is through the comparison of a model with what is happening in the world. SSM is well-developed in this regard through its grounding in Vickers' concept of an appreciative system, as illustrated by models of the intertwining of inquiry into the world and the world doing its thing (see for example Checkland and Casar (1986)). With HPM, action is motivated by discrepancies between judgment, or actual measures, of process performance and the '*ideal*' needed to enact purpose. The current approach of focussing on the red process '*intervening*'

15 Of course, any general-purpose drawing package can be used to draw an activity system model (e.g. see Pidd (2009, pp. 89–108), or indeed Figures 5.1, 5.2 and 5.3 on SSM) but I was meaning here in the sense of *specifically* designed to support activity systems modelling in SSM.
16 This is discussed further in Section 16.3

and the white process '*enacting*' or '*finding out*' is perfectly workable in practice (see the examples in Chapter 10). A process model to achieve a purpose that was judged, or measured, to have all its constituent processes '*green*' would require no action; purpose is being achieved.

The 7 ± 2 heuristic in SSM arising from Miller (1956) can also be applied to HPM for the same reasons. However, it also arose as a consequence of the original way that evidence was combined through calculation. It was found that the default necessity and sufficiency conditions, arrived at through trial and error in consulting engagements, actually force model structure with not too many subprocess. Very broad process models not only broke the algorithm (Marashi, 2006) they were difficult to understand for the same reasons. This is a useful heuristic that is shared between the two modelling approaches.

HPM has been used quite successfully by groups. Both in traditional same-time/same-place workshops, with participants clustered around flip charts and whiteboards using Post-it® notes and also online using Group Support Systems (see Chapter 14). Whereas, group model building and SSM have mainly come together where there has been some reference to multimethodology – for example in PartiSim (Tako, Kotiadis, & Vasilakis, 2010).

There is no reason why either HPM or SSM need any form of software tool support to be applied. HPM has a software package called PERIMETA, mainly because of its origins to calculate system performance, which requires implementation by an algorithm. Chapter 10 illustrates the repurposing of the Strategyfinder platform to implement hierarchical process modelling.

9.6 Assembling into a Methodology

Adding the generic constitutive rules for an interpretive methodology from Section 5.2, with the epistemology and affordance of modelling approach for HPM presented above creates the CR/E/A framework for HPM used as a PSM. The four elements of HPM can be assembled into a problem structuring methodology by the addition of an initial problematisation, agreeing an initial, anchoring, purpose. The overall process contains the following four phases

1. Problematising: leading to an initial purpose
2. Group model building
3. Evaluating conceptual system performance
4. Identifying actionable options

These phases can be conceptualised into a cyclical process as shown in Figure 9.6.

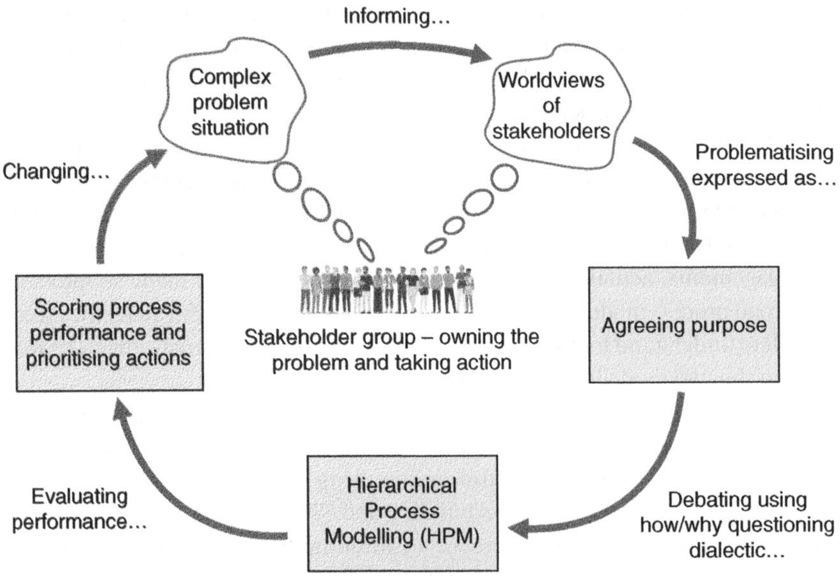

Figure 9.6 HPM as a PSM. Source: Based on an original PSM teaching idea adapted from Hindle (2011).

10

Practical Aspects for Using HPM as a PSM

> *'Experience has shown that many professionals are used to process flow mapping, and that this familiarity can cause confusion when developing a model such as that which is produced using HPM. It is important to move people away from a temporal orientation (i.e. this happens then this happens etc.) towards a purpose-driven perspective. Questions such as "How can I do this" and "Why am I doing this" are useful tools in navigating up and down the hierarchy.'*
>
> (Davis, MacDonald, & White, 2010, p. 902)

Chapter 9 introduced the basic principles of hierarchical process modelling (HPM). Its development to support problem structuring – HPM as a problem structuring method (PSM) – was instigated by a number of industry case studies and consultancies and was first brought to the wider attention of an Operational Research (OR) audience by Davis, MacDonald and White (2010). The Managing Agency Contracts (MACs) project addressed the problem of assessing the performance of contracts between the English Highways Agency and third-party providers of services, such as gritting roads in wintry weather, to maintain the availability of the Strategic Road Network (SRN). In the sense of the meaning of the top-level process established in Section 9.3.2, the overall purpose was *contracting*. This chapter presents insights into how HPM has been used in a number of projects, industry engagements, and teaching exercises across a variety of purposes.

10.1 Case Study I – STEEP and REPLICATE Projects

The STEEP (Systems Thinking for comprehensive city Efficient Energy Planning) project was an EU-funded project collaboration between the cities of Bristol, San Sebastián (Donostia), and Florence between October 2013 and September

Problem Structuring: Methodology in Practice, First Edition. Mike Yearworth.

2015 (Fomento de San Sebastian et al., 2013). The University of Bristol was a partner in the project and tasked with providing a project-wide methodology to assist each city with their energy master planning. The '*STEEP methodology*' was essentially '*HPM as a PSM*' as described in Chapter 9. A significant challenge for the implementation of any methodology to be useful was that the project needed an approach that could be largely self-facilitating in each city, mainly because there was not the budget or time available to facilitate workshops in the three cities on an ongoing basis. An additional requirement aligned with this was to enable the partners in Florence and San Sebastián to host multi-organisational group workshops in Italian and Basque respectively. To achieve the goal of self-facilitation, project partners convened for a one-day training course in the use of HPM as a PSM, which was video recorded, edited, and hosted on the STEEP project website from 2013 to 2021. The methodology, as presented to the workshop participants, is summarised in Yearworth, Schien, et al. (2015). HPM for the energy master planning in each city was developed in a series of group workshops using traditional facilitation tools. However, each city had a license to use the PERIMETA software to capture the models from the workshops. Similar to Checkland and Scholes (1990, pp. 290–297), the process of <Problem structuring using HPM> can also be modelled in HPM (Yearworth, Schien, et al., 2015, p. 9).

The use of HPM as a PSM was carried forward into the subsequent EU-funded Lighthouse project REPLICATE (Renaissance of Places with Innovative Citizenship and Technology) from February 2016 to January 2021 (REPLICATE Project Consortium, 2015). An example of HPM in use by the Comune di Firenze is shown in Figure 10.1.[1] This shows the process for the migration of the taxi fleet in the city to electric vehicles, covering such processes as procuring e-vehicles, licensing, and building-out the charging and monitoring infrastructure (SPES Consulting & Comune di Firenze, 2017). All of the modelling carried out by the Comune di Firenze with their local stakeholder groups for their smart city transition was self-facilitated.

Another example from a facilitated workshop held with Fomento San Sebastián and a local stakeholder group is shown in Figure 10.2. The purpose being modelled in this workshop was <Making the transition towards self-funding of smart city developments>. The workshop format was based on a traditional approach using Post-it® notes, Sharpie® pens, and flipcharts, and the processes and hand-drawn containment links were captured using the PERIMETA software tool after the workshop. PERIMETA enables the combination of the process structure with the measurement model as set out in Section 9.3.4. However, to complete the measurement model, one must have some knowledge of interval numbers and

1 Despite the presentation of models developed during workshops in this chapter, these models are not outcomes from the problem structuring, although they represent the problem *structured*. Outcomes were the actions to taken achieve the purpose.

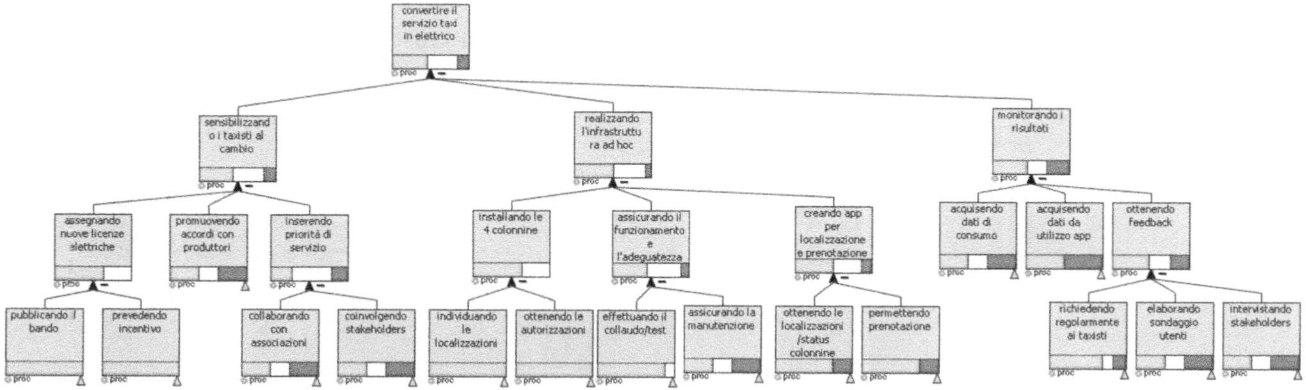

Figure 10.1 HPM for enacting the purpose <Migrating the taxi fleet to e-vehicles> in the City of Florence. Self-facilitated problem structuring was conducted by the Comune di Firenze as part of the REPLICATE project.

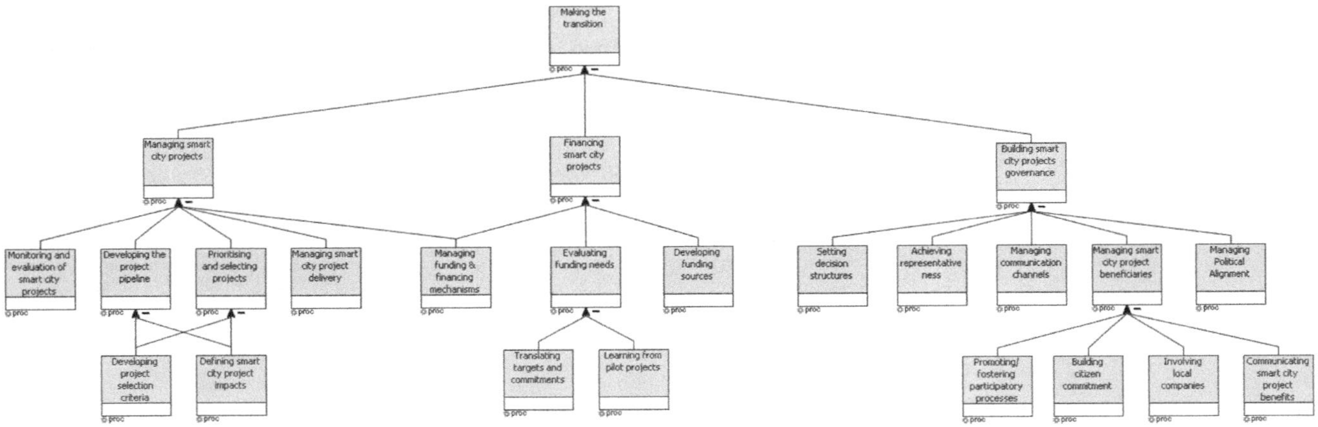

Figure 10.2 Capture of the model developed by the group in the San Sebastian workshop using the PERIMETA software tool.

their graphical representation as Italian flags. To achieve this evaluation of process performance, participants' self-judgements of process performance were captured after the workshop by using an online survey tool. In order to simplify the process of codifying process performance, the Italian flag scoring schema for assessing the performance of a process was modified by combining a judgement of process performance with an assessment of the certainty of judgement and is shown in Figure 10.3 based on an original concept for a mapping from linguistic descriptions by Hall et al. (2004, p. 62) and further refined by Deakin Crick et al. (2017).

The horizontal axis in Figure 10.3 is basically the judgement of process performance, ranging from very poor performance (in red, to the right) to very good performance (in green, to the left)[2]. The vertical axis is the uncertainty in the evaluation of judgement. In the top row there is very low uncertainty, so there is no white in any of the numbers. Going down to the bottom row, we have very high uncertainty where the interval number is essentially all white. There are various ways this chart could have been weighted. The example shown in Figure 10.3 is taken from Lowe, Espinosa and Yearworth (2020, p. 1021).

As show in the top half of Figure 10.4 the '*green*' processes are judged by the workshop participants to be performing well. The bottom half of the Figure shows the processes that have been judged to be performing badly (the '*red*' processes'), or there is uncertainty about process performance/enactment (the '*white*' processes). Actions arising from the workshop are centred on concerns for the following processes (i) Developing methods for project selection, (ii) Translating targets into commitments, (iii) Developing funding sources, (iv) Managing communication channels, (v) Promoting/fostering participatory processes, (vi) Building citizen commitment, and (vii) Communicating smart city project benefits.

10.2 Updating the Tool Support for the Modelling Approach

The model shown in Figure 10.4 has been converted to load into the Strategyfinder platform by interpreting the interval numbers for process performance as preference scores using the simple mapping as follows

$$N_{\text{green}} = \text{int}(N_{\text{pref}}\, S_n)$$

$$N_{\text{white}} = \text{int}(N_{\text{pref}}\, (S_p - S_n))$$

$$N_{\text{red}} = \text{int}(N_{\text{pref}}\, (1 - S_p))$$

---

2 As the reproduction of figures in the book is in monochrome, green will be the lighter shading to the left and red will be the darker shading to the right. Colour versions are available in the online version of the book.

	Judgement of Process Performance				
The process performance has been evaluated with …	Very poor performance	Poor performance	Neutral performance	Good performance	Very good performance
Very low uncertainty	[0.00, 0.00]	[0.33, 0.33]	[0.50, 0.50]	[0.67, 0.67]	[1.00, 1.00]
Minor uncertainty	[0.00, 0.25]	[0.25, 0.50]	[0.42, 0.58]	[0.50, 0.75]	[0.75, 1.00]
Moderate uncertainty	[0.00, 0.50]	[0.17, 0.67]	[0.25, 0.75]	[0.33, 0.83]	[0.50, 1.00]
High uncertainty	[0.00, 0.75]	[0.08, 0.83]	[0.17, 0.83]	[0.17, 0.92]	[0.25, 1.00]
Very high uncertainty	[0.00, 0.83]	[0.00, 0.92]	[0.00, 1.00]	[0.08, 1.00]	[0.17, 1.00]

Figure 10.3 Italian flag scoring schema for assessing the performance of a process – combining judgement of process performance with an assessment of the certainty of judgement.

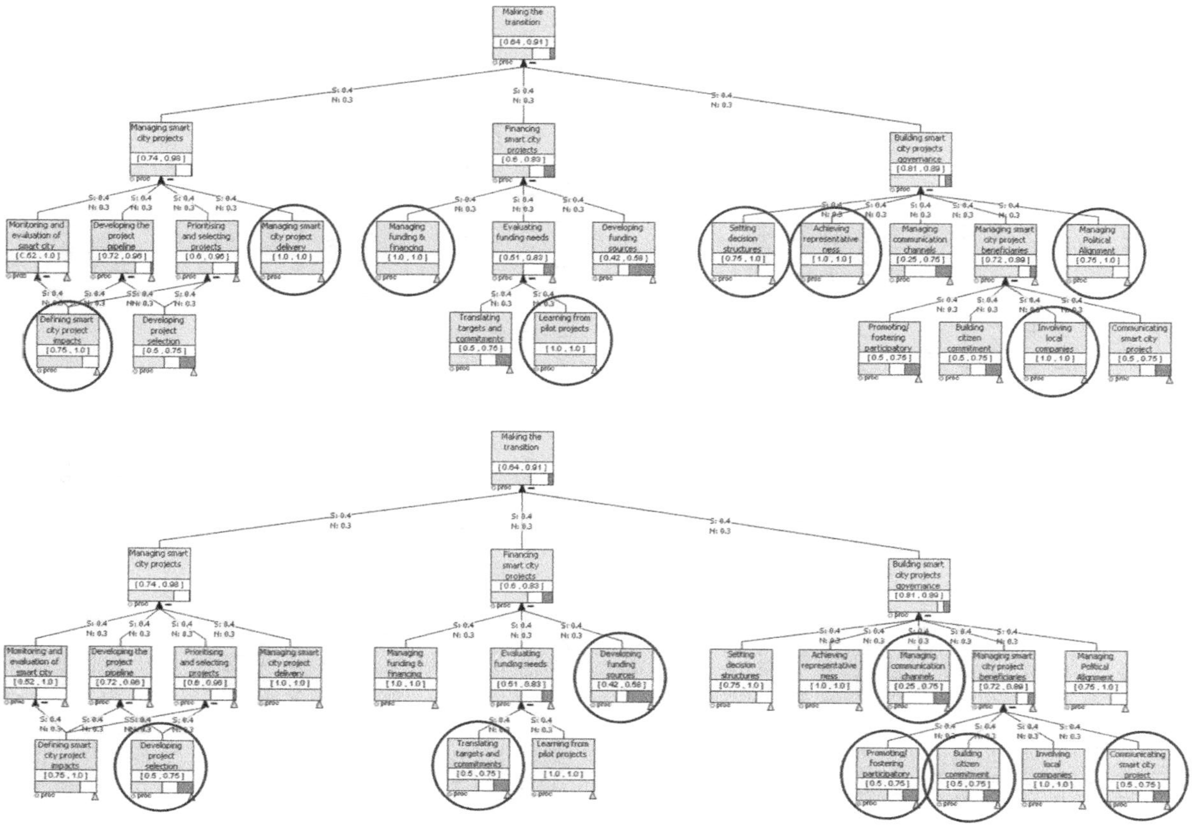

Figure 10.4 San Sebastian workshop model with evidence of process performance – processes judged to be performing well (above) and processes judged to be not performing well and/or there is uncertainty about process performance (below).

For example, the evaluation of process performance '*poor performance with high uncertainty*' (as shown in Figure 10.3) would correspond to the interval number [0.08, 0.83] and get mapped to the preferencing score (8)(75)(17) for $N_{\text{pref}} = 100$ as shown in Figure 10.5.

Issues, Options, and Arguments can also be expressed naturally in Strategyfinder[3] and are translated '*as is*' from PERIMETA.[4] Here, by way of an example taken from a teaching exercise, the process <Tracking customer interactions with the organisation> is evaluated with poor performance and no uncertainty as a consequence of an Issue of enactment – there is simply not the right sort of IT system available to the organisation that can implement this process (Figure 10.6). A number of Options have been elicited to resolve this Issue – (i) Acquire a Customer Relationship Management (CRM) system, (ii) Enhance existing systems with CRM capabilities, and (iii) Build a custom CRM system (Figure 10.7). The Arguments for and against each Option are shown in Figure 10.8. Arguments in favour of the Option are indicated using normal links in Strategyfinder, Arguments that are against the Option are indicated with negative links (shown by the minus sign decoration) (Figure 10.9).

10.3 Teaching Example I – '*SuperCo*'

An example of problem structuring is provided from an MBA Operations Management class group exercise. This was designed to teach an approach to problem structuring using HPM. Each group was asked to structure a transformational challenge facing a senior management team using the following brief.

> '*The "traditional" retail supermarket "SuperCo"*[5] *currently has no home delivery service. Your team has been tasked by the CEO on behalf of the Board to lead a transformation to acquire the "Acme Platform" as a managed service and integrate it into the existing retail business in order to launch a new home delivery service. The Acme Platform consists of technologies, especially robotics and optimisation algorithms, that are likely to be new to the SuperCo*

---

3 Strategyfinder allows for the arbitrary labelling (and formatting) of statements, following Decision Explorer.

4 If Strategyfinder were to be used for hierarchical process modelling then the preferencing would be elicited directly from the workshop participants, not via this convoluted method of translation (albeit by some straightforward python code that converts the XML in the PERIMETA model file to a JSON representation that can imported into Strategyfinder).

5 '*SuperCo*' is nationwide, but single country, and includes 500 stores, 5 distribution centres, 500 tractor units, and 2000 trailers. The business employs 50 000 people. The underpinning case for this teaching example is actually the '*Acme Platform*', examples of which can be found as commercial service offerings. '*SuperCo*' is just a fiction.

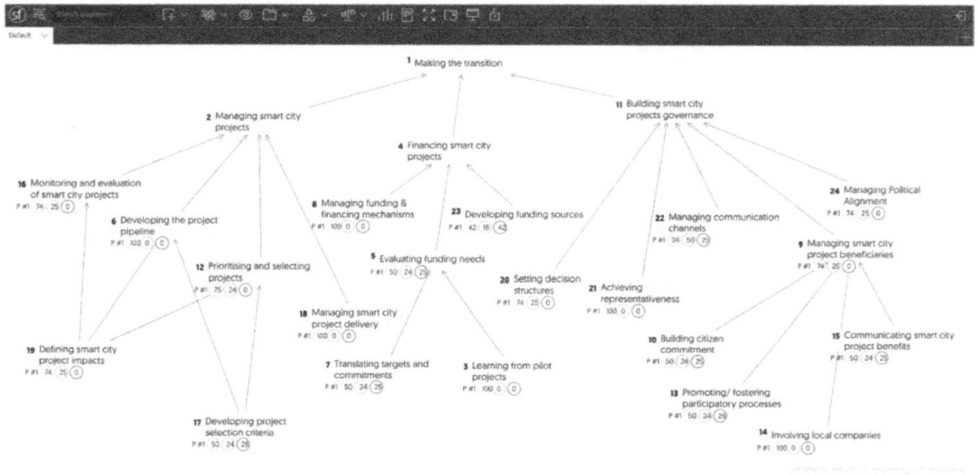

Figure 10.5 The HPM shown in Figure 10.4 using the PERIMETA software translated for representation in strategyfinder.

Figure 10.6 Introduction of an Issue. The performance of the process '*Tracking customer interactions*' has been evaluated as very poor with no uncertainty. The Issue is that there is no IT System to do this.

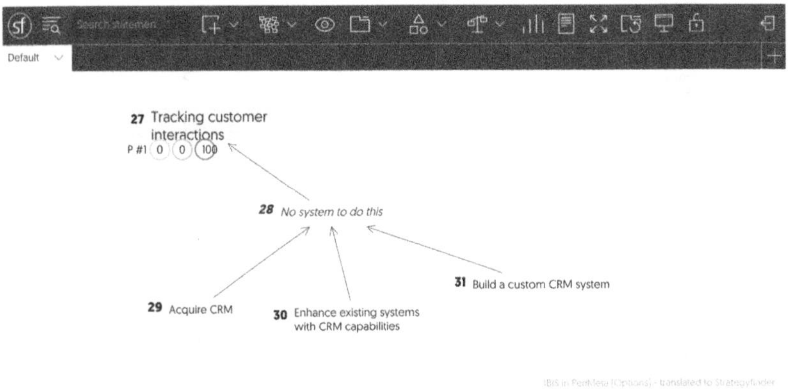

Figure 10.7 Options have been elicited. System has, in this case, been interpreted as an IT System and the Options are all variations of a Customer Relationship Management (CRM) System. The interpretation could have been different.

management team. The introduction of robotics is likely to have an impact on staffing. How will this transformation impact the existing retail store footprint i.e., as customers switch from in-store to home shopping? Remember to think about the 4 V's and where the current business sits on the line of fit and its flexibility versus cost focus.[6] *Does the proposed transformation change this? Think about the resilience of the new Acme Platform if there was an event such that order rates and volumes suddenly increased by a factor of 2 or more, or that availability of staff becomes a problem.'*

6 See a standard textbook on Operations Management (e.g. Slack and Brandon-Jones (2019) or Brown, Bessant and Jia (2018)).

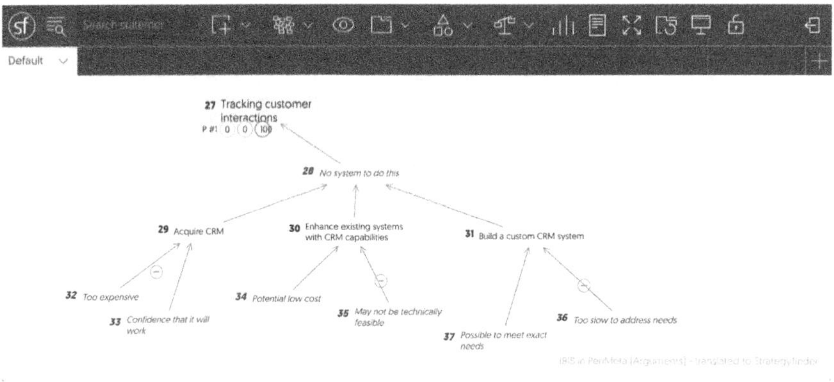

Figure 10.8 Arguments for and against each Option have now been introduced into the model.

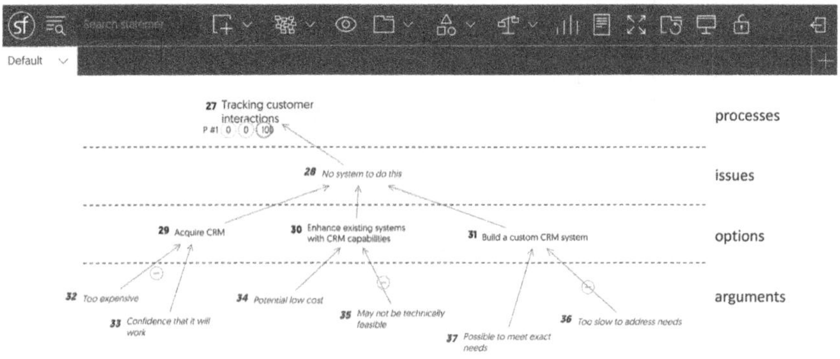

Figure 10.9 Highlighting the key components of argumentation at the Process to Issue boundary.

Groups were expected to explore the problematisation by using HPM as a PSM and questioning of purpose through the How/Why dialectic inherent in the modelling approach.

The starting point for modelling embodies, as its ostensible purpose, the transformation of a supermarket with no home delivery service to one with a home delivery service. This is the design of an operational change set for the operations team in SuperCo. However, in addition to the ostensible purpose, there are, in fact, a number of other transformations that will take place at the same time. These are some of the consequences of the decision-making inside SuperCo on wider stakeholders.[7] As shown in Table 10.1, some of these might be intended, but others are

7 The methods for identification of such stakeholders are discussed in Section 6.4. The approach described by Gregory et al. (2020) would be of particular relevance.

Table 10.1 Messiness spectrum for the SuperCo class exercise.

Less Messy	Messy	More Messy
Looking into the platform, we see ample scope for quite traditional hard OR. There is plenty to optimise here. The routing of robots in the picking process to avoid collisions, reduce the distance travelled, allow for battery charging time and avoid having underutilised resources, i.e. robots not doing anything, space unoccupied. Needs picking algorithms, routing, etc. Consider issues such as delivery scheduling, shelf-picking/warehousing, supply chain management, predictive analytics, logistics optimisations requiring vehicle routing, charging, fleet sizing, customer service level agreements.	The context provides an initial problem statement that would be considered firmly within the realm of strategising or transforming, e.g. managing the supplier relationship, dealing with staffing issues, managing the transformation team, managing customer expectations, marketing, deciding warehouse location, etc.	Looking at the wider context, we see the following additional transformations as possible in addition to the ostensible transformation: • Local workforce employment → Local workforce more/less employed • Local economy → Local economy buoyed/depressed • Road congestion → Roads more/less congested • Local pollution levels → Local pollution levels higher/lower • Civic amenities → Civic amenities enhanced/reduced • In-work poverty level → In-work poverty level raised/lowered • Housing stock affordability → Housing stock affordability raised/lowered Local water capacity → improved/challenged

clearly undesirable. Local employment levels will change; perhaps there will be an increase in unemployment as jobs are replaced by robotics. Perhaps there will be new jobs supervising the robots or keeping them working? The establishment of new customer fulfilment centres might stimulate the local economy. Roads and transport infrastructure could be affected. There may be impacts on civic amenities, local pollution levels, an increase in water load, and social conditions may change – perhaps there will be increased levels of in-work poverty arising from downward-skilling or impacts on the affordability of local housing.

10.4 Teaching Example II – Servitisation and the Circular Economy

An example is shown in Figure 10.10 of a hierarchical process model developed by an Operations Management student group during a period when the MBA programme at the University of Exeter was being delivered online as a consequence of the COVID-19 pandemic lock-downs and travel restrictions (Yearworth & White, 2021). This was a case study addressing the servitisation (Neely, 2009) of a retail furniture company's business model and the transformation towards furniture rentals and the circular economy (Geissdoerfer et al., 2017; Hopkinson et al., 2018). This case example is discussed in detail in Chapter 14.

10.5 Teaching Example III – Heritage Railway MEng Consulting Project

The introduction of a problem-structuring approach and teaching of HPM as a systems-modelling technique in the final-year M-Level Civil Engineering Systems Module at Bristol (see Chapter 7) meant that students were in a position to use this approach to structure the problematic situation of their capstone consultancy project for their MEng qualification. In this case the consultancy project was an engagement between the four-person student team and the client to produce detailed designs for extending part of a heritage steam railway. While the detailed design elements[8] were the ostensible purpose of the assessment exercise for their Engineering Council-accredited MEng qualification in Civil Engineering, the consultancy project needed to provide advice to the client on all aspects

8 For a proposed extension of the railway line, these were designs for (i) a new drainage system, (ii) a footbridge, (iii) a road bridge, and (iv) a slope stability remediation scheme. The hyperlinks in the model shown in Figure 10.11 point to the design documents.

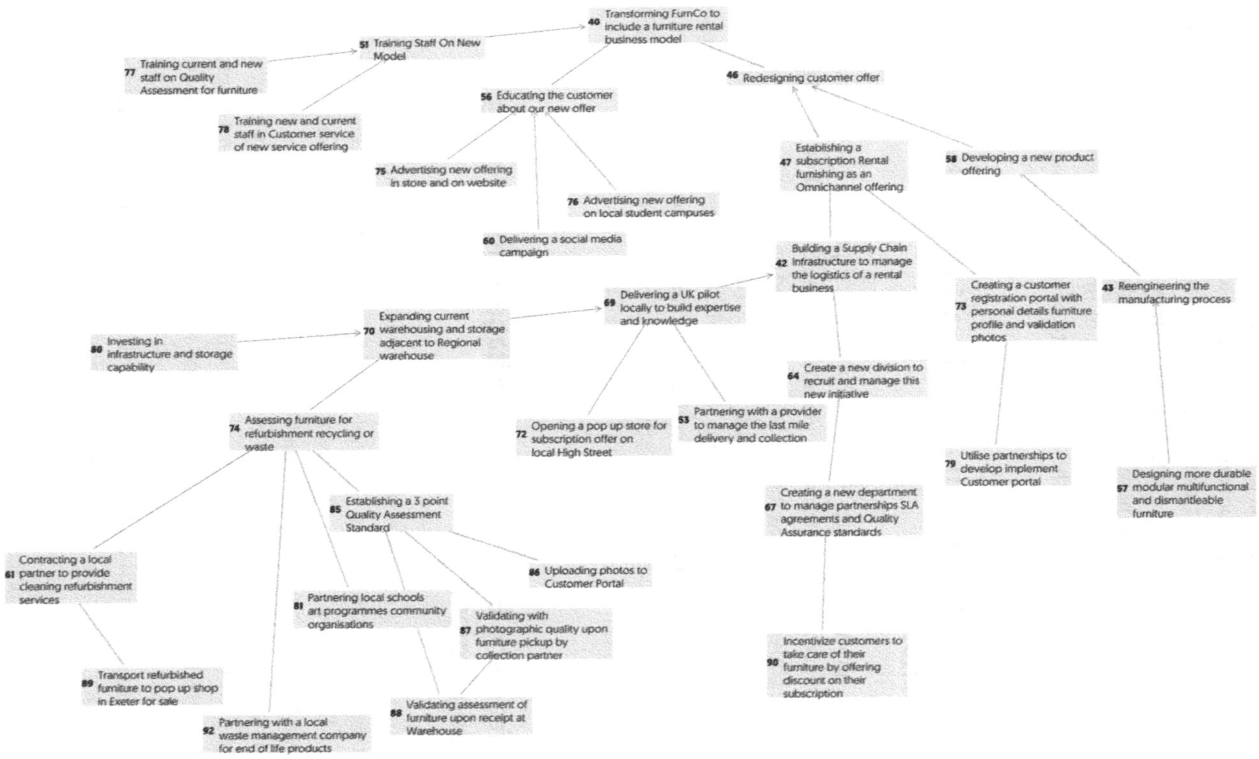

Figure 10.10 Example model from a student group of the purpose <Transforming FurnCo to include a furniture rental business model>.

of the problem context, not least broadening the funding basis and addressing issues of sustainability and accessibility. The problem context and the structuring of the problem, therefore, provided additional requirements and constraints on the designs that were produced that were not apparent in the original brief. The process model used to structure the problem, including some evaluation of process performance, was also used to support the writing of the report to the client and is shown in Figure 10.11.

10.6 Case Study II – '*Continuous Process Manufacturing Company*'

A continuous process manufacturing company ('*CPMC*') was looking to transform its business model from an installation of plant plus break/fix service to a fully servitised (Neely, 2009) availability guarantee service.[9] Moving customers away from thinking in terms of capital expenditure on plant and towards purchasing a continuous process manufacturing service. I was invited to spend a short time with the senior leadership team to conduct a workshop exploring this question. The opportunity arose as a consequence of a prior meeting with engineering colleagues who were offering potential technical and research services. I facilitated the team through a modelling exercise that covered the transformation they would have to enact to deliver the new service. Whilst it was clear that quantitative modelling would be required for predictive maintenance – i.e. expert mathematical modelling of vibration in elements of the process machinery – this was only part of the overall picture. It was a necessary element, but not sufficient to achieve the overall purpose. I was looking at the whole transformation, and not just from within the company but also what they would have to achieve in their industry. For example, customers would need to be reoriented towards buying a service rather than capital equipment. The company would also have to show evidence of being a thought leader in this space by demonstrating that they possessed an appropriate analytics capability.

This case study illustrates the speed with which it is possible to achieve meaningful results. In this case, we used an opportunity where the senior leadership team of CPMC was in the same place and had some (limited) time available. Getting additional agenda time such as this is often all the time we have to work with busy teams; organising any time with a senior leadership team or anyone with the

9 Similar in idea to the business transformation that led to introduction of the Rolls-Royce '*TotalCare*' service.

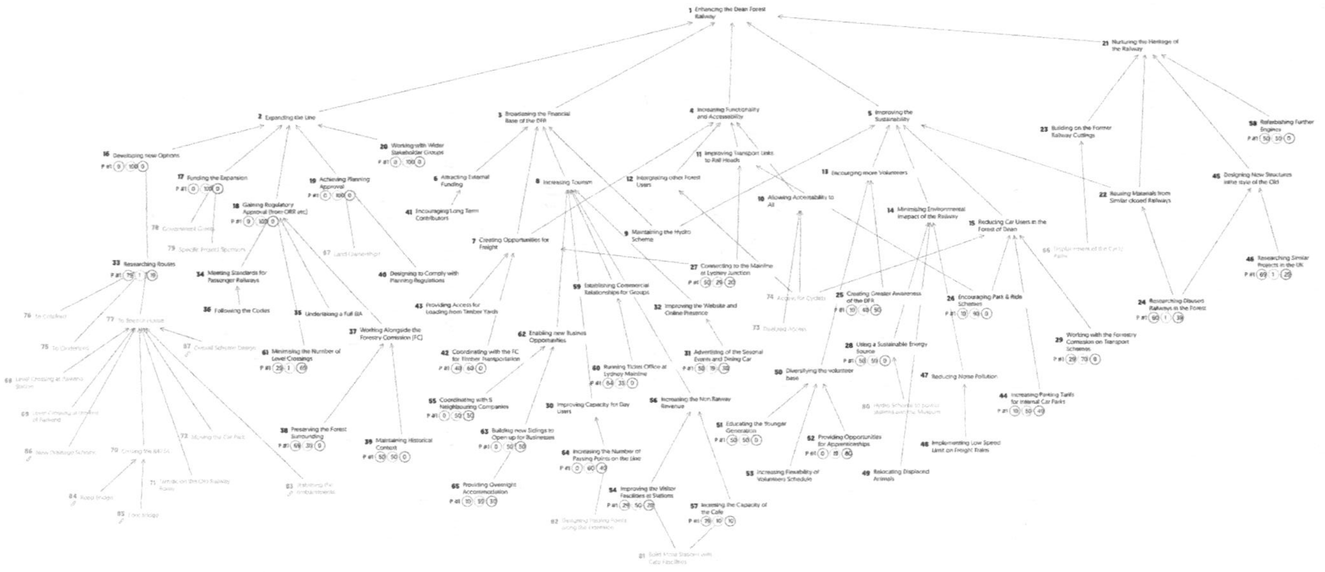

Figure 10.11 Enhancing the heritage steam railway. Note that statements 83 to 87 illustrate the facility in Strategyfinder to include hyperlinks. In this case, links to the relevant design documents.

power to commit resources to change is hard. One of the things about PSMs is that they can be deployed and used quickly. In this case, I used hierarchical process modelling as a PSM as described in Chapter 9, and the actual modelling was done as a group using Post-it® notes on a whiteboard. The How/Why questioning dialectic and use of gerunds in the modelling process were explained quickly, and a shared purpose was agreed on the fly to start with the initial top-level process as <Moving from cost-plus to selling performance>. The case study also illustrates how quantitative modelling is incorporated into the conceptual model of the transformation as <Maintaining mathematical/software models of our process> and <Predicting reliability> with associated sub-processes – see Figures 10.12 and 10.13. The delayed reification of the mathematical modelling meant that valuable time with the senior leadership team could be spent looking at the overall transformation. From the point of view of assessing the performance of the overall transformation, it is also clear that the results from such modelling can be represented at this stage not just by their presence as a process in the model but also by the associated uncertainty, in this case representing complete uncertainty in enactment.

10.7 Case Study III – Healthy Resilient Cities

The Healthy Resilient Cities project was a short study funded by the United Kingdom Natural Environment Research Council (NERC) and intended to address questions about the resilience of healthcare provision in Bristol in the face of potential extreme climate events (Yearworth, 2015a). The systemic problem structuring approach for the project was based on using HPM for the process modelling.

Limited availability with stakeholders meant that initial models were developed from a number of semi-structured interviews with key actors before the single group workshop was held, in a similar intent to cognitive mapping in strategic options development and analysis (SODA). The subsequent group workshop consisted of three distinct phases: pre-workshop preparation, workshop delivery, and analysis and reporting back to stakeholders. The problem structuring with stakeholders was quite distinct from a separate quantitative modelling activity that was needed to develop and predict measures of the resilience of the healthcare system. A purpose for the project team to work on was derived using a root definition from soft systems methodology (SSM).

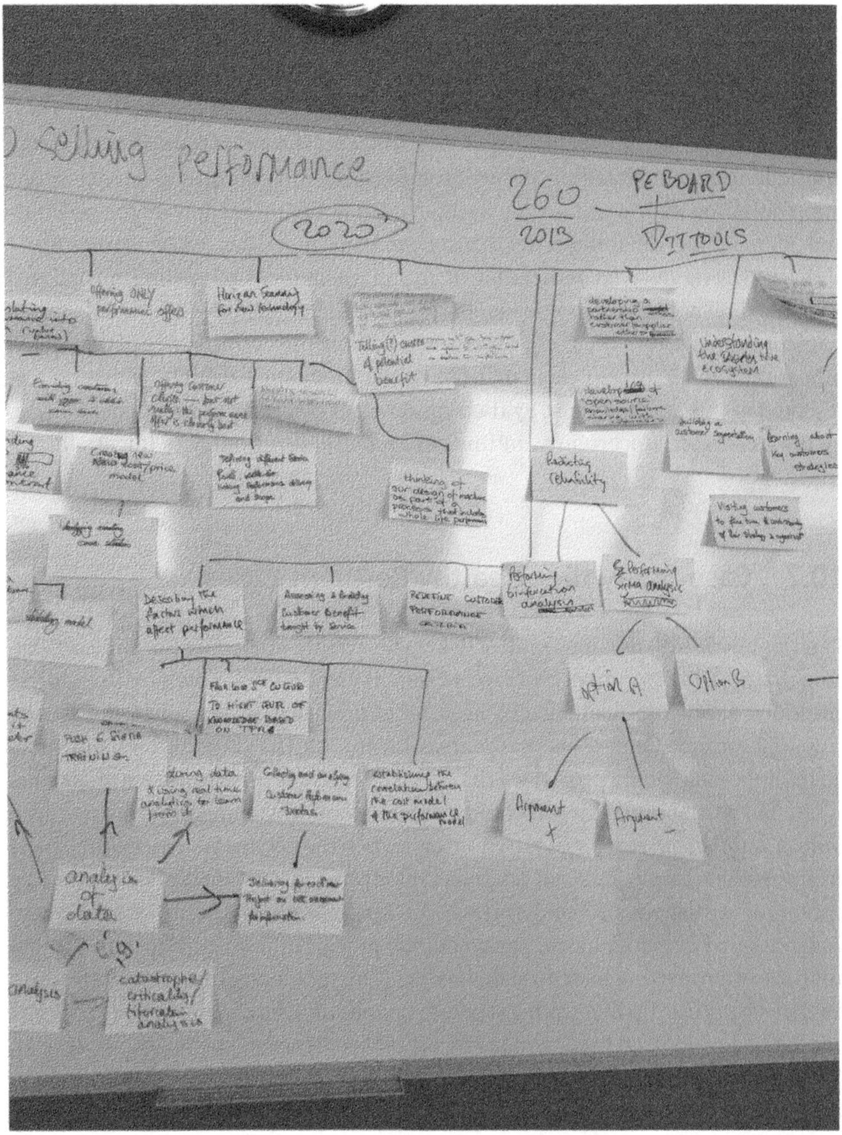

Figure 10.12 Section of the HPM developed CPMC. Note the two '*hard*' analysis projects required for the predicting reliability process.

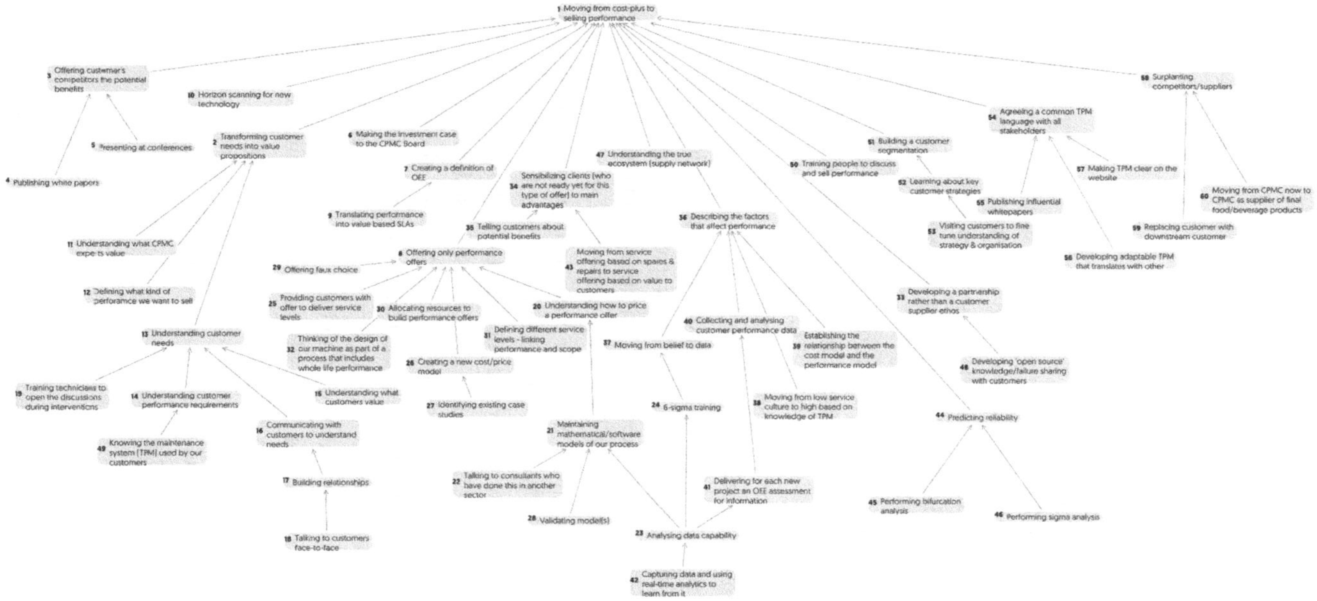

Figure 10.13 Workshop model transferred to Strategyfinder (via PERIMETA) for clarity.

> *'A system for the Healthy Resilient Cities project team (A) funded by Innovate UK (O) to improve the resilience of healthcare provision in Bristol to extreme climate events (T) for the Bristol community at large (C) based on the view that compelling business cases for taking action can be developed (W) and which is seen as essential activity to deal with the predicted effects of climate change (E)'*

This led to the purpose expressed as the process <Improving the resilience of healthcare provision in Bristol to extreme climate events>. The prevailing worldview from the preliminary interviews with key stakeholders in the healthcare trusts involved in the project was that any action to improve the resilience of healthcare provisioning in Bristol could only proceed if '*compelling business cases*' could be developed.[10]

The problem structuring approach required a clearly defined shared ownership of the transformation to be achieved. In setting the scope of the transformation process, the project team had to ensure that participants invited to the workshop had the necessary power to implement any actions identified in the workshop to achieve the transformation. The project team also had to take responsibility for recruiting the stakeholder group and ensuring participation in the workshop. The facilitator took responsibility for preparing a preliminary model based on the project proposal document and the initial stakeholder mapping meetings. The model was used to initiate the group workshop, similar to causal mapping from merged cognitive maps in SODA or preliminary models in the Group Model Building (Vennix, 1996). We recognised that this problem structuring reflected our understanding of how the project needed to be delivered, but also that we needed to be sensitive to issues and concerns that stakeholders were likely to have and that by working with these we could ensure the effectiveness of the project and discover possible synergies. The overall process for the project was also modelled by the project team and is shown in Figure 10.14. In effect, the problem to be addressed by the project and the project itself were both viewed as sub-processes of the original purpose <Improving the resilience …>. Part of the motivation for doing this was a consequence of a stipulation of the funding from NERC that the project should also demonstrate how its methodology could be operationalised elsewhere and delivered as a commercial consulting offering as a possible mechanism for scale-up.

---

10 This was not too dissimilar from the worldview prevalent in the REPLICATE project. Action was predicated on developing feasible business cases.

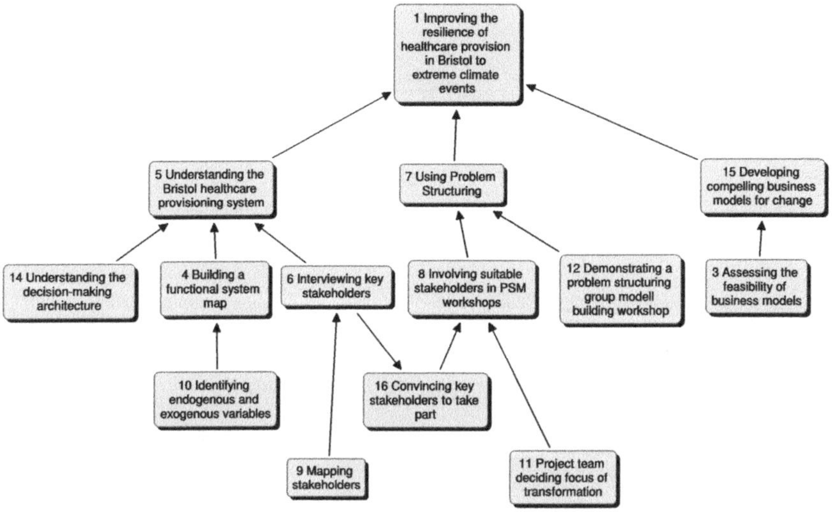

Figure 10.14 Modelling the Healthy Resilient Cities project itself as a sub process of the overall purpose.

10.7.1 Workshop

The participants were introduced to the purpose of problem structuring, where it was explained that problem structuring methods have a long tradition of being used to help stakeholders develop a shared understanding of a messy, problematic situation, learn about it, and build commitment to taking action to improve the situation. The workshop was initially run with three sub-groups, each tackling one aspect of the preliminary models developed in the stakeholder interviews. The models were constructed in the sub-groups traditionally using Post-it® notes, Sharpie® pens and flipcharts. The models were transcribed using Decision Explorer after the workshop. The first model, shown in Figure 10.15, reflected the group's focus on the project achieving a breakthrough whilst recognising that much work had already been carried out in this space, and it was therefore important for the project to build on this and not alienate people who have already invested a lot of effort in their own attempts to achieve similar outcomes.

The second group explored the difference between short- and long-term decision-making in NHS trusts and, crucially, the need to reconnect the process of developing long-term strategy with the annual budgeting cycle, shown in Figure 10.16.

The third group, see Figure 10.17, mainly focused on how to influence an NHS trust board and introduced the idea that, because of organisational differences, business cases would have to be tailored for specific boards. This group also

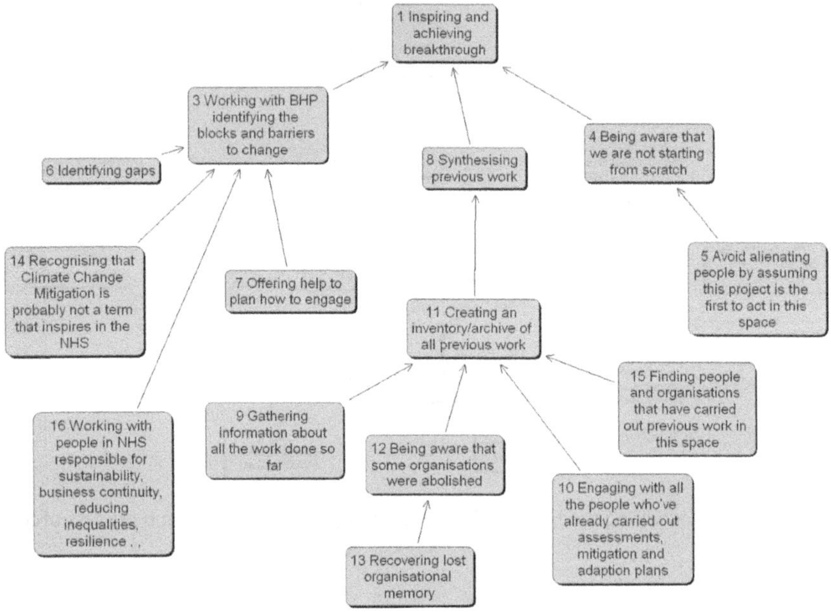

Figure 10.15 Inspiring and achieving breakthrough.

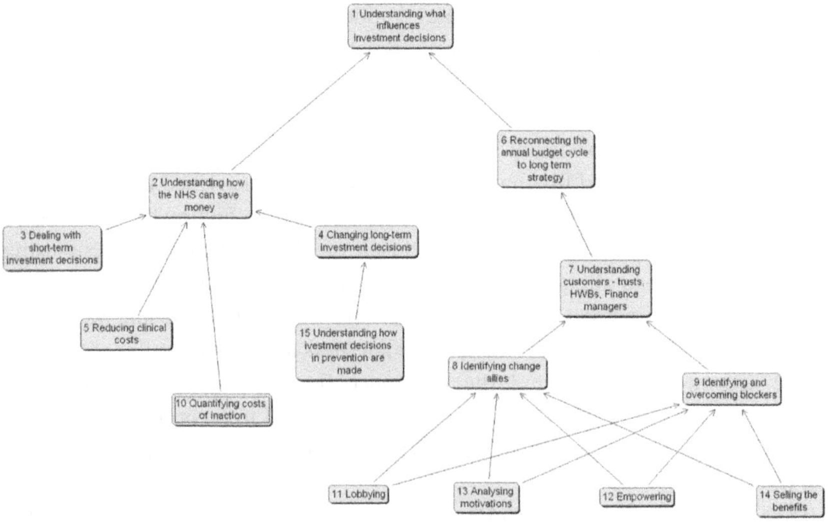

Figure 10.16 Understanding what influences investment decisions.

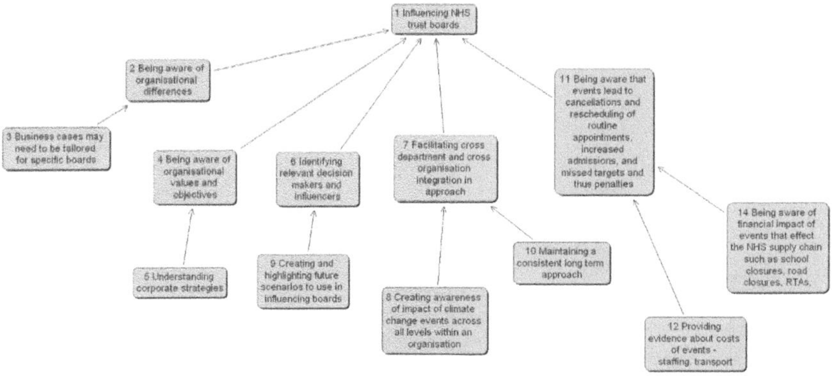

Figure 10.17 Influencing NHS Trust Boards.

explored the problem of (extreme) climate events leading to missed targets, either directly through the need to reschedule booked appointments and dealing with admissions arising from the event or due to events having an impact on the NHS service delivery supply chain, such as school and road closures, which would degrade service delivery performance by preventing staff from getting to work.

The work from the three groups was then consolidated for a final session of the whole group, as shown in Figure 10.18.

Preliminary analysis of the outcomes from the problem structuring workshop modelling suggested the following areas for improving effectiveness, achieving synergies, and operationalising the findings that would lead to actions to achieve the purpose.

1. Piggybacking or inserting business models into existing planning focused on business continuity, resilience, and addressing health inequalities (achieving synergies)
2. Identifying where business models developed in the past have not led to funding and implementation or failed to deliver returns as expected (being effective)
3. Presenting new business models with clear reference to and acknowledgement of previous work (being effective)
4. Working out how to reconnect the process of developing long-term strategy with the annual budget cycle (operationalising)
5. Connecting new business models developed in this project to current efforts to save money in the NHS (achieving synergies)
6. Being aware that different NHS trust boards will evaluate business cases differently and, therefore, that new business cases emerging from the project will have to be tailored (being effective)

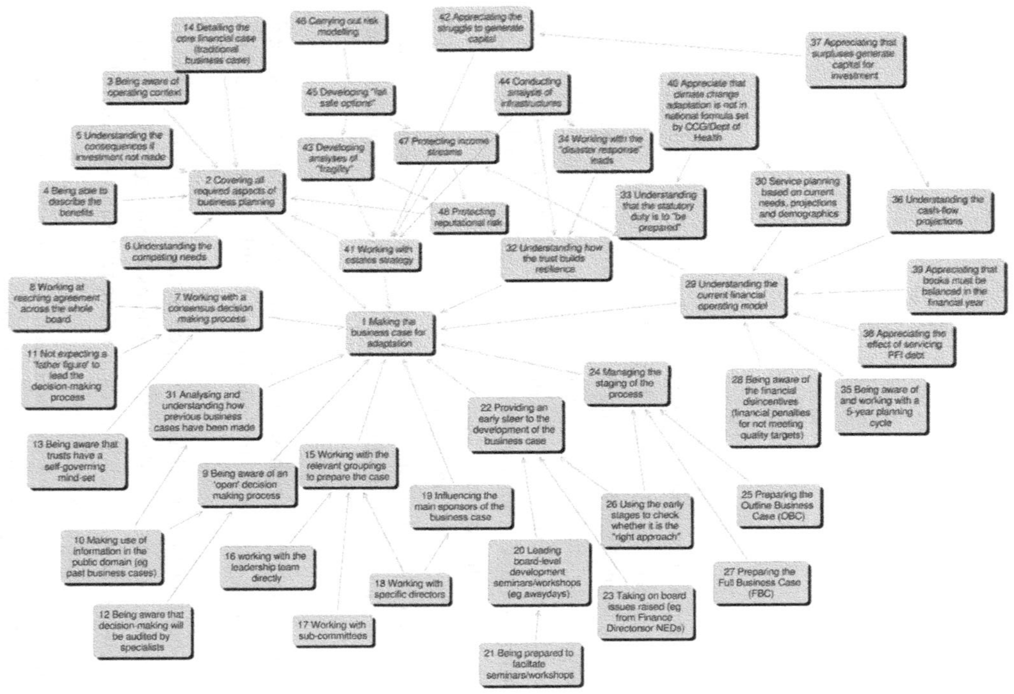

Figure 10.18 Healthy Resilient Cities: Building a Business Case for Adaption (NERC grant NE/N007638/1). Source: Adapted from Yearworth (2015).

7. Appreciating that developing new business cases is not enough. Part of making them *compelling* is the process of identifying and removing blockers (being effective)
8. The nature of a compelling business plan is unlikely to be purely economic. Boards will need to be influenced by targeting key individual decision makers with suitably convincing material, e.g. scenarios, and their organisations primed to be receptive (achieving synergies)
9. Understanding both the direct impact of events on costs and the indirect impact arising from impairment in the NHS resource supply chain will be a necessary component of developing new business models (operationalising)

10.7.2 Key Findings from the Project

Business cases for mitigating actions needed to fit into the current budget decision-making process in trusts. The scale of interventions, and hence project scope, is bounded by the financial limits of decision-making of the NHS trust board. Existing risk and business continuity plans can be enhanced by the use of infrastructure interdependency workshops and by using a broad interpretation of infrastructures (and interdependencies, see Section 10.7.4). The economic valuation work carried out in the project provided the necessary supporting data for estimating consequential losses. However, the nature of a *compelling* business plan is unlikely to be purely economic, which requires identifying where business models developed in the past have not led to funding and implementation or failed to deliver returns as expected. Presenting new business models needs to be done with clear reference to and acknowledgement of previous work. Being aware that different NHS trust boards will evaluate business cases differently and, therefore, that new business cases emerging from the project will have to be tailored. Appreciating that developing new business cases is not enough, part of making them compelling is the process of identifying and removing blockers. Boards will therefore need to be influenced by targeting key individual decision-makers with suitably convincing material, e.g. scenarios, and their organisations primed to be receptive to the climate change agenda. Strategies to meet this would require '*piggybacking*' or inserting business models into existing planning focused on business continuity, resilience, addressing health inequalities, and connecting new business models developed in the project to current efforts to save money in the NHS. Understanding how both the direct impact of events on costs and the indirect impact arising from impairment in the NHS resource supply chain will be a necessary component. Other, non project, interventions could be focused on making trusts more effective at dealing with climate change-induced events, e.g. by developing strategic and organisational capabilities – although such interventions would fall outside the business case justification approach of the project. Whilst the project raised the visibility of the risk to healthcare provision of extreme weather events arising from climate change, system-wide changes, such as changes to funding allocation models at a national level, are required to substantially improve the resilience of healthcare provision.

10.7.3 Selling the Approach as a Commercial Service

The issues involved in translating the methodology from this project into a commercial offering have already been discussed in relation to the value paradox of PSMs discussed in Section 6.3. Outcomes from using this approach more generally, i.e. specific business cases or organisational changes arising from subsequent commercial engagement are unpredictable and therefore unknown at the outset. Placing any sort of value, especially financial, on the engagement is therefore problematic. The question of value has to focus on the *process* that enacts the engagement rather than the outcomes. Consequently, the client needs to trust the provider of the process since it represents a journey for the client into the literally unknown. Value is therefore associated with the *quality of trust* in the relationship. Management consultancies know this and is a key feature of their brand. Trust needs to be established early in a potential commercial relationship, i.e. in the pre-sales phase due to the uncertain nature of problematisation, which may mean on occasion that translation into an engagement does not happen. New entrants will have to build trust relationships on the basis of specialist, niche knowledge; in the case of this project, the economic valuation approach is key, followed by (i) insights from interviews and the first workshop on how NHS trusts develop and decide on business cases, and (ii) bringing about re-appraisal of risks using the interdependency method.

The service offering is likely to have three main phases

1. Raising awareness of risks with potential clients
2. Developing strategic responses
3. Constructing business cases for specific project-based adaptations

Delivery requires specific and, what seems to be, at present, niche capability in

1. Being able to place an economic valuation on the consequences of extreme weather events,
2. Problem structuring, i.e. being able to translate competencies developed in this project into alignment with specific needs in an NHS trust and thus constructing meaningful business cases,
3. Infrastructure interdependency analysis: re-evaluating the risk register in a trust to reflect extreme weather events.

10.7.4 Infrastructure Interdependency

An additional workshop was held to enhance the impact of the project by taking a structured, systemic, and rigorous approach to understanding '*infrastructure*' interdependencies with key stakeholders. It attempted to address the problem of hindsight, i.e. '*Who would have thought that…?*' events when it comes to the impact on healthcare service delivery in Bristol of extreme weather events and climate change. The approach was based on the '*N-Squared*' technique developed by Rosenberg et al. (2014) and used successfully with HM Treasury (2015). A simple example emerging from the workshop is presented in Figure 10.19.

The causal map illustrates the deleterious effects of (i) loss of availability of a data centre, (ii) decrease in quality of road gritting, (iii) loss of availability of local schools, or (iv) lack of available fuel supplies. These either lead to a decrease in the availability of NHS staff or an increase in '*bed blocking*', i.e. the inability to discharge patients from hospitals, and leading to an overall reduction in the overall quality of healthcare service delivery.

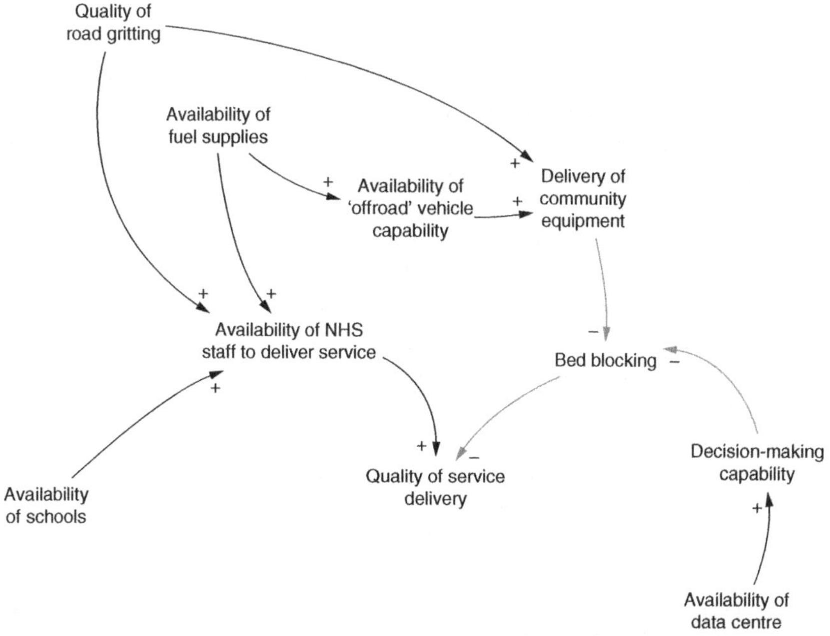

Figure 10.19 Example of a model extracted from the infrastructure interdependency analysis.

10.8 Conclusions

One of the unique contributions of HPM is the possible integration of hard and soft in the same model, not in the paradigmatic sense or as multimethodology,[11] but in the way in which the conceptual and physical (objects, and quantities such as measurements or outputs from predictive models) can be mixed together in a single coherent whole model of processes. In this way, it bears some relationship to the aims of modelling a socio-technical system within a variety of application domains/discipline traditions.[12] As demonstrated at the start of the chapter, HPM has been used to model a wide range of smart city/low carbon interventions in Florence as part of the REPLICATE project (SPES Consulting & Comune di Firenze, 2017), and these models made extensive use of this delayed reification.

The affordance of the modelling approach that enables mixing the conceptual and the physical in a single model arises from treating every element in the model as a process (Blockley, 2010). Whilst this is easier to follow for conceptual ideas (i.e. it is similar to SSM), the notion of physical things being a process is highly challenging. The argument presented above for delayed reification offers many advantages when modelling, especially when making use of the limited time available to participants in a workshop, where laborious debates about implementation of actions can be deferred until (much) later. The mathematical education for most engineers should equip them with sufficient regard for the use of *abstractions* and to work with them in everyday practice. Reification, or at least delayed reification, should be making use of a familiar skillset.[13] Reification also seems related, perhaps distantly, to the notion of solutionism. Quips about the tendency for engineers to jump into debates with ready solutions to complex problems perhaps have a degree of truth. Maintaining the modelling at the process level for as long as possible prevents problem exploration from getting bogged down into implementation issues, which may never actually be required.

It is possible to develop these conclusions much further, and with implications for the practice of OR, generally. The '*everything is a process*' mantra of Blockley (2010) was inspired by systems thinking, but perhaps missed signalling a more fundamental grounding in process philosophy. A completely processual philosophy of engineering could be constructed along similar lines as a manifesto for such in biology (Nicholson & Dupré, 2018).[14] The breadth and depth of work introduced

11 Which implies either in sequence or parallel instances of two different kinds of methodology.
12 HPM would in fact be well-suited to modelling in the following examples; resilience of infrastructure systems (Francis & Bekera, 2014), sustainability transitions (Geels, 2005; Seyfang & Haxeltine, 2012), and healthcare (Harrison, Koppel, & Bar-Lev, 2007).
13 See Sfard (1991) for an example of how this notion is employed in mathematics education.
14 Nicholson and Dupré acknowledge the primacy of Whitehead's contribution (e.g. Whitehead (1929)) but also critique his opacity of language and introduction of neologisms, and comment

by Nicholson and Dupré certainly reflects a coherency of viewpoint arising from a single EU-funded project (ERC 324186). In contrast, the piecing together of an underpinning processual philosophy for engineering requires a synthesis across all the Bristol-based sources, and breadth of the underpinning work, cited in this and Chapters 7 to 9. However, the impact should be no less as a consequence. The availability of a ready-made modelling approach in Hierarchical Process Modelling, its use in bridging between hard and soft without getting tangled up in paradigm matters, and the simple linguistic trick of forcing process descriptions through the gerund form of verbs certainly make (the beginnings of) a broader processual philosophy for engineering within reach for teaching basic principles and have certainly been proven ready for use in practice.[15]

One of the biggest changes in thinking would be the way in which engineers view time when modelling. The processual view models each individual process with a sense of *ongoingness*, processes that continue over time without the event-driven, synchronic notion of processes starting and stopping. This orientation to a *diachronic* view, together with movement away from a substance ontology, would represent a major shift in viewpoint for engineering thinking (Forbes, 2023). For example, the stability of form of constructed objects and our reference to them as nouns, e.g. a house, would be seen as a special case and require explanation. Instead, thinking of entities enacting processes, e.g. providing a home, would become the normal way of addressing problems and developing interventions. This shift would also help in a better alignment of engineering thinking, ab initio, with sustainability thinking in areas such as Life Cycle Assessment (see, for example, Finnveden et al. (2009)) and the Circular Economy (see, for example, Geissdoerfer et al. (2017)). The more widespread use of HPM to support problem structuring, as well as integrating the results of quantitative modelling and ongoing measurement (as demonstrated in Section 10.6), would certainly be a valuable asset in repairing some of the entrenched divides that exist between hard and soft OR.

---

on the possible harm this has caused to the wider uptake of process philosophy. In their manifesto they openly declare that '*re-inventing the wheel*' was a preferrable path to take. The discovery of processual thinking in engineering has a similar flavour of reinvention.

15 In this regard, perhaps engineering has taken a lead here in that there have already been examples of commercial consultancy completely based on this processual interpretation.

11

Problem Structuring in Engineering Practice

> *'When we think of making a decision, we generally think that a responsible decision can be made only on the basis of complete information. We must know the facts of the matter, collect all of the information, and then, in the face of the force of information, we will know in good conscience which way to decide. But Derrida insists that, at the same time as we must collect information and use reason, if there is to be a decision that does not simply involve following rules then there must be at least some degree of indecision. So we could say that indecision is a condition of possibility of decision. In order for there to be a decision, I must have an experience of not knowing what to do, otherwise there is no decision'*
>
> (Jones, 2003, p. 231)

> *'To make a decision is to choose from among alternative courses of action or interaction in situations in which the decision maker's choice can make a difference in the value of the outcome to him. Our common interest tends to focus on decision situations in which the decision maker is in a state of doubt about what to do. Such doubt converts the possibility of choice into a problem'.*
>
> (Ackoff, 1974a, p. 237)

> *'...every human problem is fundamentally elusive, and that the attempt to clarify its meaning and to identify a solution distorts the problem and destroys its real significance. Put otherwise, uncertainty is the key concept, but not the clear uncertainty of probability or game theory. Furthermore, uncertainty and ambiguity are the highest values...'*
>
> (Churchman, 1970, pp. B-39)

Problem Structuring: Methodology in Practice, First Edition. Mike Yearworth.

11.1 Introduction

The use of problem structuring methodology is a reasonable practice for engineers to adopt. The recognition of messy problem contexts and responding appropriately should be a *deliberative* action. For this to happen, it requires both awareness and training to become part of normal engineering practice. I see the process of *problematisation* as crucial to being an engineer and *doing* engineering. However, how should this process be enacted? What is the practical way of answering the question what *should* we be doing as engineers, which is not the same as what our knowledge and technology *allow* us to do? I see PSMs as the way forward for engineering practice, but these need some degree of interpretation from how they have been developed and presented in the Soft Operational Research (OR)/problem structuring method (PSM) literature.

Related to the idea of problematisation is the notion of engineering '*know why*' as a deliberate counterpoint to how we conventionally think in terms of engineering '*know-how*'. This is engineering practice with awareness of the processes by which engineering activity commences, rather than just considering what is produced by way of outcomes from engineering, i.e. solutions to problems. This refocusing of attention on the *process* of engineering, rather than the representational forms of its knowledge (i.e. content), naturally leads to an ethical engagement in problems, simultaneously questioning *why* we are doing this and *how* we do it. This engineering starts from the perspective of engaging in and questioning *purpose* rather than the simple justification of action through the equivalent of the situational logics in OR practice introduced in Section I.

11.2 Anticipation, Injunctions and Translations

The notion of injunction (Adams, Murphy, & Clarke, 2009) is useful since it captures the idea of the *motive* that drives translations (Latour, 1987). Engineering practice naturally occupies a state of thinking about anticipated futures, choices that can be made in anticipation of future states of the world, e.g. imagining a future world where certain requirements will exist and building for that need today.

> *'One of the valences of anticipation is "injunction". Anticipation is not only an epistemic orientation toward the future, it is also a moral imperative, a will to anticipate. From climate change, to emergent disease and biosafety, there is a moral injunction to anticipate as an act in which life, death, identity and prosperity are at stake personally and collectively. Anticipation calls for a heralding of the emergent "almost" as an ethicized state of being. Being ready*

> *for, being poised awaiting the predicted inevitable keeps one in a perpetual ethicized state of imperfect knowing that must always be attended to, modified, updated. The obligation to "stay informed" about possible futures has become mandatory for good citizenship and morality, engendering alertness and vigilance as normative affective states'.*
>
> (Adams, Murphy, & Clarke, 2009, p. 254)

Such deliberation occurs between the perception or anticipation of a problematic situation and the *will* to take action to intervene to bring about some change in the current state of affairs. It is in this process of translation that the opportunity to reflect on what we should be doing, as opposed to what we could do, breaks the purely logical, rules-based link between situation and action (i.e. avoiding situational logic). It is an opportunity to reflect on what our engineering practice should be. The idea of injunction allows us to examine the '*moral imperative*' for moving from problematisation to performative action. I use reflective questioning to examine this moral imperative in my own practice. Are our motives for action based on a *weak moral injunction*, i.e. without much thought or reflection, or are they the result of our deeply held beliefs? The notion of a *strong moral injunction* can only be a motive to act built on personal convictions, not borrowed or assumed. I emphasise this point because we must be vigilant to avoid falling into weak moral injunction for what might seem to be very good reasons, for example, signing-up to '*grand challenges*' that vie for our attention as engineers (Cech, 2012; Riley, 2012).

Chapter 9 presented the epistemology and affordance of a modelling approach that, when enacted by the constitutive rules for a generic interpretivist process of enquiry (Table 5.1) defines a new PSM well-suited to meeting the needs of problem structuring in engineering practice. The how/why questioning dialectic of the modelling approach enables the examination of purpose and is a strong foundation on which to base an ethical and reflective practice. How these might be turned into operational strategies is entirely up to the individual.

I have attached the idea of a strong moral injunction to practicing ethically and reflectively. This is not the same as *not-mal*practice. The double negative construction here is intentional and follows the formulation of Watzlawick et al. (1974). Engineers are already governed by rules of professional practice and must abide by them to avoid the imputation of malpractice, but this is not enough. Drawing on debates about ethical practice in OR, Rauschmayer et al. (2009) rightly acknowledge that '*following rules of good OR practice is necessary, but not sufficient for ethical OR*'. What these rules might be seems linked to the notion of situational-logics (see the discussion in Section I). Bellenguez, Brauner and Tsoukiàs (2023, p. 5) note that '*most mainstream OR methods underestimate the importance of an accurate problem structuring before using any method*'. Whilst pointing out that the moral dilemmas raised by this ethical engagement may be unsettling, they do not approach the darker conclusions of Churchman (1970) in dealing with the ambiguity and uncertainty of social problems. The OR profession has had 50-plus years to embrace this challenge yet still significantly avoids direct engagement on these matters, particularly in its scholarly output.

If we can just apply '*best practice*' to a problem, rely on situational logic and deliver solutions, then this is a sure sign that, as practitioners, we are not engaging in the messiness of the world. Decisions in such cases are, in effect, just programming, i.e. they become '*practitioner free*'. Drawing on the concept of an *aporia* from Derrida, '*I must have an experience of not knowing what to do, otherwise there is no decision*' (Jones, 2003, p. 231). Different from a paradox, an aporia captures the idea that the experience of *not knowing* what to do actually creates the possibility for a practitioner to make a decision.[1] The alternative, knowing what to do, just replicates the situational logic/practitioner-free approach to problem *solving*. Ackoff (1974a, p. 237) makes a similar observation in his address to the American Institute for Decision Sciences – when the '*decision maker is in a state of doubt about what to do. Such doubt converts the possibility of choice into a problem.*'

11.3 Widening Participation and Broadening the Scope

The prospect of widening participation and engagement in decision-making is likely to be uncomfortable for any expert, not just engineers. The use of PSMs or approaches that have emerged from other research traditions relating to the co-production of knowledge models with citizen groups[2] represents new ways of working that seek to break down the barriers that have divided expert practitioners from the publics affected by their decision-making. Hybrid forums are also likely to arise spontaneously due to the way in which social media affords the means with which '*concerned groups*' can arise and organise. Here, experts are likely to be responding to communities rather than using more conventional ways of working, such as the PSM workshop from Soft OR, or public engagement and debate.[3]

Engineers will also need to meet the challenges of decision-making in messy problem situations within their own organisations or amongst expert groups. Here, the recourse is likely to be the use of PSMs as in the Soft OR community.

1 Checking for the existence of an aporic state is perhaps a reasonable way of discriminating between the presence of automated decision making and the exercise of human judgement.
2 For example, Hybrid Forums (Callon, Lascoumes, & Barthe, 2009b) and discussed further in chapter 13.
3 In the co-production of knowledge model (CKM), Callon (1999) places the role of publics central to the process of knowledge creation and contrasts this new way of working with the more traditional and familiar public engagement model (PEM) and public debate model (PDM).

However, exigencies of time and budget are likely to lead to situations where decision-making processes are enacted informally in the absence of established methods. Yearworth and White (2014) argue that this is likely to be common in engineering organisations and that problem structuring would thus be *ad hoc*. These dynamic '*non-codified*' responses to messy problem situations are likely to be characterised by the generic constitutive definition of PSMs, as confirmed by the work of Reynolds (2021). Engineering organisations might call in outside experts, likely to be management consultants, or attempt to build internal capability by learning more about PSMs and developing engineers in this role. This dual role of facilitating groups through problem structuring, as well as being an expert in a particular aspect of engineering is likely to become more common. The ability to achieve strong injunctive translation is dependent on widening the participation in problematisation and decision-making, as well as taking on different roles and modes of working; both as facilitator and expert, and in appreciating the different modes with which scientists and engineers have engaged with society (Callon, 1999).

As a counterexample to this way of thinking, the 14 Grand Challenges set out by the US National Academy of Engineering (2017)[4] were formulated largely to promote engineering education and inspire engineering students to come up with '*solutions*'. Whilst some of these challenges have some parallels or passing similarities with the strategic challenges set out by other bodies,[5] the framing of these grand challenges clearly reflects the narrow constituency of the committee convened to draw them up. These challenges mostly ignore wider social issues and are narrowly focussed on technical interests and solutions. In fact, there is a strong flavour of what Morozov (2013) calls *solutionism*, which he defines as the tendency or viewpoint to recast all complex social situations as neatly defined problems with definite, computable solutions or as transparent and self-evident processes that can be easily optimised – if only the right algorithms can be found. However, this quest is likely to have unintended consequences that could eventually cause more damage than the problem they seek to address. Morozov goes a lot further by viewing this as more than just a tendency but as a legitimising ideology. Recent calls for a move away from this way of thinking in the training of engineers are encouraging (e.g. see Sorby, Fortenberry and Bertoline, 2021). Ackermann (2024) makes a similar call in her review of the grand challenge literature oriented towards a Soft OR/PSM perspective.

---

4 A component entity within the US National Academies of Sciences, Engineering, and Medicine.

5 For example, see the challenges set out by the United Kingdom Economic and Social Research Council (2019) Delivery Plan.

11.4 Moving Away from Singular Outcomes

Changing the engineering perspective away from singular outcomes represents breaking away from the assumption of a privileged viewpoint in a problematic situation. This habit, acquired from engineering education, tends to locate much undergraduate learning in the laboratory. This is emulating the situation of the natural scientist in the use of '*secluded research*' (Callon, Lascoumes, & Barthe, 2009d) to isolate phenomena from the messiness of the world in order to study them. Engineering education accordingly produces engineers who naturally identify as being scientific in their approach to problems.[6] A consequence of education oriented towards natural-science is that the habits of secluded research are brought into everyday practice in a messy world. As a result, there is a temptation to try and carve off the tractable parts of a problem that have known solutions whilst ignoring the wider mess – the '*opt out*' described by Ackoff (1977). Ideas of '*optimum*', '*best*', '*right*', and so on in relation to engineering '*solutions*' are easily bandied about without much thought about the *privileging* of viewpoint this entails. To test this idea, a simple language game can be played in meetings whenever these words are uttered – challenge them! Leaving them uncontested can only ever lead to weak injunctive translation; as engineers, we must get into the habit of asking the question '*best/optimal/right/...*[7] *for whom*?' or start to dig into what is actually meant by best/optimum/right/..., not what is assumed.[8]

11.5 Working with Multiple Viewpoints

In the process of widening participation and moving away from singular outcomes, we are required to recognise that multiple viewpoints, conflicts and power exist. Rather than retreating into passive, instrumental behaviour, I believe it is legitimate for engineers to engage in these conflicts and struggles in decision-making. In the problem structuring community, the concept of worldviews and working within an interpretivist framing are well understood (see for example Checkland (1999, pp. 215–221) and Jackson (2000, pp. 211–290)). The essence of problem structuring methodology is to provide the practitioner with the practical means

6 This is based on my experience of teaching postgraduate engineers. See my account in Chapter 8 of including the work of a social scientist (Vaughan, 2006) in a critical reading exercise and the push-back this generated.

7 This is only a indicative list of qualifying terms that are suggestive of a privileged viewpoint and should be contested. The point is to hone a critical facility for words that lend an easy injunction from problematisation to decision-making.

8 Be prepared for the disruption this may cause. The idea is to surface privileged viewpoints and bring them to the attention of the meeting.

to bring about agreement on what action to take. Grand challenges are seductive, but they require scrutiny to avoid becoming instrumental in possibly questionable outcomes (Cech, 2012).

Building on the original case study analysis conducted by Yearworth and White (2014), a teaching device was introduced to communicate the range of positionings that could be taken with respect to messy problem contexts, as shown in Table 11.1. Where traditional engineering problems and solutions are presented at the '*not so messy*' end of the spectrum, a re-problematisation is presented in the middle column, which answers the '*why*' questions that arise. The third column presents possible consequences of inadequate problematisation at the outset and are themselves messy problems. The references are to cases that represent various positions along this continuum.

11.6 Being Aware of Conceptual Limitations

Drawing on the constitutive definition of PSMs, we can ask the question of any approach to problematisation and structuring used by engineers – has it been recognised that the stakeholders of the problem situation are not '*divorced from the problem*' and that they could not act as objective '*outsiders*' (Yearworth & White, 2014, p. 939). Continuing with the theme of the ways in which engineers can use modelling in problem structuring, it is possible to see this acknowledgement of subjectivity as a further argument for the '*epistemic shift*' of the soft systems paradigm, introduced in the context of teaching engineers in Section 7.4. However, there are other examples of where this acknowledgement has been made without the need to go to a full open adoption of interpretivism. The first five of the '*ten "commandments" for a socio-environmental modelling agenda*' of Voinov et al. (2014) are entirely consistent with the co-production of knowledge model of Callon (1999), acknowledging not just the role that subjectivities play in modelling, but also of values and assumptions. There are other conceptual limitations that are relevant to our use of problem structuring, including difficulties in mastering technique and convincing participants that the methods are rigorous and valid (see Chapter 6), as well as addressing questions of evaluation (see Chapter 12).

11.7 Valuing Process Knowledge

Valuing process knowledge on an equal basis with representational knowledge is essential to realising a problem structuring sensibility in engineering practice. Much of what is appreciated in engineering practice is knowledge about *how* to do things (i.e. this is engineering '*know-how*'), and engineering education is

Table 11.1 Mapping the messiness of problematisations in engineering practice. Some examples extending from the original cases analysed by Yearworth and White (2014).

Not so messy **Why? →→→**		**Messier** **←←← How?**	**Messy – consequences for inadequate appreciation of messiness**
Delivering a vehicle shuttle for the channel tunnel as a sub-contractor	...	Delivering the wider TML[a] project	Dealing with[b] litigation arising from the specification and delay in providing approval to design documents (Williams et al., 1995; Ackermann, Eden, & Williams, 1997)
Designing a levee for a specific location to withstand certain standardised storm criteria	...	Deciding the Nationalised Standardised Storm Criteria	Learning from the Hurricane Katrina levee failure in New Orleans (Kysar & McGarity, 2006; Little, 2011, p. 274)
Building coastal flood defences	...	Condition monitoring and asset management of flood and coastal defence systems (Hall, 2003)	Dealing with coastal flooding (Government Office for Science, 2004)
Delivering an IT system to a certain requirement specification	...	Understanding what enhanced healthcare services might be through new IT services for the NHS (National Programme for IT in the NHS) (Harrop, Wood-Harper, & Gillies, 2006)	Inquiring into inadequately conceived Government procurement processes for IT (House of Commons Committee of Public Accounts, 2013)
Designing, Building and Operating an onshore wind farm for a specific location	...	Engaging in spatial planning approval hurdles for onshore wind farms (Josimović, Cvjetić, & Furundžić, 2021) Ensuring environmental, social, and inter-generational justice in planning processes (Lombardi, 2023)	Dealing with an inadequate understanding of energy security (Chester, 2010; Winzer, 2012; Jenkins et al., 2016)

Table 11.1 (Continued)

Not so messy **Why? →→→**		**Messier** **←←← How?**	**Messy – consequences for inadequate appreciation of messiness**
Redesigning a gas turbine engine component to meet a certain set of requirements	...	Improving Systems Engineering practices in an engineering organisation (Dunford et al., 2013)	Dealing with poor profitability
Delivering diesel powered vehicles that meet requisite emissions regulations	...	(Not) cheating the emissions testing process (Contag et al., 2017)	Dealing with corporate malfeasance (Rhodes, 2016)
Deciding critical infrastructure locations	...	Understanding infrastructure interdependencies (Carhart & Rosenberg, 2016)	Dealing with wasted opportunities, exposure to natural hazards
Providing a Strategic Road Network (SRN) maintenance service	...	Managing Agency contracts to deliver services (Davis, MacDonald, & White, 2010)	Dealing with failures in the safe maintenance and operation of the SRN

a) The TransManche Link construction consortium responsible for delivering what is now known as the Channel Tunnel between England and France.

b) The gerund '*dealing with*' becomes somewhat repetitive but does provide a perfectly adequate label for the process.

largely directed towards this outcome. In the university education of engineers, this is achieved using methods and techniques that mimic the way in which a scientist learns, i.e. through laboratory-based experiments. However, messy problem contexts are, by definition, unique, and here, the valuable knowledge is in the *approach* that is required to dealing with the mess. Engineering know-how is still essential to any endeavour, clearly, but this knowledge needs to be embedded/contextualised within the knowledge of processes that work in these contexts. Much work remains to be done to get across the message that this *process* knowledge must also be valued.

Section IV

Evolution

The presentation of hierarchical process modelling (HPM) as a problem structuring method (PSM) in Section III illustrates how it is possible to position a new (to OR) process modelling approach as a PSM despite its origins in engineering practice with quite different original motivation compared to the core methods, i.e. *engineering* as opposed to organising, strategising or planning. This flexibility of positioning also reflects adaptivity in applications and developments of PSMs to meet future needs.

Chapter 12 provides fairly concise overview of why the evaluation of PSMs is hard.

Chapter 13 presents a survey of PSM applications across a range of domains. The intention is to showcase not just the range of application domains but also the fact that a lot of what we know about these applications is published by authors working outside of OR – by both discipline and journal. Uptake is increasing the degree of intersection with other disciplines and really highlights the need for new conversations. Rather than (re-)describing method, the practice-theory-based evaluation framework of Ormerod, Yearworth and White (2023) is used to 'zoom-in' and 'zoom-out' of practice issues to highlight methodological lessons, i.e. issues of context/choice and implications for wider practice in the application areas. Intersections with communities of practice built on different research traditions to PSMs are discussed.

Chapter 14 explores the opportunities for the use of Group Support Systems (GSS) to enable online 'same time/different places' problem structuring workshops to take place. Extension to large-scale group working is also discussed as well the potential for problem structuring platforms – where procedural scaffolding for a problem structuring methodology is built in to the GSS.

Problem Structuring: Methodology in Practice, First Edition. Mike Yearworth.

Chapter 15 describes two strands of technical development in machine learning with relevance to PSMs and implementation in the GSS as discussed in Chapter 14 – augmented qualitative analysis based on the use of probabilistic topic models and prospects for the use of large language models for ensuring ideals of procedural rationality and justice in stakeholder workshops.

Chapter 16 concludes by setting out an agenda for Soft OR with a focus on bringing the Soft OR project back into a better alignment with mainstream OR, whilst at the same time building links with new communities of practice revealed from the survey of PSMs in Chapter 13. Secondly, some aspects of education are discussed in the context of the OR Specialist Degree Apprenticeship specification. Finally, the implications of the process epistemology of HPM described in Section III are discussed in relation to a possible processual turn in OR.

12

Evaluation

> *'It is an irony of work of the kind describe here that the obvious question posed by the very existence of an approach to bringing about improvements in a problem situation cannot be answered! When people hear for the first time about SSM… or any such methodology they naturally ask: Is it any good? Does it work? And they are very frustrated when they develop an understanding of the nature of methodology sufficient to realize that the question is unanswerable, that methodology is in fact undecidable.'*
>
> (Checkland & Scholes, 1990, p. 299)

12.1 Why Wait Until Now?

It may seem strange to defer the discussion of the evaluation of problem structuring methods (PSMs) to now, but it is only here that sufficient ground has been covered to show that evaluation, and the related issue of replication, can be discussed with the full range of material to support the argument presented here. The realm of PSM application, like engineering and any applied social science, is not the laboratory. In line with the thesis of Callon, Lascoumes and Barthe (2009a), we are dealing with '*research in the wild*'. There is no laboratory in which we can conduct experiments in organising, strategising, and planning, where we can pick off selected problems, or parts of problems, of practice to study in undisturbed isolation.

Even if we are optimising (in operational research practice), or building, making, constructing (in engineering practice) etc., our actions as practitioners will have broad effect on the world. There will be winners and losers from any intervention. As Checkland so clearly articulated through the notion of transformation of a plurality of entities, any intervention will have its *ostensible* purpose, with a single entity as the focus. However, the world is inherently systemic; our intervention

Problem Structuring: Methodology in Practice, First Edition. Mike Yearworth.

will perforce transform other entities. Some will have been considered through the representation of involved stakeholders. Other entities will have been missed, or deliberately ignored, and might get referred to as externalities.

Since the purpose of PSMs is the structuring of wicked/messy/swampy problems, which are by definition unique contexts, then measurement has no baseline for relative comparison. Attempts to deconstruct elements of a PSM technique and perform quantitative measures in laboratory type environments for comparison are likely to generate misleading results (e.g. see Cunha and Morais, 2019). Even if we do decide to minimise the effect of a phenomenon in decision making, e.g. the removal of a bias, then there is the difficulty of translating the results from laboratory experiments into the wild. Even for well-crafted laboratory experiments focussed on decision problems, there is the issue of the reproducibility of the results – the replicability crisis – leading to a growing industry in this sort of work; the unearthing of biases, the fixing of replicability problems.[1] A good example is how behavioural corrections are added into the game-theoretic modelling of platforms and supply chains. Whilst scientific in the sense of deepening our understanding of how to model decision-making processes, the domain of applicability still seems to reside within a positivist framing of an improved algorithm to, ultimately, optimise something.

From the point of view of a consultant attempting to justify the value of a PSM intervention to a potential customer then this has been discussed extensively in Chapter 6 as the value paradox (Tully, White, & Yearworth, 2019). Also, as established in Chapter 4, PSMs emerged within a phenomenological paradigm framing and evaluation should follow appropriate techniques e.g. realist evaluation (Pawson & Tilley, 1997). Their CMO framework was used as the basis for the investigation by Tully, White and Yearworth (2019) that led to the empirical construction of the value paradox. We can also look at specific evaluation frameworks that were specifically designed for PSMs e.g. White (2006) and Midgley et al. (2013). Broader considerations of the complexities of PSM evaluation are discussed by Franco and Rouwette (2022). However, despite these attempts to position practice against evaluable outcomes, we will always lack an objective baseline against which to judge the success, or otherwise, of an intervention; and a counterfactual exercise is unlikely to provide anything fruitful. What would have happened had we not planned, organised, or strategised using a problem structuring approach? Or what would have happened had we used a different approach, perhaps soft systems methodology (SSM) instead of strategic choice approach (SCA)? Or used an ad hoc approach – because planning, organising, and strategising were still required and something needed to be done. The Generic Constitutive Definition

1 This concern has been elevated to the establishment of specific task forces to address such replicability issues e.g. see Simchi-Levi (2022, pp. 3–4).

(GCD) work of Yearworth and White (2014) showed that there was a way of looking at how organisations deal with messy problems as non-codified problem structuring. They posited the idea that problem structuring could be an implicit activity within any organisation and Reynolds (2021) provided evidence for such *everyday* problem structuring through an extensive ethnographic study.

Adopting a processual view sees organisations in a state of constant organising, i.e. they are constantly strategising, planning, and transforming and the use of PSMs will be rather singular activities within this much broader flux of doings. Isolating the PSM intervention for special evaluative treatment then becomes problematic, especially when there is everyday problem structuring happening anyway. Studying and evaluating change, with or without PSM interventions, really requires '*process studies of organizing by narrating emergent actions and activities by which collective endeavours unfold*' (de Ven & Poole, 2005). This would seem to provide a more suitable context for discussing the evaluation of PSMs, but there are no simple answers. The more recent developments of the strategic options development and analysis (SODA) project (SODA-3) towards strategy-making and the detecting of emerging strategy blurs the line between studying organisations and actively engaging with them *processually*[2] and further complicates the problem of evaluation.

To make matters worse, there exist broader criticisms about Soft OR/PSMs that do not seem to be of the kind that can be resolved by evaluation techniques. These, in my experience, have fallen into two main strands

1. Soft OR/PSMs do not belong in the discipline of OR because the models are not quantified or quantifiable. To be considered as a scientific approach we must produce quantifiable and reproducible findings.
2. Objections to codification of practices as methodologies, because problem structuring is viewed as common sense. Experts, such as analysts and engineers, should just do this sort of thing *naturally* as part of their everyday work.

The first strand of critique cannot be resolved rationally; this is a question of how the discipline collectively chooses to define itself and is a social problem that I return to in Chapter 16. The second strand provides a useful avenue for moving forward with discussions about evaluation, mainly because it correctly grounds problem structuring in everyday professional practice.

The theoretical work of Yearworth and White (2014) on non-codified PSMs and subsequent empirical work by Reynolds (2021, 2023b) provides sufficient motivation to view problem structuring practice on a continuum that at one end consists

2 Detecting emergent strategy in SODA-3 and '*narrating emergent actions*' (de Ven & Poole, 2005) would seem to be activities of a similar kind.

of problem structuring-like practices that do not refer to known PSMs i.e. none of the elements of practice match the epistemologies or affordances of modelling approach of known methods. On the other, we have the endless varieties of PSM practice that emerge from the fact that problem structuring methodologies are inherently generative of a wide range of behaviours because of the operation of the constitutive rules.

To address the second point fully, we should turn to practice theories. Ultimately, the practices that we engage in as operational researchers (or as engineers or applied social scientists) are so much more than the methodologies and processes employed in any specific engagement. Ormerod, Yearworth and White (2023) conducted an extensive review of the relevance of (social) practice theories to operational research. Their findings are expressed as a research agenda for firmly turning the gaze of evaluation towards understanding the practices of OR to improve them, rather than trying to measure the value of specific interventions or the utility of specific methods or elements. They propose an evaluation framework that goes well beyond the confines of any individual engagement and look at the question from a viewpoint that would be relevant for an ethical examination of practice, continual professional development and learning. This extends the notion of an enquiring system from the immediate problem at hand to an enquiring system that encompasses practice, generally, including individual specific engagements or applications. This would lead to reflexively determining what better practice would be from a humanist perspective rather than from some, spurious, objectification.

The Ormerod Yearworth White framework – henceforth the OYW-Framework – is accessed through the following eight questions derived from original work by Nicolini (2013) on the notion of zooming-in and zooming-out and elements from the '*palettes*'[3] in Nicolini's conception of a theory-methods package.

The OYW framework attempts to broaden the framing for the analysis of PSM applications and can be used to step outside discipline boundaries. For the survey of PSM applications conducted in Chapter 13, all instances of '*OR study*' and '*OR practitioner*' in Table 12.1 can be replaced by study and practitioner with no loss of applicability. As observed by Ranyard, Fildes and Hu (2015), in their survey conducted on behalf of the International Federation of OR Societies (IFORS), PSMs have extended the scope of OR practice and can be viewed in contrast to the '*dominance of mathematical OR disassociated from practice*'. They comment on the gaps opening up between academic research and practice and draw on (Abbott, 1988) to characterise the dangers that this poses to the profession of OR. I come back to this point in Chapter 16 where I consider another theoretical position from Abbott, this

3 In the sense of a composition drawn from materials on an artist's palette.

Table 12.1 The Ormerod Yearworth White (OYW) framework.

Performative elements (zooming-in)	
Q1	Is there a sequential record of the flow of participant-to-participant and participants-to-OR practitioner/scholar interactions over time? For example, audio or video recordings of workshops (especially video), computer logs from Group Support Systems, field notes, interviews, email records... (*'sayings and doings'*, *'interactional order'*, *'timing and tempo'*)
Q2	Has the report of the OR study been situated within the everyday activities of the OR practitioner/scholar? Why did they get involved? What was their motivation? Was their contribution successful? What happened next? (*'practical concerns'*)
Q3	How were meetings and/or workshops between the OR practitioner and participants/clients organised e.g. onsite, offsite, online... (*'bodily choreography'*)
Q4	Have the results, recommendations, action plans, models, boundary objects of the OR study been recorded and analysed for their intermediating role and function? (*'tools, artefacts and mediation work'*)
Q5	How did the OR study, as performed by its actors, align with, or challenge, its ownership? How did the interplay of organisational roles and status constrain or expand the OR study? To what degree did constitutive rules of methods lead to generative practices? To what extent was the OR study a recognisable and reproducible (by others) method (*'tension between creativity and normativity'*, *'processes of legitimation and stabilisation'*)
Intersectional elements (zooming-out)	
Q6	Have the performative elements of the OR study (Q1–Q5 above) been analysed in terms of their cultural, social, political, and material context?
Q7	How has the OR study impacted wider practices within the organisation? Has it supported existing ways of working, created new socio-technical configurations/assemblages (translations), and/or led to tensions or conflicts?
Q8	What possibilities have the OR study opened-up? Where are we now? What is different as a consequence of the intervention?

time the dichotomous splitting within disciplines (Abbott, 2001), and argue that the hard/soft division in OR is in fact drawing PSM practice closer to a new constituency of academic research and practice originating outside of OR. Chapter 13 begins to shed some light on this new community of practice.

13

Survey of PSM Applications

'An Abundance of Problematisations'

(Callon, 1980)

13.1 Introduction

What is the breadth of problem structuring method (PSM) application? The question can be approached through the use of citation analysis and the comparison with some details of alternative methodological approaches from different traditions. At the very least these suggest promising avenues for conversations to share experience by scholars and practitioners alike, but also to prompt a re-evaluation of the scope of applicability of PSMs.

13.2 Method of Selection

There is an important caveat to be aware of in the selection of papers for this survey. It is the case that journal Editors are highly sensitive to the selection of articles that are making a new contribution. In the case of PSMs, from the perspective of an Editor, there are only so many applications of a specific methodology to address a problem context before the work ceases to be an '*innovative application*'. There is little to be gained by the journal from countenancing the acceptance of a new paper; it is, at best, only likely to be making an incremental contribution. In the case of theoretical and methodological innovations the bar is set even higher, the contribution needs to be significant before there is any chance of getting published, although such contributions are highly prized. The publication dates in many cases reflect something akin to the priority date in patent applications – reflecting the first time that methodology X has been articulated, or

Problem Structuring: Methodology in Practice, First Edition. Mike Yearworth.

the first time that methodology X has been applied in problem context Y. Furthermore, having a survey of PSM applications solely based on academic publishing is highly limiting. However, despite these caveats, the intention of the survey is to take a practice theory lens to the reading of each application and draw attention to evidence of specific elements of practice that correspond to the 8-question zooming-in/zooming-out framework of Ormerod, Yearworth and White (2023) introduced in the Chapter 12. A simple processual interpretation has been introduced for each of the questions and is shown in Table 13.1.

By selecting text using this framework the '*analysis*', as such, is following the exact opposite process of how the framework was derived.[1] The focus on the enacted processes provides insight into how other researchers have applied PSMs in their problem context – in their own words[2]. The intention is to identify what Checkland (1972, p. 93) regarded as the taken for granted and understated activities that constitute the methodological aspects of the interventions that have been surveyed.

As can be seen from Figure 13.1, the term System Dynamics is quite dominant even though it is not explicitly mentioned in the search query. There are many

Table 13.1 OYW framework questions as a simultaneous process of zooming-in and zooming-out.

Question	Zooming out	Zooming in	Interpretation
Q1		↑	Making practices visible
Q2			Justifying the methods
Q3			Organising workshops
Q4			Recording the process
Q5			Legitimising participation (stakeholding)
Q6			Contextualising practice
Q7			Achieving translations
Q8	↓		Opening-up possibilities for change

Source: Adapted from Ormerod et al. (2023).

1 The same *'reverse qualitative'* approach could have been used to select quotations from the papers in the survey by using different frameworks e.g. looking for evidence of a pragmatic framing of practice (Ormerod, 2006) using the process model in Figure 3.1, or focussing on facilitation (Ackermann, 1996) using the process model in Figure 6.2.

2 The amount of quoted text would be prohibitive to reproduce in a book. Therefore, the full tables are available at https://problemstructuring.com/Book/Chapter13/Table-13.2.html. For each paper that is referenced, page numbers point to the text that illustrates the specific element of practice.

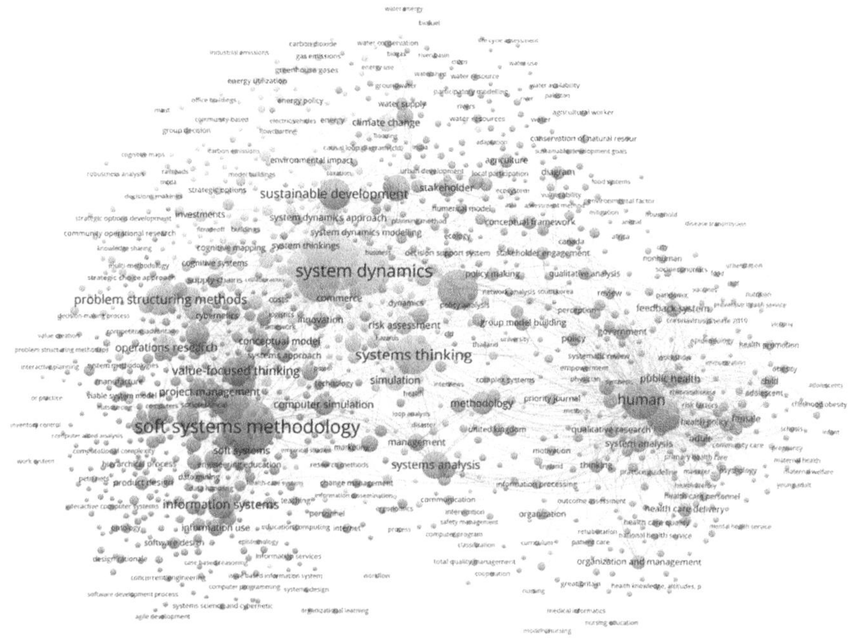

Figure 13.1 A visualisation of the problem structuring methodology landscape as a keyword co-occurrence analysis of 3637 article references obtained from Scopus (Note: Search query = TITLE-ABS-KEY (*'problem structuring method'* OR *'soft operational research'* OR *'soft systems method'* OR *'strategic options development and analysis'* OR *'strategic choice approach'* OR 'hierarchical process model' OR *'causal loop diagram'* OR *'value focused thinking'* OR *'issue based information system'* OR *'drivers*pressures*state*impact* response'*). An interactive version of this map can be explored in detail at https://problemstructuring.com/Book/Chapter13/Figure-13.1.html.

published applications of System Dynamics that follow a recognised approach to the development of a dynamical model that can be simulated (e.g. see Sterman (2000, pp. 83–105) and Morecroft (2007, pp. 59–85)), which will naturally include the development of causal loop diagrams (CLDs) (e.g. see Sterman (2000, pp. 137–190) and Morecroft (2007, pp. 31–57)) as part of the methodology.[3] However, where the purpose of the modelling seems to be the development of a quantitative System Dynamics model rather than the use of CLDs expressly for problem structuring the paper has been excluded from the survey.

3 In his book, Pidd (2009) makes use of this bridge between CLDs and quantitative System Dynamics modelling as the segue between the sections on interpretive modelling and mathematical and logical modelling.

13.3 Classification of PSM Applications

A simple classification scheme has been used as follows:

1. *Methodology Classification*
   Either a single methodology, labelled as *Single*, or a multi-methodology labelled as follows
   Multi Type A: $PSM_a + PSM_b$
   Multi Type B: PSM + Quantitative method (i.e. Soft OR then Hard OR)
   Multi Type C: Qualitative analysis + PSM
2. *PSM Type*

The PSM type is defined by the CR/E/A framework (see Chapter 5) as follows

$$\text{PSM type} = \{\text{SSM}, \text{SODA}, \text{SCA}, \text{DPSIR}, \text{VFT}, \text{HPM}, \text{IBIS}, \text{CLD}\}$$

The advantage of breaking down the analysis of PSMs into application domains is that we get to see how different methodologies are adapted to similar problem contexts. The examples illustrate something of the how the relationship between constitutive rules, epistemologies, and affordances of the modelling approach plays out in an application context. Also, how multimethodologies have been employed to suit certain contexts. There does seem to be the question of whether the application is firmly grounded in a PSM '*tradition*' or whether it is just '*dipping-in*' or borrowing ideas for use in an established field. On the whole, the applications selected for presentation showed evidence of being reasonably grounded in the PSM literature in that key papers were cited. All the examples provide a bridge that connects the survey of PSMs and the nature of wicked/messy problems with a specific problem context.

13.4 Survey of PSM Applications by Subject Area

Most of the articles found in this survey are clearly an application of an identifiable methodology, singly or in combination, and where the epistemology and affordance of the modelling approach follow closely the analysis presented in Chapter 5 and are offering some particular advantage, or matched, to the problem context. For example, it is no surprise to see Strategic Choice Approach (SCA) more widely used than other methodologies in urban planning applications. Where the approach is more generally following the constitutive rules of an interpretivist methodology rather than a named methodology, the CR/E/A framework is simply labelled as Soft. The summary of PSM applications based on the practice-oriented perspective using the zooming-in/zooming-out 8-question framework of Ormerod, Yearworth and White (2023) is shown in Table 13.2. The

Table 13.2 Summary of PSM applications surveyed from a practice-oriented perspective using the zooming-in/zooming-out framework of Ormerod, Yearworth and White (2023). Further details about the analysis can be explored at https://problemstructuring.com/Book/Chapter13/Table-13.2.html

Article	Classification	CR/E/A framework
Water Resource Management/Flooding		
'*The implications of complexity for integrated resources management*' (Pahl-Wostl, 2007)	Multi	Soft OR
'*Using causal loop diagrams for the initialization of stakeholder engagement in soil salinity management in agricultural watersheds in developing countries: A case study in the Rechna Doab watershed, Pakistan*' (Inam et al., 2015)		CLD QSD
'*Integrating conflict analysis and consensus reaching in a decision support system for water resource management*' (Giordano et al., 2007)	Multi	SODA
'*An approach for supporting problem structuring in water resources management and planning*' (Schramm & Schramm, 2018)	Single	SODA
'*The strategic choice approach to the maintenance management of a water distribution system*' (de Sousa Pereira & Costa Morais, 2020)	Single	SCA
See also '*Assessment of actions to tackle the shortages of water in La Paz, Bolivia*' (Rivero, Morais, & De Sousa Pereira, 2020)		
'*A participatory system dynamics modeling approach to facilitate collaborative flood risk management: A case study in the Bradano River (Italy)*' (Perrone et al., 2020)	Single	CLD
Urban Planning/Public Administration/Sustainable Built Environment		
'*Sustainability focused decision-making in building renovation*' (Kamari, Corrao, & Kirkegaard, 2017)	Multi Type A	SSM VFT
'*Applying futures studies and the strategic choice approach in urban planning*' (Khakee & Strömberg, 1993)	Single	SCA
'*A computer-aided strategic approach to decision-making in urban planning: an exploratory case study in Thailand*' (Kammeier, 1998)	Single	SCA
'*Environmental Technology, Materials Science, Architectural Design, and Real Estate Market Evaluation: A Multidisciplinary Approach for Energy-Efficient Buildings*' (Fregonara et al., 2013)	Single	SCA
See also '*Integrating Environmental and Economic Sustainability in New Building Construction and Retrofits*' (Fregonara et al., 2016)		

(Continued)

Table 13.2 (Continued)

Article	Classification	CR/E/A framework
'*Strategic visualization: the (real) usefulness of cognitive mapping in smart city conceptualization*' (Vaz et al., 2022)	Single	SODA
'*Experimental Use of Strategic Choice Approach (SCA) by Individuals as an Architectural Design Tool*' (Todella, Lami, & Armando, 2018) See also '*Negotiating perspectives and values through soft OR in the context of urban renewal*' (Tavella & Lami, 2019)	Single	SCA
'*A multi-methodological combination of the strategic choice approach and the analytic network process: From facts to values and vice versa*' (Lami & Todella, 2022)	Multi Type B	SCA ANP
'*A multicriteria decision aid perspective that guides an incremental development of knowledge and fosters relationships and decisions*' (Curto, Norese, & Rolando, 2022)	Multi Type A	SCA MCDA (ELEC-TRE)
'*A participative research strategy for propagating soft methodologies in value management practice*' (Green, 1999)	Single	SODA SSM SCA
Ecological/Environmental/Ecosystem/Waste Management (Including Circular Economy)		
'*Participatory modeling and analysis for sustainable forest management: Overview of soft system dynamics models and applications*' (Mendoza & Prabhu, 2006)	Multi Type B	Soft Systems CM QSD Fuzzy Cognitive Mapping (FCM)
'*A problem structuring method for ecosystem-based management: The DPSIR modelling process*' (Gregory et al., 2013). See also… '*Managing the Marine Environment: Is the DPSIR Framework Holistic Enough?*' (Atkins et al., 2011) '*DPSIR = A Problem Structuring Method? An exploration from the "Imagine" approach*' (Bell, 2012)	Single	DPSIR

Table 13.2 (Continued)

Article	Classification	CR/E/A framework
'*Analysis of electronic waste reverse logistics decisions using Strategic Options Development Analysis methodology: A Brazilian case*' (Guarnieri, e Silva, & Levino, 2016)	Single	SODA
'*Enhancing public participation in natural resource management using Soft OR––an application of strategic option development and analysis in tactical forest planning*' (Hjortsø, 2004)	Single	SODA
'*The Use of Cognitive Mapping in Analysing Sustainable Tourism Policy: Methodological Implications*' (Farsari, Butler, & Szivas, 2010)	Single	SODA
'*Integrative multi-attribute negotiation model to define stakeholders' responsibilities in the reverse flow channel*' (Silva & Fontana, 2021)	Multi Type A/B	SODA SSM MOLP[f]
'*Life cycle sustainability impact categories for sustainable procurement*' (Oliveira & Souza, 2023)	Single	CM
Multi Organisation Collaboration and Project Coordination		
'*Project review and learning in the construction industry: Embedding a problem structuring method within a partnership context*' (Franco, Cushman, & Rosenhead, 2004)	Single	SCA
'*Guiding interventions in a multi-organisational context: combining the Viable System Model and Hierarchical Process Modelling for use as a Problem Structuring Method*' (Lowe, Martingale, & Yearworth, 2016) See also… (Lowe, Espinosa, & Yearworth, 2020) for a deep dive into the use of the VSM as a PSM.	Multi Type A	VSM HPM
'*Understanding stakeholders' perspective of cost estimation in project management*' (Doloi, 2011) See also… (Burger, White, & Yearworth, 2019)	Multi Type C	Semi-Structured Interviews SSM
'*Problem-structuring methods and project management: an example of stakeholder involvement using Hierarchical Process Modelling methodology*' (Davis, MacDonald, & White, 2010)	Single	HPM
'*Managing soft change projects in the public sector*' (Crawford et al., 2003)	Single	SSM
'*Improving understanding on degrowth pathways: An exploratory study using collaborative causal models*' (Videira et al., 2014)	Single	CLD

(Continued)

Table 13.2 (Continued)

Article	Classification	CR/E/A framework
Requirements Engineering/Capture for Information Systems		
'*Improving the quality of conceptual modeling using cognitive mapping techniques*' (Siau & Tan, 2005)	Multi Type A	Cognitive Mapping SSM (Rich Pictures)
'*Conceptual model of a future farm management information system*' (Sørensen et al., 2010)	Single	SSM
'*A knowledge centred framework for collaborative business process modelling*' (Adamides & Karacapilidis, 2006)	Single	Soft OR
'*Putting soft OR methods to work: the case of IS strategy development for the UK Parliament*' (Ormerod, 2005)	Multi	Soft OR
Healthcare		
'*PartiSim: A multi-methodology framework to support facilitated simulation modelling in healthcare*' (Tako & Kotiadis, 2015)	Multi Type B	SSM DES
'*Depression as a systemic syndrome: mapping the feedback loops of major depressive disorder*' (Wittenborn et al., 2016)	Single	Qual SD
'*The Use of Soft Systems Methodology in the Development of a Simulation of Out-patient Services at Watford General Hospital*' (Lehaney & Paul, 1996)	Multi Type B	SSM DES
'*A Community Based Systems Diagram of Obesity Causes*' (Allender et al., 2015)	Single	CLD
'*Using systems science to understand the determinants of inequities in healthy eating*' (Friel et al., 2017)	Single	CLD
'*Care and Flow: Using Soft Systems Methodology to understand tensions in the patient discharge process*' (Emes et al., 2017)	Single	SSM
Safety Culture/Major Incidents		
'*Applying systems thinking concepts in the analysis of major incidents and safety culture*' (Goh, Brown, & Spickett, 2010) See also '*The Use of Group Model Building for Analysing Event Causality within the Nuclear Industry*' (Carhart & Yearworth, 2010)	Single	Qual SD
'*Dynamics of safety performance and culture: A group model building approach*' (Goh et al., 2012)	Single	CLD
'*A study of the precursors leading to "organisational" accidents in complex industrial settings*' (Taylor et al., 2015)	Multi Type C	Qual. HPM CLD

Table 13.2 (Continued)

Article	Classification	CR/E/A framework
Energy Planning/Efficiency		
'Structuring an MCDA model using SSM: A case study in energy efficiency' (Neves et al., 2009) See also… *'Using SSM for structuring decision support in urban energy planning'* (Coelho, Antunes, & Martins, 2010)	Multi Type A/B	SSM VFT MCDA
'Understanding behaviour in problem structuring methods interventions with activity theory' (White, Burger, & Yearworth, 2016) See also… *'STEEP Project Deliverable D2.1 – Energy Master Plan Process Modelling'* (Yearworth, Schien, et al., 2015) *'Why so Serious? Theorising Playful Model-Driven Group Decision Support with Situated Affectivity'* (Burger, White, & Yearworth, 2018) *'Climate change and cities: problem structuring methods and critical perspectives on low-carbon districts'* (Freeman & Yearworth, 2017)	Single	HPM
'A holistic multi-methodology for sustainable renovation' (Kamari et al., 2019)	Multi Type A	SSM MCDM
Transport Planning		
'Maximising stakeholder learning by looping again through the simulation life-cycle: A case study in public transport' (Jones, Kotiadis, & O'Hanley, 2021)	Multi	HPM DES ABM
'An overview of air delay: A case study of the Brazilian scenario' (Pamplona & Alves, 2020)	Single	VFT
'Changes in transport planning policy: Changes in transport planning methodology?' (Kane & Del Mistro, 2003)	Single	SSM
'Integrating multicriteria evaluation and data visualization as a problem structuring approach to support territorial transformation projects' (Lami et al., 2014)	Multi Type B	MCDA (ANP) Other visual tools
'Soft-systems methodology as learning and management tool' (Jotin Khisty, 1995)	Single	SSM

(Continued)

Table 13.2 (Continued)

Article	Classification	CR/E/A framework
Public Policy – Explicitly Addressing the use of PSMs in a Policy Setting Context		
'Applying Large-Group Interaction Methods in the Planning and Implementation of Major Change Efforts' (Bryson & Anderson, 2000)	Multi?	SODA SCA LGIMs
'Into the Swamp: The Analysis of Social Issues' (Rosenhead, 1992)		Soft OR
'Voices and values: Linking values with participation in OR/MS in public policy making' (White & Bourne, 2007)	Single	Soft OR
'The importance of eliciting stakeholders' system boundary perceptions for problem structuring and decision-making' (Pluchinotta, Salvia, & Zimmermann, 2022)	Multi Type C	QSD CLDs CMs Qual
'Blending systems thinking approaches for organisational analysis: Reviewing child protection in England' (Lane, Munro, & Husemann, 2016)	Single	QSD

application domain labels are suggested by the contents of the material surveyed. The allocation of papers to subject area has in some places been a matter of choice.

13.5 Observations

Ideally, the quotations from the cited works[4] speak for themselves in terms of illustrating how the PSM engagement was conducted (broadly, zooming-in), and the implications for the stakeholders of the intervention and potentially on practice (broadly, zooming out). The selection process meant that there was no attempt to ensure coverage of any specific methodologies or particular multimethodology combinations were represented. Coverage was intended to be indicative of what can be achieved with PSMs. The process of zooming-in highlighted where the authors were reporting on how they justified their selection of methods, organised workshops, and identified and involved stakeholders. Where there was useful content on making the problem structuring process visible and

4 See https://problemstructuring.com/Book/Chapter13/Table-13.2.html

repeatable by other researchers this was highlighted too. Zooming-out focussed attention on how translations were achieved. Further quotations were included where there was useful reflection by the authors on how their practice was contextualised within the application domain and the impact that PSMs could make for achieving wider change. One immediately apparent commonality was that the zooming-in/zooming-out process of social theory analysis very much mirrors the stages of soft systems methodology (SSM) e.g. as set out in order by Jotin Khisty (1995), and very clearly in the analysis of the case study reported by Sørensen et al. (2010). Emes et al. (2017) can be read as a particularly good example. As the PSM most closely associated with an interpretivist sociology, this is perhaps unsurprising.

From the range of interventions surveyed, it is possible to suggest that PSMs afford an accessible methodological approach to intervention in messy problem contexts. If the purpose of a PSM-based intervention is broadly labelled as planning, strategising or transforming and the range of intervention purposes in this survey are interpreted processually, then it is possible to see many as some variation on planning, strategising or transforming. Applications in urban planning, energy planning, water resource planning, flood resilience planning... obviously share similar planning specific concerns. However, if the initial generic set of purposes (i.e. planning, strategising, or transforming) is extended by including policy making, coordinating (e.g. multiple projects), ensuring (e.g. safety), contracting, ... then we see the potential scope for modelling purpose more broadly as introduced in Section 9.3 in the context of using HPM as a PSM.

From a *methodological* perspective, there is clearly no barrier to the uptake and then re-dissemination of PSM knowledge outside the narrow confines of the OR journals. The articles referenced in Section 13.4 were published in a wide range of journals. This comes as no surprise; a simple exclusion of OR (and closely related) journals from the query (shown in note of Figure 13.1) used to select papers still leaves 86% of the total number of original number of articles, and this ratio has never fallen below 57% on a year-by-year comparison. Prior to 1989 and the publication of RAfaPW there was not widespread awareness of the term PSM, although the core methodologies themselves were known. However, since then, whilst the diffusion of PSM knowledge outside of OR as a discipline has been encouraging it has also introduced problems. As noted by Paucar-Caceres (2010), '*the proliferation of secondary "soft" OR literature is so extensive that [it] has distorted the real message of methodologies such as Checkland's SSM*' – which complicates the interpretation of the data. As Harwood (2019) has observed, and as shown by the analysis of the published record, the centre of gravity for PSM application is outside OR. However, the caveat I stated at the start of this chapter concerning the publication selection effect in OR journals needs to be considered. Masked by the steady rise in published outputs outside OR journals there are still significant theory and methodology contributions being made to the field being published in the OR journals.

13.6 Interdisciplinarity

As quoted from Popper (1963) in the Preface '*[w]e are not students of some subject matter, but students of problems. And problems may cut right across the borders of any subject matter or discipline*'. Popper's remark clearly elevates the epistemic status of problems to be higher or more important than that of any discipline. Would this make PSMs of particular relevance and/or appropriateness for interdisciplinarity? As discussed in Chapter 4, Dando and Eden (1977) drew attention to the divergent nature of Soft OR with respect to the mainstream from very early on in its emergence and declared the need for a...

> '...*truly interdisciplinary OR which recognizes the complex and potentially partisan nature of what we do. Related work in the UK is noted in order to emphasise the beginning of a widespread concern over such issues. The remainder of the Congress demonstrated a distinctly different view of OR, which was characterized by a disregard for the nature of human behaviour and of power in organizations and society. We suggest that the reflections made by the Invited Speakers demonstrate again the need for a fundamental reorientation of OR.*'
>
> (Dando & Eden, 1977)

Ackoff (1979, p. 194) also argued that OR should consider its claim to interdisciplinarity be taken seriously. However, what this interdisciplinary OR might look like has largely gone unexamined – or at least the one we see through academic publication. As I discuss later in Chapter 16, it seems apparent that Soft OR has independently developed as an applied social science in its own right rather than being part of an OR *plus* interdisciplinary activity. The survey of applications in this chapter suggests that knowledge of PSMs has mostly *diffused* into other areas of applied social science rather than being developed as the result of any specific interdisciplinary partnership. I believe this causes OR, and Soft OR, in particular, some challenges in the interdisciplinary landscape. Is Soft OR a *discipline* that can be partnered with other disciplines? White (2025) certainly argues strongly for treating Soft OR as a distinct discipline. It is also the case that Soft OR is already in an interdisciplinary partnership with Hard OR. This is evident from (i) Checkland's observation that '*any piece of work based on phenomenology can always encompass within itself "hard" studies based on positivism so that hard and soft OR are complementary to each other*' (Checkland, 1989b, p. 40) (see also Pidd (2004)), (ii) a general consideration of multimethodologies (see Mingers and Brocklesby, 1997; Mingers and Gill, 1997; Mingers, 2001; Pollack, 2009), and (iii) examples of specific hard/soft integration such as PSM plus multi-criteria decision analysis (MCDA) (see Marttunen, Lienert and Belton, 2017 for a review) and the other varied multimethodology examples in the survey.

This notion of problem-oriented or challenge-led research seems to be the primary driver for interdisciplinarity. The British Academy (2016) draws attention to the need to be a specialist and an expert in at least one discipline in order to be an effective collaborator in any project that crosses disciplines. And that expertise lies primarily in knowledge – and methods of its production – *grounded* in a specific discipline, without which it is difficult for a researcher to make a robust contribution. The United Kingdom Economic and Social Research Council (2019) (ESRC) delivery plan is clearly aligned to funding challenge-led research for delivering economic, social, and cultural impact. It highlights the intention to strengthen the United Kingdom's interdisciplinary, cross-sectoral research partnerships by investing in building new interdisciplinary research and investing in research methods training to identify future methodological needs. However, it is not clear how methodologies for problem structuring fit within such priorities, as they are primarily concerned with designing interventions at an appropriate scale to match the problem context; i.e. in the case of the ESRC delivery plan, discipline knowledge is valued and the methods for its production (process) are acknowledged, but this seems to deprecate the opposite case. Yet the ESRC specifically calls out research challenges in climate change, innovation in health and social care, and housing that intersect with the application areas for PSMs reviewed above. Here, action-oriented *process* knowledge is the starting point of an intervention in these problem contexts.

There is also the vexed question of how disciplines might approach problematisation differently depending on how close the discipline is to practice and characterisation along the hard/soft dimension (see Section 16.1). What I mean by this is captured in the following questions. Firstly, what is the nature of a real-world problem or challenge that sets it apart from purely theoretical problems – and as a consequence requires an interdisciplinary approach? Secondly, does the nature or type of challenge depend on the discipline – or do all disciplines agree on roughly the same set of societal problems as research challenges? Thirdly, do different disciplines perceive and approach these challenges in ways that make it hard for disciplines to work together? Also, framing interdisciplinarity solely in terms of disciplines is somewhat narrow when a practice orientation might see the challenge as additional to having to work across diverse agencies or organisations e.g.

> *'Unstructured problems, such as the grand challenges, are characterised by the existence of multiple actors with different perspectives and conflicting interests, and they require a coordinated effort from multiple organisations.'*
>
> (Pluchinotta, Salvia, & Zimmermann, 2022)

A coordinated effort seems to require some form of extra-disciplinary expertise to be employed, i.e. knowledge about what might constitute a *process of interdisciplinarity* does not itself sit within a discipline. Therefore, in this process space *orthogonal* to discipline content, expertise must either come into existence as some emergent phenomenon arising from bringing discipline content knowledge together, or already exist as a form of cross-cutting or integrating approach. The former is assumed, and whilst the latter are also fields of research, in areas of process expertise, they seem to sit differently with our traditional notions of academic structuring i.e. schools, faculties, and departments. I also consider them as contributing essential concepts, frameworks, and tools to interdisciplinarity in different ways. In addition to PSMs, these integrating perspectives would be (amongst others) complexity science, responsible innovation, systems thinking, citizen science and co-production, and data science (Frodeman, Klein, & Pacheco, 2017). None of these cross-cutting perspectives have clearly defined disciplinary homes, nor are they mission- or challenge-led as is the current fashion for structuring the academy, and all are vast subjects in their own right. Complexity science provides us with a lens on emergent phenomena that are characteristic of the wicked and messy problems and central to the essence of challenge led research. Responsible innovation considers the wider societal impacts of research and hence its relevance to interdisciplinarity and challenge-led research. Systems thinking and systems approaches are often invoked in response to dealing with complexity and wicked and messy problems (see Strijbos (2017) for an alternative perspective). Citizen science and co-production are instrumental to public participation in research (see Section 13.7). Finally, the rise of data science has been inexorable, and any interdisciplinary research project that is likely to acquire very large data sets will need to draw on a data science contribution – although as a field it also considers itself to be interdisciplinary.

Calhoun presents a position originally set out by Bourdieu that social science disciplines have often resisted putting problem-oriented research front and centre

> *'As Pierre Bourdieu (1988) argued, academic fields value what they regard as distinctive and essential to them; transactions at their margins may be required to secure resources but they are commonly denigrated—So each discipline values both that which is interior to it and that which is more pure or fundamental. This hierarchy imposes distinctions and divisions where problem-oriented research might better be served by collaboration*

and communication. The problem here is not just the prestige hierarchy; it is also the very notion that pure science always precedes application in a one-directional linear process. In fact, the relationship may be much messier and more iterative; application may also be experiment, and the observation of poorly informed practical action may be the basis for better knowledge. This often involves a process of cocreation in which nonacademics are linked into the knowledge-producing networks and processes. The nonacademics may be researchers working in other, more practically oriented settings. They may be "practitioners" who bring knowledge from their experience. Among their contributions may be to help academic researchers ask better questions, so it is important that they be engaged earlier in the planning of research.'

(Calhoun, 2017, p. 125)

It is clear in this viewpoint that the knowledge that Soft OR/PSMs bring to such collaborations is firmly at the practitioner (lower) end of this '*prestige hierarchy*', hence White's attempt to establish a Soft OR ontology (2025).

13.7 Other Traditions

All of the examples in this chapter so far have been grounded in the core PSMs of SSM, Strategic Options Development and Analysis (SODA), and SCA with the addition of CLDs, Value Focussed Thinking (VFT), HPM, and Drivers-Pressures-State-Impact-Response (DPSIR). However, there are other approaches in applied social science that follow a different tradition that nonetheless share similarities with PSMs (Yearworth, 2016). Callon's work in actor-network theory (ANT) in particular provides theoretical grounding for approaches to problem structuring wicked/messy problems, or as Callon, Lascoumes and Barthe (2009a) put it, '*acting in an uncertain world*'. There is a coherent set of themes in this body of work encompassing (i) deciding what is and what is not problematic and the notion of there existing an *abundance* of problematisations facing experts (Callon, 1980), (ii) analysing the micro-processes of translation (Callon, 1984), (iii) working with citizen groups and the co-creation of knowledge model (CKM) (Callon, 1999), and (iv) navigating Hybrid Forums (Callon, Lascoumes, & Barthe, 2009b,2009c).

The Pickering flood-risk case study described by Lane et al. (2011)[5] is clearly an operationalisation of these themes into a practical intervention. The researchers were acting as practitioners in leading the intervention, not just using ANT to

---

5 And related details about participatory modelling and knowledge co-production (Landström et al., 2011).

theorise about other experts intervening in a problematic situation. It is in this respect that the interpretation of Callon's work can be viewed as methodological in the same action research/action learning sense as using a PSM. This case study should be contrasted to the water resource management/flood management examples using PSMs summarised in Section 13.4.1. What is immediately apparent is that despite completely independent developments paths in ANT and Soft OR we have methodologies that fulfil a similar purpose. However, finding other examples of this type of work is complicated by the absence of clear methodological labels (unlike with PSMs) and requires looking for action-oriented accounts where there has been a process of co-production of knowledge leading to action including descriptions of participatory modelling. For example, using Scopus queries like TITLE-ABS-KEY ("co*production" OR "participatory model*") AND (REFAUTH ("Callon")) are useful for finding similar engagement approaches to Lane et al, but what of other theoretical strands that would also lead to similar approaches? For example, engagement methodologies constructed around stakeholder theories would seem relevant and there are potential overlaps to be made between thinking in OR (e.g. see (Müller, Groesser, & Ulli-Beer, 2012) and (de Gooyert et al., 2017)) and elsewhere (see (Mitchell, Agle, & Wood, 1997; Maskrey et al., 2016)). Conversely, focussing on the engagement method then precludes finding new theorising emerging from the engagement e.g. Donaldson et al. (2013). This area remains underexplored but suggests new linkages between OR and applied social scientists, with Soft OR/PSMs functioning as the link.

13.8 Conclusions

Sitting within a Soft OR/PSM tradition and looking out towards practice grounded in the application domains surveyed in this chapter, or grounded in different methodological traditions, it is difficult to achieve perspective on how we are viewed from the outside. As an example of what we need, Leong et al. (2023) review the use of Soft OR methods in climate actions through a general review of the literature and analysis of two case studies, including the STEEP project introduced in Chapter 10 as an application of HPM as a PSM. They use the Generic Constitutive Definition of PSMs (Yearworth & White, 2014) to provide criteria for assessing the extent with which the interventions analysed conformed to a Soft OR engagement. Whilst the limitations of the application encountered during the STEEP project are fully accounted for by Freeman and Yearworth (2017), the Leong et al. review does provide a useful outside perspective and their conclusions are encouraging. They state that '*...notwithstanding the two cases reveal in their application, Soft OR nevertheless has much to offer policymakers* *in the arena of climate change policymaking and deserve more attention and use*' (p. 1).

The evidence presented in this chapter provides ample support for the application of PSMs across a wide range of application domains that exhibit characteristics of wicked/messy problems. This in itself is not surprising, but the overall number of published examples is not encouraging. Of course, not all applications of PSMs are published, and not all problem structuring is rigorously codified, so it is not possible to really know the prevalence of problem structuring activities '*in the wild*'. However, within the scope of the problem context set out in Chapter 2, and the range of available problem structuring approaches there seems sufficient evidence of efficacy to conclude that their wider use should be promoted and encouraged.

It is reasonable to continue with the assumption that the disciplinary '*centre of gravity*' for this Soft OR practice should be OR. It would be useful for theoretical and methodological advances of problem structuring to be critiqued with respect to the existing OR literature, but this is not strictly necessary (for example, see Freeman and Yearworth (2017)). On the other hand, in the review of wider PSM applications Paucar-Caceres (2010) observes that the secondary literature shows a sort of methodological '*distortion*' in the use of PSMs, which clearly does not help with maintaining coherent communities of practice. Whilst the constitutive rules are generative of problem structuring behaviours, some of this '*distortion*' is probably arising from misunderstandings of epistemologies and affordances of modelling approaches, which *are* prescriptive.[6] The ongoing '*homing*' of Soft OR in the wider OR community at least ensures continuity of disciplinary coherence through the '*normal*' operation of journal peer review and conference engagement. However, the argument to the wider OR community to re-engage with the Soft OR project really comes back to the critiques presented in Section I – for OR to retain its policy relevance to societal challenges its means of connection to the messiness of these problem contexts is through processes that ensure adequate problematisations, and specifically through the use of PSMs. This discussion is continued in Chapter 16.

---

6 This is still an unexplored area for investigation. For example, given the relatively high volume of published accounts of using CLDs compared to cognitive/causal maps in SODA, the diagramming of SCA, and use of HPM, then it is probably more likely to find instances of CLDs that do not make sense or are misleading.

14

Digital Transformation Through Group Support Systems (GSS)

'Problem structuring methods (PSMs) have been widely and successfully used in many organizations, but the reliance on face-to-face meetings and workshops makes a typical PSM project difficult and time-consuming to organize, and means that the process may only involve a narrow cross-section of the organization. Yet much interaction in organizations is neither face-to-face nor even synchronous.'

(Morton, Ackermann, & Belton, 2007, p. 547)

14.1 Motivation

Although I had some previous experience of hosting distributed participant workshops at Hewlett-Packard Research Labs using the Halo video conferencing system,[1] my original motivation for the use of problem structuring method (PSM) workshops hosted via an online platform was based on working with a charity with staff based in multiple countries (Doswell et al., 2016). This work was originally based on the idea of using Delphi techniques with PSMs – see Rowe and Wright (2011) and specifically following the approach developed by Morton, Ackermann and Belton (2007). However, on encountering the Group Explorer/Decision Explorer System, attention switched to deploying a version of this system to the MS-Azure cloud platform as described by Yearworth and White (2017).

1 This was a Joint Venture between Hewlett-Packard and Dreamworks to develop a high-quality video conferencing product for *'C-suite'* customers consisting of a board-room studio installation divided in two by a wall of video screens. A great deal of attention was placed on designing a user experience that was as close as possible to both *'studios'* appearing as if they were actually part of a single board-room. This included matching up the positioning of sound from each participant with their apparent position in the screens and integration between studios over a dedicated low-latency network (Hewlett-Packard, 2007; Paay, Kjeldskov, & O'Hara, 2011).

Problem Structuring: Methodology in Practice, First Edition. Mike Yearworth.

The original work of Morton, Ackermann and Belton (2007) mixed a Delphi-style approach (Dalkey & Helmer, 1963) with cognitive mapping in a multimethodology.

> *'A characteristic of PSMs is their extensive use of workshops, where participants will create a model "on-the-fly". Because of this reliance on workshops, while delivering significant value to the organization overall, PSMs can be very costly in terms of participant time, as well as in financial terms if the organization is not physically collocated. Moreover, because involvement is restricted to workshop participants, what is often a key goal of PSMs—to involve a broad constituency of opinion in organizational and policy decision making—may be compromised.'*
>
> (Morton, Ackermann, & Belton, 2007, p. 547)

Whilst their approach was not grounded in the use of an online group support system (GSS), their motivation was based on the requirement to overcome some key limitations of PSM workshops. First, the time investment required from participants, both for travel and for participation in the workshop itself; and second, limited workshop places acting to restrict representation from a broader range of stakeholders. The need to overcome these restrictions has only become more important over time. The problem with the time commitment necessary for assembling a group of stakeholders into the same place at the same time for a workshop has only become compounded by concern about an organisation's CO_2 emissions. The effect of the COVID-19 pandemic, the need for home working, and now hybrid forms of working add to this requirement. To these, a new requirement can also be added, the need for very short lead times. The time-penalty associated with the process of assembling a workshop of participants at a given time and place may be too slow for certain types of organisational decision making.

The remainder of this chapter is a reflection on the findings arising from the implementation of a problem structuring approach using a GSS platform to replace the onsite group work for the fulltime MBA Programme at the University of Exeter during the lockdowns and travel restrictions associated with the COVID-19 pandemic. Teaching was switched to online delivery at very short notice during the first lockdown imposed in March 2020. Ethical approval[2] was sought from the University to collect data from the online delivery of my core module (Managing Strategic Resources and Operations) and individual consent was obtained from students for a survey administered during the delivery of the course. The primary question driving the research was concerned with rapidly identifying '*best practices*' that might be transferrable across the MBA teaching

2 University of Exeter ethics approval eUEBS000731 v2.0 23/03/2020.

team as we dealt with the unique circumstances of the pandemic lockdown and travel restrictions. However, given that some initial work had already been done to convert an existing same-time/same-place GSS (for SODA-4 – see Chapter 5) to a cloud-hosted environment to support same-time/different-places workshops (Yearworth & White, 2017) and the timely availability of the new Strategyfinder platform, the research question naturally enlarged to include developing an understanding of the effectiveness of distributed working using online group workshops for problem structuring and strategy making. The initial reflections on findings from the first round of data collection were moderated by a second online delivery of the course later in 2020 and the simultaneous preparation of a chapter for the Handbook of Group Decision and Negotiation (GDN) on experiences of same-time/different-places working (Yearworth & White, 2020). As background, Eden and Ackermann (2020) present an overview of GSS principles, and Ackermann (2020) a review of GSS developments.

This was a limited and qualitative only study. It was not based on the singular purpose of letting theory emerge from data (Strauss & Corbin, 1998) and there were some implicit assumptions that I brought to the study. First, I did not believe there would be any widespread '*best*' practices' that would emerge, but that some practices would prove useful according to specific purpose. Secondly, I was assuming that the students would find online same-time/different-places working just not as good (effective, preferable, enjoyable, satisfying, politically acceptable…) as the in-person group work they were originally expecting (and had paid for) when they joined the Programme in September 2019. On the other hand, I thought the emerging needs that demand online working, such as enabling faster workshop setup times, reducing carbon emissions, and supporting emerging post-pandemic working patterns (particularly hybrid-working) would be of value to the students in their post MBA employment. The experience of seeing another colleague scale a design exercise for a very large cohort of undergraduate classes also led to a further question – how could problem structuring approaches using GSS platforms like Strategyfinder scale to support very large group participation? For example, would it be possible to get a cohort of hundreds of undergraduate students trained-up in using one of the core PSMs using only online tools?

14.2 Experimental Setting

The purpose of using an online GSS in conjunction with a conferencing system such as Microsoft Teams or Zoom, is to support a formal problem structuring approach that in a traditional setting would have been achieved using basic workshop tools such as whiteboards or flip charts, Post-it® notes, and Sharpie® pens – and of course all the participants assembled in the same room. For many facilitators, this '*traditional*' same-time/same-place workshop with participants clustered around flipcharts or whiteboards is still the easiest and most effective way of getting a stakeholder group or leadership team to work on a problem. It is also an approach that I would normally have used with my students working in break-out groups on a transformation task set for them during the delivery of my course on the MBA programme. Of course, strategic options development and analysis (SODA) workshops have long benefited from the replacement of the traditional facilitator tools with two distinct software elements, Decision Explorer (supporting the cognitive and causal mapping and projecting the model onto a screen) and Group Explorer (providing the '*Chauffeur*' interface on the tablets/laptops sitting in front of the participants). Whilst the Decision Explorer/Group Explorer combination has now been replaced by the web-based Strategyfinder platform the basic difference between the two can be summarised in Figures 14.1 and 14.2.

In the '*conventional*' workshop setup, participants in the room are sat in front of a tablet or PC with the '*Chauffeur*' interface presented on their screens. This interface allows statements to be entered, links to be made between statements, and preferencing and voting scores to be entered – all under the sequencing control of the facilitators. The model is projected onto a screen from Decision Explorer running on the '*Public*' server such that it is visible to all participants. Two facilitators are shown – one focussed on facilitating the workshop participants through their task and the second facilitator focussed on managing the layout of the model, such as presenting sub-views, conducting graph-based analyses, and controlling the sequencing of Group Explorer through its different phases, gathering, preferencing and voting (Yearworth & White, 2020).

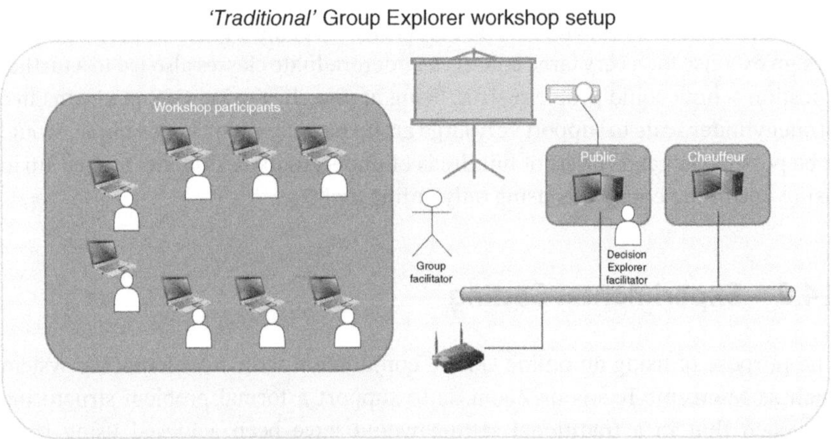

Figure 14.1 Conventional Group Explorer workshop setup. Two facilitators are shown here, one in effect facilitating the workshop participants through the use of the methodology and the second is managing Decision Explorer. Source: Yearworth et al. (2020)/with permission of Springer Nature.

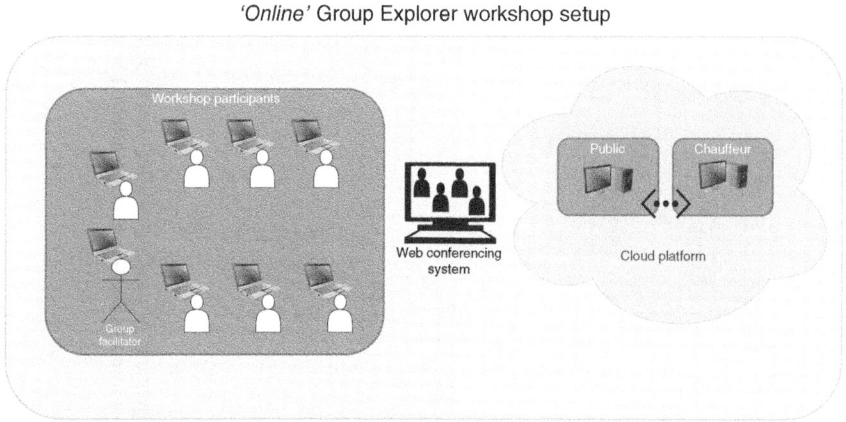

Figure 14.2 Online Group Explorer workshop setup. A web conferencing system is required to connect the participants in the online meeting and the Public and Chauffeur computers are hosted in a cloud computing environment. Source: Yearworth et al. (2020)/with permission of Springer Nature.

In the original port of Group Explorer/Decision Explorer to a cloud-based hosting environment for same-time/different-places workshops, the projection screen was replaced by a conferencing system that could share the Public screen from the server and the participants were connected to the Chauffeur interface via the internet. Note that amongst the many changes that this cloud-based configuration enforced was the appearance of the facilitator as just another participant in the system. It was this de-centring of the facilitator role that motivated a new area of research. In this configuration, we started to address some of the questions concerning the role of facilitation in such settings (Yearworth & White, 2017, 2020). This is illustrated in the hierarchical process model (HPM) to address the purpose <Making Group Explorer usable in a same-time/different-places mode> shown in Figure 14.3. The analysis techniques used are discussed in Section 6.5. The development of the Strategyfinder platform replaces this brute-force porting to a cloud-based hosting environment with a web-based interface that seamlessly combines the Public view of the Decision Explorer map with the Chauffeur interface. Additionally, this new platform encapsulates learning from the evolution of the SODA methodologies (Eden et al., 2020).

14.3 Teaching Requirements

The rapid response to the shift to remote teaching in response to the first COVID-19 lockdown in the UK generated an immediate set of requirements to replace the originally planned group work using HPM as the problem structuring approach. These were

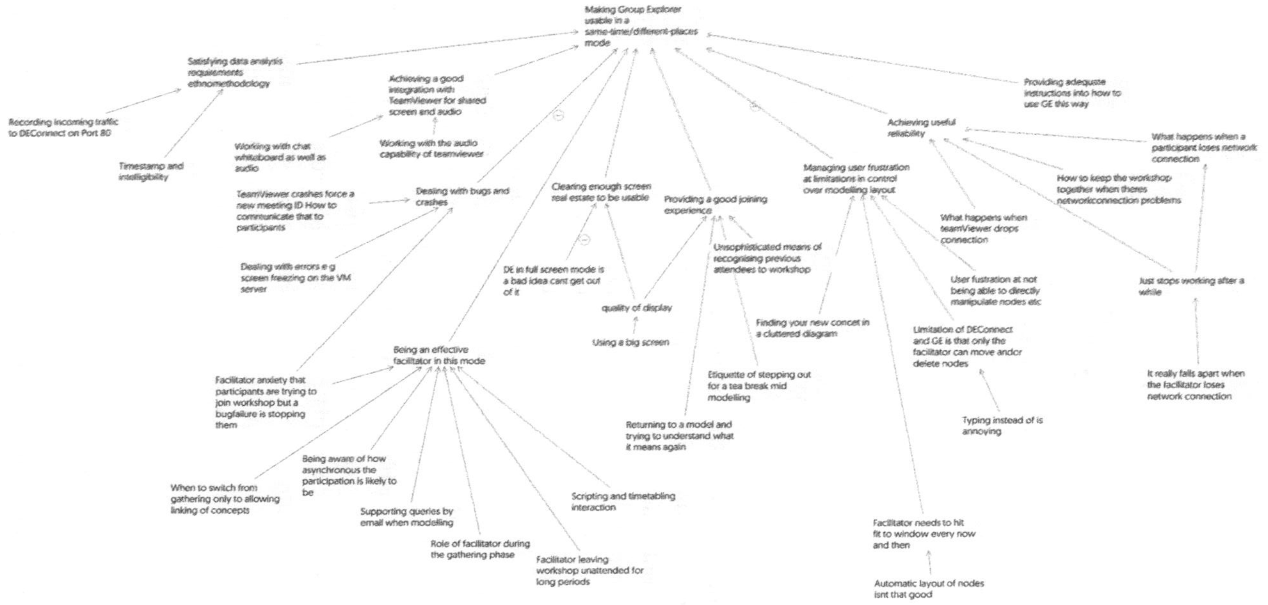

Figure 14.3 The original model from Yearworth and White (2017) converted for use with Strategyfinder. (Note: Original Decision Explorer models in XML format were parsed and converted to an equivalent JSON representation for import into the Strategyfinder platform.) Source: Yearworth et al. (2017)/with permission of Springer Nature.

1. Organising small ($N_{members} \leq 7$) groups of geographically distributed students with mutually convenient time zones
2. The problem structuring approach would need to be self-taught by using a combination of presentation materials and a video recording of a lecture teaching its use; both stored in the University's standard Virtual Learning Environment (VLE) i.e. Moodle in this case
3. The problem structuring approach would also need to be self-facilitated and self-scheduling to manage differences in individual workloads and time zones[3]
4. The problem structuring approach would need to be supported by a GSS that could be used without the input or control of a facilitator
5. Use of the GSS would also have to be self-taught in the same way as the problem structuring approach
6. Group work would need to be supported by conventional online conferencing and team collaboration systems i.e. MS-Teams in this case
7. Presenting the HPM of the business transformation as part of the final '*pitch*' was a requirement to show their working

Whilst the overall cohort size was not large, these requirements reflected the needs of students that were widely geographically separated, covering −8 to +10 hours of time difference from GMT; although each group, of no more than 7 students, was created to be relatively close in time zone. The requirement for groups to self-facilitate emerged simply from the impossibility of being able to facilitate groups over such a wide time zone difference and the need for multiple groups to self-schedule group working sessions at mutually convenient times.

The exercise was constructed as a management simulation. Each group was presented with the situation that they represented a consultancy invited to pitch to the board of a potential customer to advise and lead an operations transformation that addressed the requirements brief that was set. Throughout the course, three different transformations were tackled

1. Transforming a traditional supermarket to a new online delivery service,
2. Servitising and circularising a retail furniture business, and
3. Integrating management analytics into a university migrating to online delivery of programmes via a content platform[4]

Presenting the HPM of the transformation was a requirement for each group to show their work. Whereas in conventional on-campus group work, where I could work with groups directly, in this online setting the time zone differences

3 Therefore, '*at-the-same-time*' modelling but not the same local time.
4 This list was enlarged in the following academic year to include; (i) platforms and the gig economy, (ii) food security and food waste, (iii) fast fashion, (iv) everything-as-a-service, (v) click and deliver – instant retail and last mile logistics, and (vi) climate risk to operations and ESG reporting.

and self-scheduling nature of the group workshops meant that this was impossible to achieve. To address the research questions stated in the ethics approval for the study it was decided to see what could be achieved solely through self-facilitation.

An example of a hierarchical process model developed by one of the groups is shown in Figure 14.4. This is from the case study looking at servitising and circularising the business model of a furniture business (Yearworth & White, 2021). One of the main findings has been that all groups in all exercises found it possible to grasp the CR/E/A framework of the problem structuring technique and use the Strategyfinder platform without any direct input from a facilitator. On the whole, for each of the three exercises, each group managed to develop a model that conformed to the rules set out in the training material. Generally, models were all acyclic graphs[5] with processes labelled using the gerund form, and could be read according to the rules of Hierarchical Process Modelling (see Section 9.3).

Whilst the modelling was designed to support problem structuring it was not the final output from the group work – this was the '*pitch*' each group made to the whole class in an attempt to '*win*' the business to lead the transformation. The class was asked to peer rate each presentation and therefore how the modelling contributed to structuring the group approach to achieving the transformation. The impression from the total of 15 models and presentations developed (that is, 5 groups and 3 exercises) was that that coverage was appreciated more than depth – participants focussed more on the breadth of concerns addressed rather than any high degree of '*how*' question decomposition – although no specific guidance was offered about how broad or deep to go with the modelling. Available time and meeting deadlines were the only limiting factors.

A lightweight evaluation was conducted based on Midgley et al. (2013), augmented by considerations of the online setting i.e. survey questions were asked against the Context, Purposes, Methods, and Outcomes of the original framework with the addition of questions under two new categories – Technical and Social Environment. The questionnaire was delivered using an online survey tool and there was a 43% completion rate ($N = 15$). It was understood that the evaluation framework was being used somewhat out of context, in that the problem structuring intervention was a manufactured problem loosely based on background case study material and the artificial construction of a management consultancy engagement, but the simulacrum of a consulting environment was sufficiently close that the use of the framework would return useful insights.

Under Context, the self-assessed cultural diversity of the groups and the breadth of experience of the participants was considered to be a strength. On the other

5 Since the Strategyfinder GSS was being repurposed away from a SODA methodology, which does support loops, to HPM where loops are not allowed it was important to see if students had discovered that loops were possible and whether they incorporated them into their models.

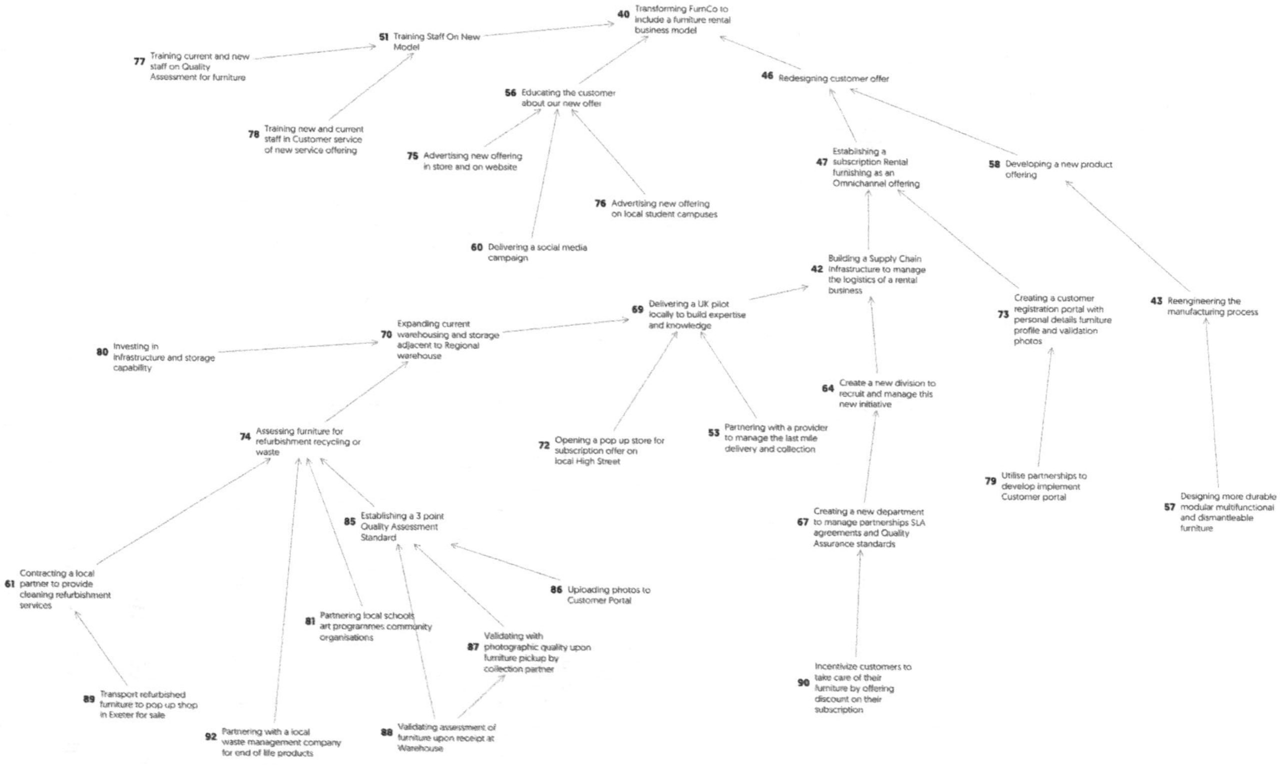

Figure 14.4 Example transformation from one of the student groups – servitising and circularising a retail furniture business.

hand, the group sizes were also considered to be too large. I believe this was a relative categorisation of group size compared to group work carried out in other modules on the programme, but also online. The groups were no larger than seven members, which was set as the maximum size based on prior experience with previous same-time/same-place workshops held using this methodology. For Purposes, the workshops had a single ostensible purpose and due to their manufactured nature, there was no real diversity of worldview or representation of broad stakeholder concerns. Questions concerning methodology highlighted the biggest participant concerns. Students struggled to disentangle the methodology from the use of the platform in the sense of user experience – which I covered outside of the original Midgley et al. framework under Technical and Social Environment considerations. There were also many expressions of concern that this way of working was just not as effective as working together in same-time/same-place groups – i.e. business as usual – and also reflecting one of my own ingoing biases. However, whether this was any different from the broad criticism levelled by students at universities for switching to online delivery rather than in-person teaching was impossible to determine from the data; the limitations of the questionnaire design were such that it was not possible to disentangle this effect. Perhaps, like everything else, a reflection of the speed with which we were working to make the transition to online. There were also some issues about scope and expectations. As mentioned already, there was no guidance offered about the breadth or depth of modelling – only a limit on the amount of time available for the activity. Outside the original evaluation framework, questions about the Technical environment surfaced difficulties with the use of the Strategyfinder platform; zooming-in and zooming-out of the model and scrolling functionality were flagged as issues. This was probably due to the limitations in the training material which did not adequately cover dragging the model around the screen and the function to re-centre models. Some participants also mentioned connection issues due to poor broadband performance. Interestingly, anonymous participation, which is a feature of Group Explorer and Strategyfinder, was considered by some participants to be annoying. Finally, Social Environment issues covered interruptions arising from working at home – from family members and neighbourhood, difficulties scheduling mutually convenient times, and dealing with the remaining time zone issues.

14.4 Reflections

My reflections focus on methodological and technical issues from this rather extreme form of non-facilitated same-time/different-places group problem structuring. They complement the more speculative discussion in Yearworth and White (2020).

Firstly, experience of teaching a modelling approach that avoids ontological commitment (delayed reification) by enforcing processual thinking has always been challenging, but the enforced lack of facilitation meant there was no possibility of any useful process guidance to be offered. On the whole, I think the students managed to build process models that avoided reification and I was satisfied, teaching wise, with what they achieved. The modelling did appear to help their group decision-making, although this was not explicitly tested. Certainly, the breadth of the issues covered by the models was encouraging. HPM has a rigorously defined epistemology and affordance of modelling approach (see Sections 9.4 and 9.5). When combined with the constitutive rules given in Section 5.2 these give the specific instantiation of the CR/E/A framework for problem structuring with HPM. The generative nature of the constitutive rules that allows expressiveness in the method and the more prescriptive epistemology and affordance of the modelling approach appeared to cause difficulty for some students. It is possible that the experience of using other online platforms, most likely gained from other modules on the MBA Programme,[6] led to a certain degree of frustration, trading-off creativity in what they wanted to express on a canvas and what was allowed in the GSS.

My second point concerns the sort of IT experience expected. Students demand IT that is easily accessible, delivers its functionality quickly, and conforms to idiom. This last expectation is driven by the ubiquity of commercial platforms. In a world that is already completely in the grip of social media platforms and with a rather loose interpretation of what *is* a GSS, then it is reasonable to think of students already being used to a type of GSS. Yearworth and White (2018) have argued that the community use of social media can be considered as a form of GSS that can support spontaneous self-organising problem structuring. Students have grown up with these platforms, which have been highly optimised for user experience with vast development budgets. They offer a user experience that is easily accessible, users do not have to work too hard to obtain desired functionality. Therefore, by idiom, I mean that the user experience of these mass platforms has set the generally expected way that a collaboration platform should work. Platforms such as Strategyfinder have origins in specialised GSS (i.e. COPE, Group Explorer/Decision Explorer) which were never meant for mass use and instead reflect theoretically informed methodology.

My final point addresses the role that self-facilitation can play in the future. Although self-facilitation in this experiment was a consequence of the requirements stated earlier, there is a promising research avenue opening up in further

---

6 Mural was definitely being used, and it is possible that Miro was too. Both these collaboration platforms are more open – i.e. they support many different types of diagramming templates – and offer web-based user interfaces that to a certain extent set the expectation for how online collaboration tools should work.

investigation into the question of *self-scaffolding* of methodology in a GSS (Yearworth & White, 2020). Self-scaffolding is a concept that describes the degree to which the process of using a methodology is built-in to the way in which the GSS platform works. A highly scaffolded platform would require little by way of training, or explanation, or facilitator intervention i.e. it could be used immediately for content and the process would be implicit in how the platform is used. This idea is discussed below, especially with regard to scaling-up to include working with large stakeholder groups.

14.5 The Effectiveness of Online Problem Structuring Practice

These reflections on the results from the experiment of designing and setting a self-facilitation problem structuring exercise raise some questions about evaluating the effectiveness of the use of GSS for problem structuring, generally. How do we disentangle content from the effects of the combination of methodology, GSS technology, and facilitation that create the process, in assessing the overall effectiveness of the intervention? In effect, how do we explore the relationship $\varepsilon = f$(Content, Methodology, Facilitation, GSS Technology), where ε is some notional measure of effectiveness. Eden (1995) points out that the evaluation of the performance of a GSS runs into the same problem as measuring the value of the PSM intervention itself as discussed in Section 6.3.

Therefore, rather than trying to understand this problem, can we instead try to influence or benefit from technical developments? For example, can automation of a methodology via the use of GSS technology reduce facilitation to its essential minimum such that we create a *problem structuring platform*? What I mean by this is this – can we imagine having a GSS available that combines the implementation of a problem structuring methodology, e.g. such as HPM or SODA, but also directs the workshop participants through the process of problem structuring? The merging of methodology, facilitation, and GSS technology into a single concept – that of platform-enabled procedural scaffolding – such that effectiveness can be reduced, more or less, to consideration of the platform and the content i.e. $\varepsilon = f$(Content, Platform)? The same conundrum of measurement exists (Eden, 1995), but the complex interaction between methodology, facilitation, and the GSS technology process has now been distilled into the single question of how the platform operates. So far, there are no online collaboration platforms that implement a problem structuring methodology with this sort of self-scaffolding capability, although it is not too difficult to imagine what it would look like. However, despite the apparent attractiveness of such a development, the emergence of a self-scaffolding problem structuring platform is likely to cause tensions with more traditional views of problem structuring that see the role of facilitation as an indispensable element.[7] Whilst acknowledging these tensions, I believe it is still worthwhile mapping out what the boundaries of feasibility might be, which leads to the following refinement

1. Pure process platforms: GSS that self-scaffold methodology sufficiently so that users can focus on content and there is no need for training or facilitation.
2. Pure content platforms: GSS that are designed to be used under the control of a facilitator e.g. Strategyfinder.

Until pure process problem structuring platforms are available (if they ever become available), hybrid positions will likely exist, much like the experimental situation discussed in this chapter, i.e. a pure content platform being pressed into use without the availability of a facilitator. In this case, it is possible to make three recommendations to improve the effectiveness of the use of GSS in such situations:

1. Pre-emptively rolling-out a cross-organisation training programme in a problem structuring methodology and the use of a specific GSS as part of the normal training diet. This would equip project teams, ad-hoc groups, senior leadership teams, with the means to self-assemble and initiate collaborative problem structuring with very short lead-times; as soon as the necessary participants are available.[8]
2. Dealing with issues arising from limitations of technology and being more prescriptive in the way in which team collaboration tools are used. Turning off video and minimising the use of chat constrains participant-to-participant interaction to happen only via the model – reinforcing its role as a boundary object in the problem structuring process (Franco, 2013).
3. Enabling new modes of working, being creative with the methodology especially using different-times/different-places modelling; i.e. allowing for asynchronous phases by making the model accessible for participants to interact with it in their own time.[9] So, for example, the gathering phase of a SODA workshop could be carried out asynchronously over a period of time prior to the actual same-time/different-places online workshop. This makes better use of participants' busy schedules in the preparation for the workshop.

7 Some of this tension is covered in the discussion of facilitation in Section 6.5. I believe the debate will continue.
8 This requirement has surfaced in questioning arising from government analysts during conference presentations. Policy contexts are very fast-moving.
9 In the original experiments carried out by Yearworth and White (2017), this was an unlooked-for affordance arising from the cloud-hosting of Group Explorer/Decision Explorer, it was not an original motivation.

4. Learning, through the use of the tools available to record workshop interactions, using the techniques discussed in Section 6.5 and analysis of the data with a specific focus on aspects of practice that are impacting effectiveness, both positive and negative.

The use of GSS for achieving rapid problematisations and problem structuring seems quite achievable so long as the organisation keeps looking for effectiveness improvements using an analysis of the micro processes. Digging into the fourth recommendation in a little more detail and making use of the analysis techniques set out by Yearworth and White (2017) it is possible to focus on the following transitions in the process

1. Process → Content. How has the methodology and/or GSS training enabled phases of workshop to flow? This is where participants want to be spending their time, working on the problem at hand. What caused the participants to stop debating process issues and get on with working on content?
2. Content → Process. What was it about methodology and/or GSS training that led to the workshop breaking down into discussing process issues?
3. … → Process → Process → … The workshop has become totally process bound, perhaps right from the start. The problem content has not been addressed at all. Was the methodology and/or GSS training completely inadequate or non-existent?
4. null → Content → Content → … Here the workshop bootstrapped into content-full activity.

The fourth point might be thought of as how a workshop immediately starts dealing with content and stays there, which is what the ideal of a self-scaffolding problem structuring platform would be. Clearly, we are not there yet but the practical implication of this framing of the use of GSS and the micro-analysis of the process is to work out how much time as a proportion of the total available to a workshop group can be spent focused on the content work.

14.6 Teaching of PSMs in an Education Setting

Implementations of self-scaffolding methodology on collaboration platforms in an education setting started to emerge during the pandemic lockdowns. An example of this – built using the Mural online collaboration platform (Tippin, Kalbach, & Chin, 2018; Mural, 2023) – orchestrated groups of students through a design exercise with minimal intervention from the module leader. The Mural page was set up for each group and left for them to self-facilitate through the exercise in a certain time period. It was also left to the group to self-organise their time and engagement with the platform. This was very successful and obviously scaled to large class sizes (hundreds of students) extremely well.

One of the primary advantages of specialist platforms like Strategyfinder and muVAM is that they hold the participants and facilitators to the methodology and prevent superfluous additions. On the other hand, collaboration platforms that came to pre-eminence during the COVID-19 pandemic are sufficiently general purpose to support many different sorts of online group work and enable different collaboration templates to be created. In order to obtain the functional implementation of a known methodology requires a general-purpose platform to be *constrained* i.e. going from the general purpose to the specific, unlike the specialist platforms that have emerged, bottom-up, from the PSM practitioners. Using the ideas of epistemology and affordance of modelling approach from the CR/E/A framework it is likely that a general-purpose platform will enable processes that are inconsistent in the use of terminology and/or symbols that are not in the methodology, and afford diagrammatic expression that is outside the visual language of a known methodology. They would allow the emergence of a process that is not rigorous in the sense with which PSMs have been defined.

14.7 Extension to Large Group Working

Despite the emergence of PSMs as a new paradigm in Operational Research (OR), as outlined in Chapter 4, they were still being used, on the whole, within the same settings as traditional OR – meeting the needs of small groups or teams in organisations. Breaking out of this traditional context – for what was still a relatively new approach at the time – White (2002) introduced the idea of using PSMs in very large group settings. The attributes of PSMs seemed well matched to the needs of large groups, especially in their cognitive accessibility, and eminently suitable for the purpose of addressing representativeness in organisations and dealing with turbulent environments. However, based on two large group case study examples, it was clear that there were problems arising from difficulties in facilitating such large groups and scaling the existing mapping approaches to deal with the very large number of issues raised. Shaw et al. (2004), in a follow-up to White's work, conducted a more in-depth analysis of the differences between using PSMs with small groups and what they called Large Group Intervention Methods (LGIMs), concluding that, as they currently stood (in 2004), PSMs would need some modification to deal with the problem of uncontrolled divergence and complexities arising from the sheer number of issues and options that might be surfaced by a large group.

There seems to have been limited further work addressing scale-up to large group working. Morton, Ackermann and Belton (2007) demonstrate scaling, but without using a GSS, mainly leveraging how the Delphi technique can be scaled to working with large groups of experts asynchronously. O'Brien and Meadows (2007) and Bryant, Darwin and Booth (2011) show how to deploy PSMs for strategy making across organisations and thus widening the participative debate (Bunker & Alban, 1996). White and Bourne (2007) show how to scale the use of PSMs in a large group Community OR application. However, whilst these examples show some degree of scaling to large groups, none took place in the era of web-based PSM platforms (available since circa 2019) and it remains to be seen how, with the emergence of new GSS for problem structuring, such as muVAM and Strategyfinder, it would be possible to re-engage with the questions raised by White (2002) and Shaw et al. (2004) with platforms that would be intrinsically better suited to dealing with scale-up issues, especially if augmented by some of the technologies discussed in Chapter 15.

The teaching of PSMs to large groups of students may be possible by arranging them into groups with a set of pre-populated platform templates for delivering a set of scalable modules or parts of modules for problem-oriented learning. A set of problems/challenges could be defined within a context that, for the subject in question, would be considered as messy and with a group-oriented organising/strategising/planning task to deliver a plan for intervention.[10] Groups would need to be self-organising in time and also in place i.e. hybrid working would be allowed and encouraged.[11] The platform would be pre-loaded with a set of scaffolds that supported the modelling approaches used by a range of PSMs e.g. SODA, soft systems methodology (SSM), strategic choice approach (SCA), HPM… constructed from their respective epistemologies and modelling affordances.[12] In addition to the self-scaffolding of the methodology, the platform would also be self-documenting i.e. the course Instructors would be able to see the transitional objects (e.g. activity models, hierarchical process models, causal maps, decision graphs…) that supported the group in their process towards agreeing an action plan.

Recent reports on the use of online GSS to support PSM workshops have been made by Lami and White (2022) on using the muVAM platform, and by Zimmermann et al. (2021) in conducting a variety of System Dynamics and Group Model Building workshops online during the pandemic. Their findings complement the work presented in this chapter and contribute to the debate about PSM practice as the use of online GSS workshops is likely to increase. The issues raised relate just as well to the larger question of how the ideas of PSMs and Soft OR practice

10 e.g. similar to the set listed in Section 14.2 above.

11 Also necessary, as was the case when teaching was largely on campus but with some members of a class attending remotely due to different local lockdown/travel restrictions.

12 This approach is of course not limited to just teaching PSMs. Templates for other group activities such as Design Thinking, Scrum/Agile, Hackathons, etc. could also be included.

could find their way into the ubiquitous platforms that dominate social and business life. Incorporating ideas from this world into the narrow domain of specialist GSS to support PSMs in '*traditional*' settings seems reasonably straightforward in comparison. Yearworth and White (2017) provide a useful summary

> '*The use of PSMs and GSS has largely been within organisations and communication about them restricted to academic and teaching texts. On the other hand, crowdsourcing platforms are open and have grown on the basis of meeting the functional requirements of their users. Development and communication have relied largely on Web 2.0 technologies and Open Source ideals and tools. Whether these two streams are ever likely to cross-share ideas is debatable. To a certain extent, the scaffolding provided by the Chauffeur is recognisable in other forms on other platforms, perhaps more as an implicit way of using them than anything that is provided by way of automation. However, the one thing that does set the two streams apart is the use of formal modelling approaches. As can be seen from the complexity of using Decision Explorer it is extremely unlikely that formal causal mapping capabilities will find their way into open platforms in the future. However, the ubiquity of open platforms means that there is potentially a ready audience for better ways of making decisions, if only a way could be found for making these more formal techniques more approachable and easier to use. As an area of further work, we suggest that heuristics could be captured from the detailed analysis of how facilitators use these modelling approaches and then used to produce highly automated versions suitable for integration into open platforms.*
>
> *Unless researchers find a way of breaking out from the confines of purely academic interests in the development of GSS and PSMs then their work will likely have little impact on the development of platforms that will be used by the majority of people in the future, even in business. Furthermore, until they find a "way in" to influencing these platforms it is possible that the quality of debate and decision making on them is always going to be less than that achievable with a well-designed GSS underpinned by an appropriate and well-theorised methodology.*'
>
> (Yearworth & White, 2017)

15

New Technologies for Augmenting PSM Practice

The developments in group support system (GSS) demonstrate how new technologies are having an impact on Soft Operational Research (OR)/problem structuring method (PSM) practice. The intersection with the broad area of platforms should not be surprising given their ubiquitous nature. Machine learning is another burgeoning technology development that has the potential to have an impact on Soft OR/PSM practice. This chapter looks at two technologies in particular

1. Augmented qualitative analysis (AQA), based on the use of probabilistic topic models, and
2. Application of Large Language Models (LLMs).

Building on Chapter 14, an obvious setting for the realisation of these augmentations is as extensions to existing GSS.

15.1 Augmented Qualitative Analysis (AQA)

If problem structuring – whether in OR or engineering practice – is viewed as a form of problem exploration then the initial process of problem exploration or '*enquiring*' into a system might well resemble a form of qualitative research. Techniques of qualitative analysis such as semi-structured interviews and the use of computer aided qualitative data analysis software (CAQDAS) would segue into the use of a problem structuring method. Some of the examples cited in the survey of PSM applications use this approach. Early uses of Strategic Options Analysis and Development (SODA) describe processes for '*helping clients talk about their problems*' (Eden, Jones, & Sims, 1983, pp. 25–38) and a qualitative researcher with experience of conducting semi-structured interviews should feel comfortable with this form of problem exploration. The joining together of qualitative analysis and problem structuring into a formal multimethodology would be a rather natural action research process (Yearworth & White, 2013).

Problem Structuring: Methodology in Practice, First Edition. Mike Yearworth.

There is some background in the integration of qualitative analysis as a preliminary stage in model development in System Dynamics. Morrison (2003), Luna-Reyes and Andersen (2003), and Kopainsky and Luna-Reyes (2008) were all an influence on the design of the multimethodology described by Yearworth and White (2013). Vennix (1996) includes qualitative analysis as a means of developing preliminary models prior to Group Model Building. More recent developments by Newberry and Carhart (2023) show a continued interest in this form of multimethodology.

In the case of organisations or multi-organisations, including project organisations, with considerable quantities of documents, in existence or are generating them as a process of ongoing documentation, then the semi-structured interviews of qualitative research could be replaced by automated topic analysis. Certainly, for any endeavour that is generating copious quantities of documents whilst needing to maintain a fast decision-making capability, then the automatic generation of potential causal maps that could be used as the starting point for a subsequent problem structuring workshop is likely to be of great benefit.

Advanced computational techniques are beginning to have an impact on social science research and have placed approaches such as automated content analysis on a path towards making a major contribution (Nelson, 2020). With the support of new computational techniques, it is possible for the qualitative researcher to move back-and-forth between the text, coding, and higher-order concepts (Mantere & Ketokivi, 2013). Probabilistic topic modelling (Blei, 2012), based on exploiting the statistical properties of the Latent Dirichlet allocation (LDA) (Blei, Ng, & Jordan, 2003) is now a fairly mature machine learning technology. Analysing a corpus to examine context and to generate themes for a theoretical model of relevance to management scholars using a probabilistic topic model is demonstrated in recent work by Croidieu and Kim (2017) and Hannigan et al. (2019). The same techniques can be used to develop preliminary models for a problem structuring workshop. The approach described here is ideally suited to data sets that sit at the '*top end*' of the scale of a conventional qualitative analysis by a small group of researchers, even if supported by current CAQDAS tools, but still very small compared to what topic modelling is able to handle (Crossley, Dascalu, & McNamara, 2017). This scale exists in a *hinterland* of qualitative analysis – not big enough for totally automated big-data analysis but too much for conventional manual techniques. Hence the suggestion that this is AQA. Analysis proceeds inductively, by interpreting the term lists within the raw topics identified by the probabilistic topic modelling as variable labels for the models.

Once a topic model has been trained on the data, it is then possible to apply the topic model as a *classifier* of the original data at the paragraph level. Once each paragraph is classified with the equivalent of a CAQDAS coding strip then it is possible to derive an adjacency matrix by counting, for each topic pair, the number of times the pair co-codes a paragraph of text. Following the method described by Yearworth and White (2013), this generated adjacency matrix can be interpreted as a putative causal loop diagram (CLD) or causal map. For a topic model of k topics, there are $k(k-1)/2$ potential causal links.[1] The reduction of all the possible pairings to a plausible number can be done by thresholding i.e. only treating co-coding of paragraphs as significant if they occur over a threshold number of times. Deciding whether this reduced number of possible causal links are actually causal is a matter for methodological debate (Luna-Reyes & Andersen, 2003; Kopainsky & Luna-Reyes, 2008), but can be supported in AQA by circling back to an inspection of the raw data, where significant co-coded paragraphs are displayed to the qualitative researcher. Yearworth and White (2013) commented that such a facility was not available in the CAQDAS tools that were available at the time. Now it is possible to proceed directly from the probabilistic topic model classification of a corpus and the automatic production of the adjacency matrix to generate candidate models in a format that can be read by a suitable application or platform.[2]

To demonstrate the technique, a corpus of 170 documents of 1.05 Million words associated with the United Kingdom Parliamentary scrutiny of the 2012 Olympic Games[3] was used to train an LDA-based probabilistic topic model with 18 topics ($k = 18$) and then to generate candidate CLDs for a range of threshold values. A visualisation of the LDA probabilistic topic model using the pyLDAvis package (Sievert & Shirley, 2014) can be seen in Figure 15.1.

To maintain the link to the method originally described by Yearworth & White (2013) the adjacency matrix generated from the topic model classification of the corpus at the paragraph level is shown in Figure 15.2.

The (raw) graph result can be seen in Figure 15.3. The nodes have been coloured in this case according to betweenness centrality. This reflects the statistical properties of the network and how it was generated rather than any indication of whether these nodes have any particular leverage (Crielaard et al., 2023). Such graphs as these are preliminary models to a group workshop and, as imported in tools such as Vensim, do not carry forward any node colouring or link polarities. However, despite these promising developments, the use of graph analysis tools as part of

1 This matrix is symmetric, we therefore have to consider both possible links between variables. However, the leading diagonal can be removed as self-loops on a variable in CLDs are meaningless. This technique bears some similarity in the $O(N^2)$ complexity as found in SODA-1 when using repertory grids, and also Interpretive Structural Modelling (ISM).
2 This has been demonstrated by generating MDL files for Vensim and JSON files for Strategyfinder. It would be possible to also create files in the now standard XML Interchange Language for System Dynamics (XMILE).
3 As published by the Select Committee of Culture Media and Sport, the Public Accounts Committee, the National Audit Office, and verbatim transcripts of the oral evidence obtained by the Select Committees.

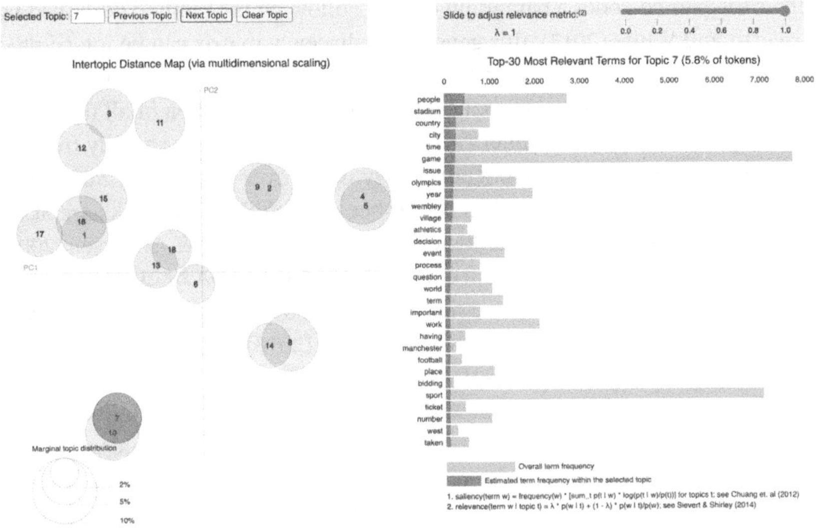

Figure 15.1 Visualisation of the LDA probabilistic topic model highlighting the term list for Topic 7, which has been labelled '*Deciding stadium location*' in the subsequent graphs.

	T_1	T_2	T_3	T_4	T_5	T_6	T_7	T_8	T_9	T_10	T_11	T_12	T_13	T_14	T_15	T_16	T_17	T_18
T_1	0	30	179	20	8	19	119	13	34	145	281	20	36	34	21	158	212	10
T_2	30	0	66	86	240	6	31	47	37	45	39	43	46	81	77	38	17	27
T_3	179	66	0	42	14	38	29	12	82	33	132	195	320	18	145	70	380	36
T_4	20	86	42	0	510	41	19	144	285	33	12	7	78	55	16	10	10	12
T_5	8	240	14	510	0	16	18	234	95	49	31	4	15	189	16	6	6	11
T_6	19	6	38	41	16	0	102	44	24	79	17	13	81	93	49	3	23	14
T_7	119	31	29	19	18	102	0	153	34	613	28	25	115	167	57	34	121	78
T_8	13	47	12	144	234	44	153	0	18	218	8	4	50	104	12	42	30	55
T_9	34	37	82	285	95	24	34	18	0	27	20	26	79	28	91	21	14	30
T_10	145	45	33	33	49	79	613	218	27	0	70	29	164	200	53	88	178	76
T_11	281	39	132	12	31	17	28	8	20	70	0	59	5	21	9	168	154	6
T_12	20	43	195	7	4	13	25	4	26	29	59	0	35	27	230	18	69	11
T_13	36	46	320	78	15	81	115	50	79	164	5	35	0	43	64	20	91	39
T_14	34	81	18	55	189	93	167	104	28	200	21	27	43	0	47	3	35	21
T_15	21	77	145	16	16	49	57	12	91	53	9	230	64	47	0	17	65	32
T_16	158	38	70	10	6	3	34	42	21	88	168	18	20	3	17	0	123	6
T_17	212	17	380	10	6	23	121	30	14	178	154	69	91	35	65	123	0	26
T_18	10	27	36	12	11	14	78	55	30	76	6	11	39	21	32	6	26	0

Figure 15.2 Adjacency matrix for the graph in Figure 15.3. Note its symmetry and setting leading diagonal values to zero. Variables looping on themselves is not meaningful in CLDs.

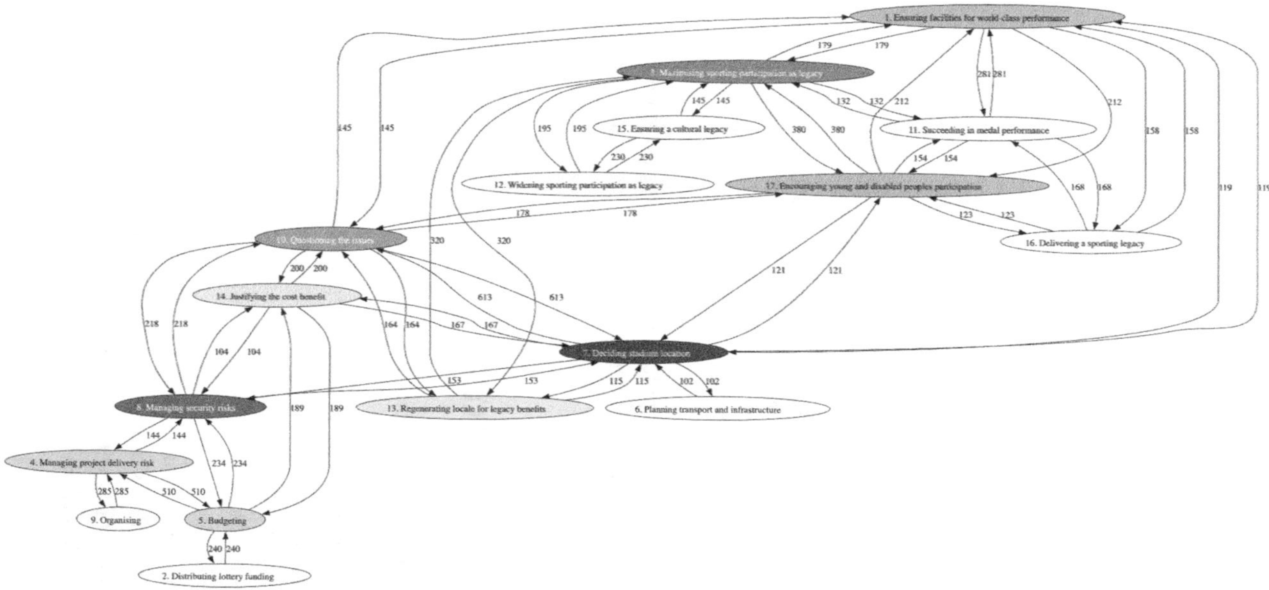

Figure 15.3 Raw graph generated from the automated probabilistic topic model classification of a corpus associated with the 2012 Olympic Games. Note that the topic numbers from the LDA model have been incorporated into the labels to show the link to Figure 15.1. The nodes have been coloured according to betweenness centrality.

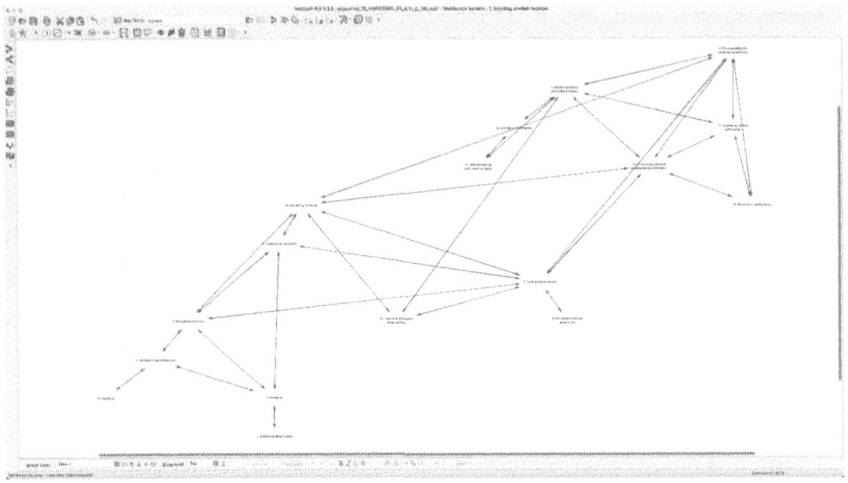

Figure 15.4 The raw graph shown in Figure 15.3 as imported into Vensim PLE. Note that the links are not bi-directional but are in fact two unidirectional links with the same start and endpoint coordinates. It is up to the modeller to decide which to keep and whether to label with +ve or −ve link polarity.

the AQA process needs more investigation, especially in regard to important questions such as the development of interventions from the CLD refined by the group workshop (Figures 15.4 and 15.5).

15.2 PSMs in the Age of LLMs

LLMs have recently become the ubiquitous manifestation of the public perception of AI. The release of a simple and intuitive interactive user interface to LLMs through the use of prompts (e.g. ChatGPT (OpenAI, 2023)) has catapulted this technology to widespread awareness. However, it is difficult to see how the process of knowledge creation through a process of *situated* enquiry/action research in a specific problem context could be more different from assembling responses to prompts from an LLM.

The prompt/response exchange with ChatGPT[4] below, represents an experiment into the relevance of this technology into the design of a problem structuring process. The exchange starts with eliciting a response grounded in the notion of abductive reasoning closely associated with the pragmatic philosophy discussed in Chapter 3.

4 March 14th, 2023 version.

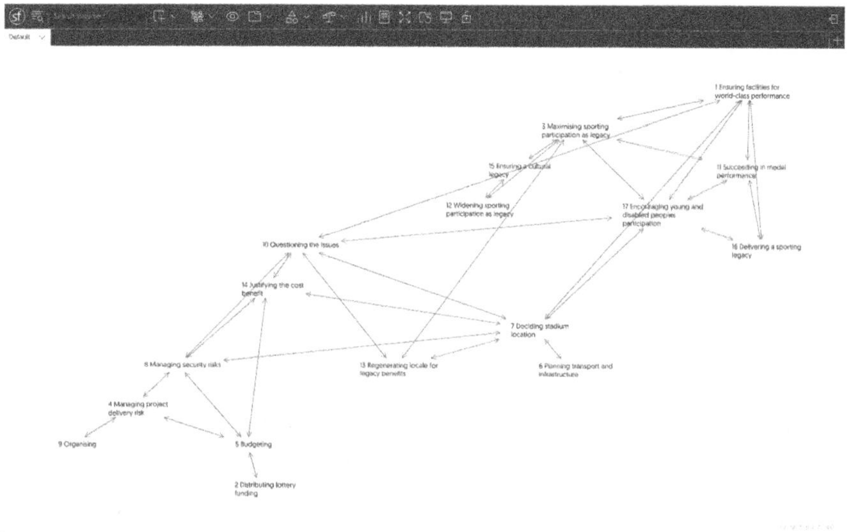

Figure 15.5 The raw graph shown in Figure 15.3 as imported into Strategyfinder. This time the links are bi-directional, but it is still up to the modeller to decide which to keep and whether to replace with unidirectional links and/or label −ve.

Prompt – '*Give me an example of how an LLM, like GPT, can be used to perform abductive reasoning to create new knowledge?*'

… a medical example was given (not shown)

Prompt – '*Give me an example from the applied social sciences. For example, addressing the needs of a community with a high flood risk*'

Response – 'Sure, here's an example:

Observation: A community is located in an area with a high risk of flooding.

Abductive reasoning: What could be done to address the needs of this community?

Hypothesis 1: Build more robust infrastructure to prevent flooding. Hypothesis 2: Develop a community-based emergency response plan. Hypothesis 3: Create public awareness campaigns to educate people about flood risks and how to prepare for them. Hypothesis 4: Implement flood insurance programs for residents.

Using an LLM like GPT, we can generate further information about each of these hypotheses and evaluate their potential effectiveness based on existing knowledge. For example, GPT may suggest that building more robust infrastructure is necessary to prevent flooding in the first place, but may also point out that this can be expensive and may not always be feasible. Similarly,

GPT may suggest that a community-based emergency response plan could be effective in managing the impacts of flooding, but may also indicate that such plans require significant resources and coordination.

By using an LLM to generate and evaluate hypotheses, we can potentially identify new strategies for addressing the needs of communities at risk of flooding, such as combining approaches or considering alternative approaches that may not have been previously considered. This can help inform decision-making and ultimately lead to more effective policies and programs for reducing flood risk and protecting vulnerable communities.'

The concept of abductive reasoning is clearly within the training material used for the model. The suggested mode of inquiry appears to be suitably pragmatic, although simplistic and technocratic (Fiorino, 1990).

Prompt – '*Give details of a complete approach that takes into account social desirability or feasibility of interventions*'

Response – 'When designing interventions in the applied social sciences, it is important to consider not only their potential effectiveness but also their social desirability and feasibility. Here is an example of a complete approach that takes into account these considerations:

Step 1: Identify the problem and goals. Identify the problem you are trying to solve and the goals you want to achieve with your intervention. For example, in the case of a community with a high flood risk, the problem might be a lack of preparedness for flooding, and the goal might be to improve community resilience and reduce the risk of harm from flooding.

Step 2: Conduct a needs assessment Conduct a needs assessment to understand the context and the needs of the community. This can involve collecting data on factors such as demographics, existing resources, and community attitudes and beliefs related to flooding. This step can help you understand what resources and strategies are already in place, as well as what gaps exist that need to be addressed.

Step 3: Develop a range of intervention options Based on the needs assessment, develop a range of intervention options that address the problem and goals you identified in step 1. For example, these could include options such as building infrastructure, creating community-based emergency response plans, implementing public awareness campaigns, or offering flood insurance programs.

Step 4: Evaluate the feasibility and social desirability of each option Using an LLM like GPT, evaluate the feasibility and social desirability of each intervention option. This can involve generating further information about each option and evaluating its potential effectiveness, as well as its potential costs, risks, and benefits. It can also involve generating information about how each option might be perceived by different stakeholders, such as community members, policymakers, and funders.

Step 5: Select and implement the most feasible and socially desirable option Based on the evaluation of each option, select the intervention that is most feasible and socially desirable. This may involve combining elements of different options or adapting them to better fit the needs and context of the community. Once a decision is made, implement the intervention and monitor its effectiveness over time, making adjustments as needed.

By taking into account social desirability and feasibility when designing and implementing interventions, we can help ensure that our efforts are not only effective, but also sustainable, equitable, and acceptable to the communities we serve.'

It is the prompting that has extracted a response that would now be considered as the beginning to address the ethical issues of an intervention. However, the language model response is still quite technocratic in flavour. Again, further prompting is needed.

Prompt – '*Who decides the most feasible and socially desirable options?*'

Response – 'The decision-making process for determining the most feasible and socially desirable options for an intervention will typically involve multiple stakeholders with different perspectives and areas of expertise.

In the case of the example I provided earlier, stakeholders might include representatives from the community at risk of flooding, local government officials, emergency response organizations, environmental groups, and other relevant organizations or experts.

The decision-making process should be transparent and participatory, with opportunities for stakeholders to provide input, ask questions, and express concerns. Ideally, the decision-making process should be based on a shared understanding of the problem and the goals of the intervention, as well as a clear and systematic evaluation of the feasibility and social desirability of each option.

Ultimately, the responsibility for making the final decision on which intervention option to implement will likely rest with a designated authority or group, such as a government agency, community organization, or steering committee. This group should be accountable to the stakeholders involved and should communicate its decision-making rationale and the criteria used to evaluate each option.'

The process is now sounding closer to something that could be realistically operationalised, but this now needs to be brought closer to the realm of Soft OR/PSMs practice.

Prompt – '*What sort of models should be used to support this process bearing in mind that not all participants have a technical background?*'

Response – 'To support a decision-making process that involves stakeholders with different backgrounds and areas of expertise, it is important to use models and tools that are accessible, easy to understand, and facilitate communication among stakeholders. Here are some examples of models and tools that could be used:

1. *Decision matrices: A decision matrix is a tool that can be used to systematically evaluate and compare different options based on a set of criteria. This can be particularly useful when trying to weigh different factors that may be important to different stakeholders. The decision matrix should be designed to be easy to use and interpret, and should include clear instructions for how to complete it.*
2. *Visual aids: Visual aids such as diagrams, charts, and maps can help to communicate complex information and facilitate understanding among stakeholders who may not have a technical background. For example, a map showing the flood risk zones in a community could be used to help stakeholders understand the risks and potential impacts of flooding in different areas.*
3. *Plain language summaries: Technical reports and documents can be difficult for stakeholders without a technical background to understand. To ensure that all stakeholders are able to participate fully in the decision-making process, it is important to provide plain language summaries of key information and recommendations.*
4. *Participatory modeling: Participatory modeling involves bringing stakeholders together to collaboratively develop and use models to support decision-making. This approach can be particularly useful when stakeholders have different perspectives or knowledge, as it can help to build understanding and consensus among participants. Participatory modeling can be facilitated by trained facilitators, who can help to ensure that all stakeholders are able to contribute and participate fully.*

Overall, the models and tools used to support the decision-making process should be tailored to the needs and preferences of the stakeholders involved, and should be designed to promote transparency, participation, and understanding.'

Prompt – '*Can you provide any examples of participatory modelling approaches?*'

Response – 'Sure, here are a few examples of participatory modeling approaches that have been used in the applied social sciences:

1. *Community-Based System Dynamics (CBSD): CBSD is an approach to participatory modeling that involves engaging community members in the development and use of system dynamics models. System dynamics models are used to represent complex systems and the relationships between different components of those systems. In CBSD, community members work together with facilitators to develop a shared understanding of the system being modeled and to identify potential interventions that could improve outcomes. CBSD has been used in a variety of contexts, including public health, environmental management, and disaster risk reduction.*
2. *Participatory GIS (PGIS): PGIS is an approach to participatory mapping that involves engaging community members in the collection and analysis of geographic data. PGIS can be used to support decision-making in a variety of contexts, including land use planning, natural resource management, and disaster response. PGIS typically involves a combination of traditional mapping techniques and digital tools, such as Geographic Information Systems (GIS) software, to create maps that reflect local knowledge and perspectives.*
3. *Participatory scenario planning: Participatory scenario planning involves engaging stakeholders in the development of alternative scenarios for the future, based on different assumptions and inputs. Scenario planning can be used to explore potential futures in a variety of contexts, including urban planning, climate change adaptation, and natural resource management. Participatory scenario planning typically involves a facilitated process of brainstorming, modeling, and testing different scenarios with stakeholders, with the goal of identifying potential risks, opportunities, and trade-offs.*

These participatory modeling approaches all share the goal of engaging stakeholders in the development and use of models to support decision-making. By involving stakeholders in the modeling process, these approaches can help to build shared understanding, generate new insights, and promote collaborative decision-making'.

Whilst the generated text here would not be out of place if it was read in a flood risk strategy document, the degree of leading that was required, via the construction of the prompts,[5] shows that this is a technology that might have a role as an *augmentation* to a PSM approach, perhaps in the directed construction of hard/soft integration methodology, and that it might improve the effectiveness of an intervention. This is perhaps not surprising – a methodologist prompting an LLM will receive process-like answers. However, this is a long way short of any actual stakeholder engagement and critical thinking that would be considered normal for a Community OR engagement (e.g. Gregory et al., 2013). The veracity of the suggestions will also need to be thoroughly checked. In this case, participatory GIS (Dunn, 2007), community-based System Dynamics (Hovmand, 2014), and participatory scenario planning (Oteros-Rozas et al., 2015) have a well-established presence in the literature and offer new perspectives on social science research that overlaps with Soft OR/PSMs, as discussed in Section 13.7.

It is clear that further investigation is required, not least in the constant monitoring and evaluation of new LLM developments. However, it is also worthwhile noting that whilst the LLM has been directed (prompted) in this example to elicit a methodological response, this is just the start of any intervention. Unless LLMs become so ubiquitous as deployable tools that it is possible to enhance their training with *situated* knowledge,[6] their role in the context of wicked/messy problems will likely remain limited to a general reflection of their broad and uncurated training material. They could become nothing more than '*stochastic parrots*' (Bender et al., 2021), offering little else than a misleading distraction to the problem context at hand. What is far more concerning is that LLM use becomes widespread without users, or indeed the LLM training material itself, attaching sufficient weight to knowledge about the characteristics of wicked/messy/swampy problems – especially their co-produced nature with the process of intervention – leading to even more solutionism (Morozov, 2013, 2023).

A recent study into the use of Chat GPT-4 at the Boston Consulting Group (BCG) by Dell'Acqua et al. (2023) attempts to quantify the effect of using LLMs on consultants' practice. In a pre-registered trial covering 758 consultants tackling a range of complex, knowledge-intensive tasks their analysis revealed a '*jagged technological frontier*' in terms of value. The more successful users of the technology, defined in terms of productivity measures, either delegated certain tasks to the LLM or completely integrated the use of the LLM into their task workflow.

5 Which highlights the emergence of '*Prompt Engineering*' as a new skill and also its difficulty for the non-AI expert (Zamfirescu-Pereira et al., 2023).

6 For example, to update their training using the data sources discussed in Section 15.1 in the context of AQA.

15.3 Future Directions

Reflecting on the small example of LLM use above and the recent BCG study, together with the use of probabilistic topic models to conduct AQA it is likely that the PSM consultant will, in time, develop a form of augmented process for their practice. A logical place for this to occur will be within the GSS platform that the consultant is using to support their model building with clients. When coupled with the idea of the self-scaffolding of methodology as something built-in to the design of a GSS (introduced and discussed in Chapter 14 as a *problem structuring platform*) there is the prospect of enabling very dynamic problem structuring capacity, as shown in Figure 15.6 (Yearworth, 2024). Individually, the use of a LLM, scaffolding, AQA, and topic labelling[7] components have all been demonstrated and shown to be feasible. However, the overall scheme is yet to be realised on any existing platform.

One further possibility for the application of LLM technology presents itself in this scheme. To date, much of the controversy and debate about the impact of LLMs has focussed on content issues. This has been either from (i) the perspective of copyright violations[8] in the training of LLMs, or (ii) concerns about job losses amongst content creators. However, there seems to have been little attention

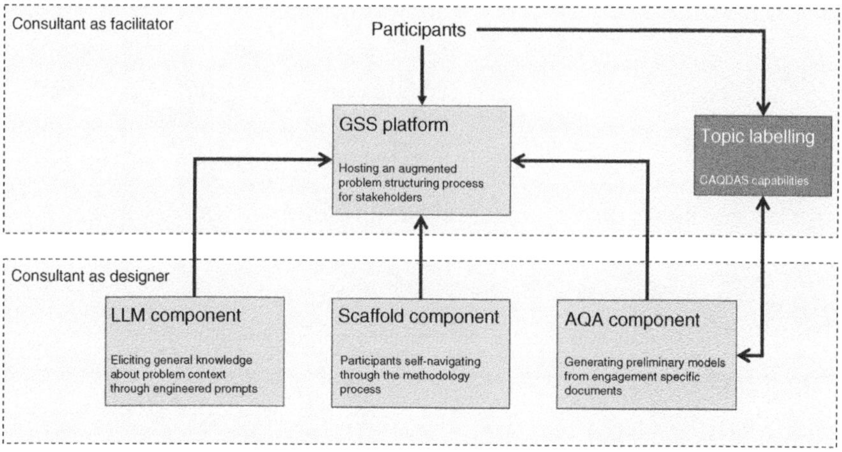

Figure 15.6 Augmented problem structuring implementing a problem structuring platform as defined in Chapter 14 (Yearworth, 2024).

---

7 Using the probabilistic topic model to classify each paragraph in the corpus and then enabling the search and inspection of the text for the purposes of inductively labelling the topics.
8 And more recently, the specific licencing of content from content providers.

paid to the role that LLMs, or other machine learning techniques, can take in supporting *process*. Concerns when facilitating workshops about ensuring ideals of procedural rationality and justice (Bryson et al., 2004; Eden et al., 2020), arbitrating and moderating participant behaviours (Ackermann et al., 2005, p. 324), or playing entirely new roles such as devil's advocate (Cunico, Zimmermann, & Videira, 2024) are only going to be exacerbated on pure process platforms, or pure content platforms with limited facilitation – especially if scaling to very large groups (White, 2002; Shaw et al., 2004), or in education settings (Section 14.6). It is possible that new LLMs could be trained to provide these types of functionality and integrated into GSS as extensions to the scaffold component shown in Figure 15.6.

16

Setting the Agenda

'the "urge to know" and the "urge to do" are different things.'

(Checkland, 1981, p. 24)

The motivation for the emergence of a Soft Operational Research (OR) and the development of problem structuring methods (PSMs) was always problem oriented – the belief, at least amongst some within OR, that it needed to retain its policy relevance, its applicability to social issues, with methodologies that were suitable for addressing wicked/messy/swampy problems. The first section of the book examined this motivation in detail and re-stated the problems of practice that were symptomatic of situational logics at play and the tendency to *'opt-out'* from dealing with the wickedness of planning, strategising, and transforming in favour of a focus on the wide variety of optimising approaches applicable to problems that had already been tamed. According to Checkland (1983), the discipline was in danger of becoming practitioner-free – if OR was becoming nothing more than the art of matching ready-made solutions to a well-known list of problems, then what was the relevance of the OR practitioner as a go-to consultant? In charting the emergence of Soft OR as a new paradigm in OR it was clear, using Abbott's notion of the fractal disciplinary landscape, that the characteristics of the new PSMs were such that the Hard/Soft OR split was an *inevitable* consequence of dichotomous choices over both the objects of study in OR, i.e. wicked/messy/swampy problems, and the methodologies required to research them, i.e. the fundamentally qualitative, interpretivist approaches of PSMs. Whilst the paradigmatic split over methodologies has largely defined the evolution of Soft OR since its emergence, it is clear that the choice – facing an individual practitioner and facing the discipline of OR as a whole – over dealing with wicked/messy/swampy problems has nothing to do with the specific methodologies of Soft OR. The argument for pragmatism in Chapter 3 was directed at the need to approach *all* problematic situations with a process of enquiry that was

Problem Structuring: Methodology in Practice, First Edition. Mike Yearworth.

equipped to deliver the practitioner, working with their client, with an adequate problematisation of the situation as the first step in any engagement.

16.1 Building Links

The agenda I set out here is motivated by the desire to keep Hard and Soft OR bound together in OR as an academic discipline and profession and is focussed on three strands (i) building links from Soft OR practice, (ii) addressing matters of OR education and maintaining links with the systems community, and (iii) looking at new ways forward for OR, building on the developments of a problem structuring sensibility in engineering practice and a new processual interpretation for modelling.

16.1.1 Towards New Disciplinary Partners in Applied Social Science

Menand (2001, pp. 209–210, 222, 364–365) highlights an interesting turning point for the American pragmatists in his account of the reactions of Dewey, Peirce, and James to the publication of Darwin's *On the Origin of Species* in 1859. Here was a new scientific theory that could replace the explanatory need for an absolute goal for life with a contingent one: a non-directed search for fitness in an environment. It was a revelatory work and fundamental to the pragmatists' project (Simpson & den Hond, 2022, p 131). It strikes me that if we go back to the United Kingdom OR Society's strapline of '*science of the better*', then a pragmatic interpretation of better is just the same as a Darwinian one. There is no higher-order, external direction for defining better, only a contingent '*what works best in this problem context*' way forward. Methodological learning becomes nothing more than the mechanism for adapting the OR '*gene pool*' to its environment – but it is not heading towards a higher, teleological goal, just evolving its fitness in problem space. Sticking with the Darwinian analogy, the type of problem context (the evolutionary niche) in which the individual operational researcher has specialised becomes the most important consideration in judging and planning for success. Pragmatism reminds us of the primacy of instrumental context/choice in the process of adaptation to the problem landscape and, as argued below, Soft OR/PSM scholars and practitioners now co-exist in the same evolutionary niche (problem context) as the applied social sciences.

The appeal for a Darwinian relevance to OR practice was just an analogy. However, as introduced earlier, Abbott (2001) provides a well-argued thesis for how disciplines develop over time, split into separate parts over dichotomising choices, and end up occupying similar *disciplinary* niches despite different starting points.

'The concept of fractal distinctions not only proves useful in understanding the external location of the social sciences generally. It also provides an essential tool for understanding relations within them. Indeed, as I shall show, both the external and the internal structures are produced by the same mechanism. We typically distinguish the various social sciences and various positions within them using a set of dichotomies. Every graduate student learns them. Some of these dichotomies concern the object of study: a focus on social structure or on culture, emphasis on the emergent versus the individual levels, beliefs about the constructed versus the real nature of social phenomena. Other dichotomies involve the aspects of social phenomena taken as problematic—choice versus constraint, conflict versus consensus. Still others differentiate methodological styles—narrative versus analysis, positivism versus interpretation. And some concern the nature of the knowledge obtained: pure versus applied knowledge, situated versus transcendent knowledge. All of these dichotomies are, ... fractal distinctions.'

(Abbott, 2001, p. 10)

The overlap with applied social sciences is not a new observation. There are pointers to the founders of the OR field using their scientific methods to address social problems (Kirby, 2011, p. 90). There have also been two conferences addressing the specific subject (Lawrence, 1966b; Jackson, Keys, & Cropper, 1989).

Biglan (1973a,b) categorised disciplines into orthogonal dimensions as the result of a multidimensional analysis of perceptions of similarities and differences obtained through surveys[1] and is shown in Figure 16.1. This provides a useful map of the disciplinary landscape on which we can plot Abbottonian trajectories.

The pure and applied dimension is very clear, at least on the right-hand side. Research in the applied fields cannot be carried out without a sufficient grounding of their nature as professional practices. I have characterised the pure/applied dimension going up the chart as a move towards practice, the application of theoretical knowledge to practical problems that in itself generates new knowledge, but of a different kind, and aligned with the emergence of regulated professions, which tend to codify practice-based knowledge as '*best-practices*'. I have reproduced the

1 I was first introduced to the Biglan classification when co-teaching a course on Advanced Interdisciplinary Research Design to PhD students in their MRes Programmes by a colleague who is a social geographer. The framework has been periodically revisited and validated (e.g. see Stoecker, 1993), and appears in different guises to fulfil a similar purpose (e.g. in characterising the disciplinary breadth of engineering in Figure 7.2). As a landscape for interpreting fractal disciplinary landscapes (Abbott, 2001) we should also think of using it recursively. Also see Peirce's classification of the sciences (Ormerod, 2006, pp. 896–897) and Schön's reflections on the vocational (professional) and research (higher learning) dichotomy (1992).

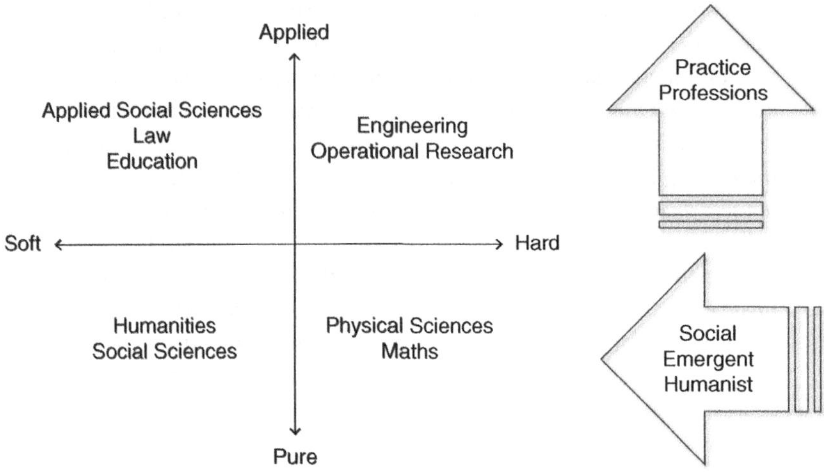

Figure 16.1 A categorisation of disciplines. Source: After Biglan (1973a,b).

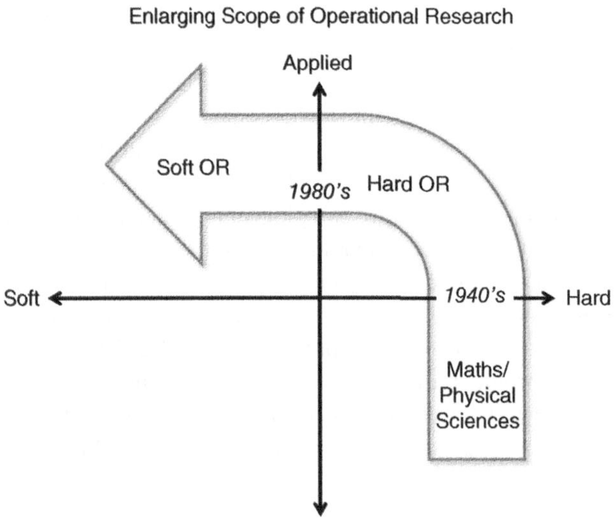

Figure 16.2 The enlarging scope of OR over time.

soft/hard labelling used by Biglan to distinguish the so called hard, rationalist sciences from the humanities and social sciences. Plotting the evolution of Soft OR on the same chart in Figure 16.2 is just a restatement of my view that Soft OR has evolved into an applied social science.

I would expect to see the core PSMs of soft systems methodology (SSM), strategic choice approach (SCA), and strategic options development and analysis (SODA) as well as qualitative system dynamics (QSD)/causal loop diagrams (CLDs), drivers-pressures-state-impact-response (DPSIR), and value focussed thinking (VFT) to find increasing uptake in the application domains represented in Chapter 13. There is the opportunity to learn methodologically, finding better ways of working with stakeholders, scaling-up to working with larger groups, and making use of online Group Support Systems for working with whole new constituencies of practitioners (Petropoulos et al., 2024, pp. 465-467). The downside is that there are risks to preserving the integrity of the PSMs that have benefited over the years from considerable investment in ensuring their rigour. Paucar-Caceres (2010) draws attention to the difficulty of learning about PSMs in action through the distorting effects on the interpretation of methodology arising from the proliferation of the secondary Soft OR literature.

16.1.2 Towards Mainstream OR

OR as a profession has been, on the whole, accretive of new approaches, methodologies, and techniques. Individual researchers will be aligned with certain communities of practice, scholars will be publishing in journals congruent with where they sit in relation to Abbottonian dichotomies of the objects of study, the aspects of these objects that are problematic, and the necessary methodologies to research them. However, OR is a catholic discipline and as a profession should be able to accommodate and work across these dichotomies. The unifying philosophy of OR is grounded in a problem orientation; i.e. the centrality of the challenge of problem formulation and individual moral choice and belief about its scope. External observers of the profession (i.e. researchers in Science and Technology Studies (STS) and the Sociology of Scientific Knowledge (SSK)) might consider Callon's work on problematisation as a useful lens for the study of operational researchers' practices with respect to problem formulation. Social theorists could well take a view on how the practice of operational research works as a profession using the perspectives of Bourdieu or Niccolini. I would expect individuals and research groupings within engineering that start to adopt an interpretivist, qualitative approach to practice – a problem structuring sensibility – are likely to face similar challenges to the pioneers of Soft OR. However, like OR, as a profession, one would hope that engineering is similarly accretive and is able to enlarge its scope rather than entrench divides.

Rosenhead's original project to draw the proponents of different methodological strands of development into a coherent agenda within the OR community was both bold and innovative (Rosenhead, 1989b). Despite problem structuring developing into a form of applied social science, its evolution from applied mathematics and science means that it, nonetheless, retains disciplinary links through its publishing venues, training, and the career paths of its practitioners. Soft OR needs to retain these disciplinary bonds so that it offers relevance to the problem formulation process in an OR intervention that may involve any amount of the mainstream Hard OR methods based on quantitative modelling. However, it must also work ethically to impose checks and balances e.g. knowing when *not* to optimise etc., avoiding the negative processes that Rosenhead refers to in setting out the new paradigm in Chapter 1. However, the Hard/Soft split in OR is a fact of dichotomising choices and maintaining links between the two strands is challenging. For example, Vidoni (2022), a Software Engineer, presents a case for 'OR engineering' in the pages of the Journal of the Operational Research Society in a strong critique of the damaging effects of such splits.

In an attempt to strengthen these links across OR, Dyson, O'Brien and Shah (2021) point out that approaches to problem structuring were inherent in its origins. They conducted a detailed study of the '*founders*' of OR, as profiled by Assad and Gass (2011), to explore the links between the emergence of the discipline and its influence on the current practice of Soft OR and PSMs.

They start with the negative point that there are still questions about the legitimacy of Soft OR/PSMs within the OR profession, especially in the United States (Ackermann et al., 2009; Mingers, 2011). They observe that the editorial policies of leading journals '*militate*' against the publication of Soft OR/PSM articles by rejecting work that does not include quantitative modelling. In a recently edited handbook of OR, the editors include the following in their preface

> *'At a more fundamental level, should "soft OR" be in scope or out of scope? We have taken the view that an inclusive stance is the most helpful, both for theory and practice. Real-world problems often require consideration from both "softer" and "harder" perspectives and need consideration of both predictive and prescriptive problems.'*
>
> (Salhi & Boylan, 2022, p. vii)

Despite the positive motivation for inclusion, the fact that the question was surfaced alludes to the ongoing question of legitimacy within the field. Recent review articles, one in the same handbook, again refer to the debate (Franco & Rouwette, 2022; Petropoulos et al., 2024, pp. 465-467).

The question, certainly from the United States perspective, seems to boil down to the simple notion that an intervention that does not involve quantitative modelling is *not* OR.[2] Of course, this can be immediately countered by the opposing

2 See the response to the Ackermann et al. (2009) letter, signed by 48 active researchers in support of Soft OR.

Table 16.1 When dealing with messy problems, problem structuring is always necessary and may be sufficient, and hard methods may be necessary but are never sufficient.

	Necessary	Sufficient
Quantitative methods	Maybe (uncertain)	Never (false)
Problem structuring methods	Always (true)	Maybe (uncertain)

Table 16.2 When dealing with puzzles, hard methods are always necessary and may be sufficient, and problem structuring may be necessary but is never sufficient.

	Necessary	Sufficient
Quantitative methods	Always (true)	Maybe (uncertain)
Problem structuring methods	Maybe (uncertain)	Never (false)

argument that any intervention that leaves out problem structuring is also *not* OR (the Ackoff '*Opt Out*').[3] The argument of Dyson, O'Brien and Shah (2021) might be summarised by two truth tables. When dealing with messy problems the logic in Table 16.1 seems to hold. On the other hand, when dealing with puzzles (see Table 1.1) the logic in Table 16.2 appears to be reasonable.

Following from these two tables, the observations made by Dyson, O'Brien and Shah (2021) concerning the definition of Soft OR and questions about its legitimacy as OR (e.g. as set out by Mingers (2011)) can perhaps be summarised by the following statements

1. In the '*Hard OR tradition*' the critique of Soft OR can be stated as – if quantitative methods are absent then it is not OR, and
2. In the '*Soft OR tradition*' the critique of Hard OR can be stated as – if problem structuring is absent then it is not OR.

Of these two statements, the first holds some importance in that it does appear to be gating what is published. The second statement is weaker. However, it also includes the contention that problem structuring can be present but not explicitly acknowledged (e.g. non-codified problem structuring methods (Yearworth & White, 2014; Reynolds, 2021, 2023b)). Evidence for non-codified PSM use can also be found in the wider corpus of OR literature using text-analytics

3 Jackson (2019a, p. 372) quotes the colourful language used by Ackoff to describe the attachment to '*mathematisation*' in OR.

techniques (Yearworth, 2022). However, none of this helps with dealing with the entrenched divide between PSMs and quantitative modelling. From the quantitative perspective, there is certainly no problem with publishing work that is purely quantitative with no reference to problem structuring, or indeed any reference to a '*customer*' of the research[4] with many examples of '*solutions*' to stylised problems or '*puzzles*' as defined in Chapter 1. The majority of work published as Soft OR/PSMs does not include quantitative modelling, although see the examples of multi-methodology in Chapter 13 and the different ways of integration of Hard and Soft modelling offered by HPM (see Chapters 9 and 10).

Of far more importance to building links, Dyson, O'Brien and Shah (2021) have unearthed in their research evidence that problem structuring was a *fundamental* activity in OR, right from the beginning. They go on to trace the influence of the founders on the originators of the PSMs by looking at the Context, Process, and Content characteristics of the methodologies. The main source of data for their analysis were the profiles in Assad and Gass (2011) listed in Table 16.3. Having established the links through a qualitative analysis of concerns, work, issues, etc. their findings can be supported quantitatively through citation analysis. Such an analysis was conducted for the authors of methodologies included in Rational Analysis for a Problematic World (RAfaPW) using data from the Scopus database[5] and is shown in Figure 16.3.

What this analysis provides is the missing piece that now links all the work presented in Chapter 5 with the founders of OR, and in Chapter 13 at one level of remove. Of course, some of the founders were also publishing contemporaneously with the PSM methodologists (e.g. Ackoff, Forrester, Beer, Saaty…), but presenting these relationships graphically is an indication of influence. Dyson, O'Brien and Shah (2021) remind us that the ideas of Soft OR/PSMs were *always* part of OR. For example, in presenting the history of Blackett's pivotal role in the emergence of OR, Kirby and Rosenhead (2011, pp. 15–16) provide an illuminating account of his involvement in a dispute about major strategic decision making; the early practitioners/scholars understood the need for problem structuring as an integral part of uncovering a customer's real needs. The criticisms of OR expressed by Ackoff (1977, 1979a), Checkland (1983), Rosenhead (1986) etc. can be read as a plea for the OR profession as a whole (i.e. scholars *and* practitioners) to pay attention to the problem of problem formulation and get back to properly addressing the issue, rather than taking the easy, '*opt-out*' route, and relying on situational

4 A reminder that this was Ackoff's reflection on his profession.
5 Citation count data were generated by using the Python-based API-wrapper pybibliometrics to access Scopus (Rose & Kitchin, 2019) and using queries of the form:
(AU-ID(scopus_id))AND(REFAUTH('*name*'))

Figure 16.3 Citation analysis of Founders listed in Table 16.3.

Table 16.3 Founders of OR as profiled by Assad and Gass (2011) in the same order as presented.

- Patrick Blackett
- Steven Vajda
- Philip McCord Morse
- John von Neumann
- Charles Frederick Goodeve
- Albert W. Tucker
- Jacinto Steinhardt
- George E. Kimball
- Ellis A. Johnson
- Leonid Vital'evich Kantorovich
- C. West Churchman
- William W. Cooper
- George B. Dantzig
- Herbert A. Simon
- Andrew Vazsonyi
- Murray Aaron Geisler
- Hugh Jordan Miser
- Abraham Charnes
- George Kozmetsky
- Jay Wright Forrester
- Russell Lincoln Ackoff
- David Bendel Hertz
- Richard E. Bellman
- Egon Balas
- E. Leonard Arnoff
- Berwyn Hugh Patrick Rivett
- Howard Raiffa
- D. Ray Fulkerson
- Harold W. Kuhn
- Saul I. Gass
- Thomas L. Saaty
- Anthony Stafford Beer
- John F. Magee
- Philip Starr Wolfe
- Harry Markowitz
- John D. C. Little
- Evelyn Martin Lansdowne Beale
- Ralph E. Gomory
- Alfred Blumstein
- Harvey M. Wagner
- Seth Bonder
- Bernard Roy
- Ronald A. Howard

Source: Adapted from Dyson, O'Brien and Shah (2021).

logics. They go on to propose two strands of work that would contribute to building bridges

> *'We are thus advocating that hard OR should embrace the rigorous and transparent approach of soft OR in problem formulation and address problem structuring in its publications where appropriate…*
> *…Implementation is often raised as a problem in enacting hard OR models, and the deep engagement with stakeholders and the transparency of soft OR may well help to resolve that issue.'*
>
> (Dyson, O'Brien, & Shah, 2021)

The first strand is a continuation of the rationale put forward by Ackermann et al. (2009) and could be considered as an ongoing project, see the means discussed below. However, unless such problem structuring leads to quantitative modelling then publication in some OR journals will remain problematic (as stated in the response to Ackermann et al. (2009) by the (then) Editor of Operations Research). The second strand could perhaps be enacted by seeking-out and highlighting case-studies from the literature of where this is already happening and enlarging the Soft OR/PSM network from these examples. The analysis advocated by Yearworth (2022)[6] would help in uncovering examples on the broad OR literature, even if not signposted by the title, abstract and keywords used by authors.[7]

Is an '*Enhanced OR*' (Jackson, 2019b) an answer to addressing the divide across OR? Whilst critical pluralism offers a theoretically coherent approach to harnessing the complementary strengths of different approaches using the system of systems methodologies (SOSMs) for orientation I am not convinced that it will heal divisions between Hard and Soft OR. I also think that Jackson too quickly dismisses the contribution of pragmatism.[8] In his history of the emergence of critical systems thinking (CST), Jackson (2019a, pp. 516–525) relates some relevant observations about the choices facing the Soft OR community with respect to critical approaches. He argues strongly that

> *'The recommendations emerging from a soft systems study are likely to be "managerialist" and regulative, it was argued, because no attempt is made to ensure that the conditions for a genuine debate involving all stakeholders are provided. The kind of open participative, discussion essential for the success*

6 Using the augmented qualitative analysis (AQA) approach based on probabilistic topic modeling described in Section 15.1.

7 As an example, see Ghosh and Troutt (2012) *Complex compound option models – Can practitioners truly operationalize them?* which was found using this AQA technique.

8 In fact, he seems to have changed position recently in a Centre for Systems Studies (2023) thought-piece and in Jackson (2023).

of the soft systems approach, and the primary justification for its outcomes, is impossible to obtain in problem situations where there is a fundamental conflict between interest groups that have access to unequal power resources. Soft systems thinking either has to walk away from these problem situations or it has to fly in the face of its own philosophical principles and acquiesce in proposed changes emerging from limited debates characterised by distorted communication.'

(Jackson, 2019a, p. 518)

Having argued in this chapter that Soft OR now finds itself positioned as an applied social science and distanced from its origins in the hard sciences, is Jackson advocating[9] that it must also abandon its interpretivism[10] and embrace a critical pluralism to become an '*Enhanced OR*'? Although pitched at the whole of OR, the argument is pointedly aimed at the Soft OR community. The argument seems to echo the same veiled prognostications made by Dando and Bennett (1981) more than 40 years ago. In just the same way as the arguments were ignored then, I do not think Hard OR practitioners today will be interested in these debates. Even if provoked by Jackson (2019b), I believe they will just carry on with business as usual. However, Soft OR practitioners, users of the core PSMs, have long been buffeted by these arguments. We could also think of PSMs and critical approaches as another example of a chaotic Abbottonian '*split*' in the fractal discipline landscape, but this time towards the right-hand side of the SOSM (Section 2.5), between interpretivism and the radical positions in the Burrell and Morgan framework (Section 2.6), or on the left-hand side of the Cynefin framework (Section 2.5) i.e. happening well away from the realm of Hard OR.

Despite reservations about critical pluralism, it is reasonable to consider it as one of four realistic approaches that are readily available for addressing the Hard/Soft OR divide, i.e. (i) returning to the roots of OR practice (Dyson, O'Brien, & Shah, 2021), (ii) achieving an '*enhanced OR*' through critical pluralism (Jackson, 2019b), (iii) adopting a grounding in pragmatism for all OR practice (Ormerod, 2006, 2020b), and (iv) a *processual turn* in OR (Yearworth). Each, in completely different ways, offers potential *means*. However, regardless of these available means (and there are probably others), the challenge to the OR profession as a whole is where it wants to stand in regard to wicked/messy/swampy problems and the problematisations broadly expressed as planning, strategising, and transforming. If OR consultants and scholars choose to be policy relevant, choose to engage with the range of social and environmental problem contexts set out in the survey of PSM

9 For example, in his 2019 Beale Medal lecture to the OR Society (Jackson, 2019b); and also earlier in Jackson (2006) and much earlier still in Jackson (1982).
10 As debated in print with the positions of Checkland, Churchman, and Ackoff (Jackson, 1982).

applications in Chapter 13, then there is much work to be done in (re-)building the links and a good place to start would be in education programmes.

I believe that the original motivations for the emergence of Soft OR/PSMs are just as valid today as they were 40 years ago, but they still have not been fully realised. Jackson was upbeat in his Beale Lecture, but the '*good health*' he refers to is not the whole of OR practice. I am concerned that the message of critical pluralism will be even harder to digest in a world that is largely managerial and financialised. It is worth considering that the prevailing global neoliberalism *is* the operating context (for Hard *and* Soft OR) in which to work and effect change.[11] In this framing of practice, managerialism is *all* there is, and the struggle to deal with messy problems needs to start from the inside. It is reasonable to consider Soft OR and PSMs as offering a way of pushing back against the management practices that are found across almost any organisational enterprise; public (e.g. '*New Public Management*' in the public sector (Hood, 1991; Diefenbach, 2009)) or private. If managerialism *is* all there is, then Hartz (2023), in a strong critique of its impact on the university sector, suggests that a possible way forward is through developing an understanding of the micro-processes of managerialism itself (pp. 14 and 15). The Soft OR project, embedded as it is within a managerialist framing, therefore provides a suitable methodological approach to an understanding of the processes of managing change, strategising, and planning that, whilst not part of the critical project, at least suggests a way forward for bringing wider stakeholder perspectives into managerial decision making (see Section 6.4). As Dyson, O'Brien and Shah (2021) make clear, it was always there in OR practice. Teaching PSMs to engineering students, business and management students, maths students, professionals on programmes of continuing professional development (CPD), to civil servants etc. would therefore become its own emancipatory project.

16.2 Education

The United Kingdom Degree Apprenticeship specification for the OR Specialist (Institute for Apprenticeships & Technical Education, 2021) does a good job of signposting the need for problem structuring as a key knowledge requirement in the syllabus. The recommendations by Dyson, O'Brien and Shah (2021) are perhaps a starting point for the introduction of PSMs into teaching materials on accredited programmes. However, I believe we can go much further. Ormerod's proposal for a pragmatic OR method provides a unifying starting point for *any*

11 Consider the business model framing of the two research project case studies presented in Section 10.1 and Section 10.7 as being perfectly normal.

OR program. Likewise, my proposal for a generic process of enquiry into problematic situations at the end of Chapter 5, provides another way in. The key difference between the two is that Ormerod is more practice-oriented and mine is more focussed on methodology. They can be viewed as complementary but compatible. Their use to support the training of OR practitioners would be viewed as the first steps in re-establishing OR's links to the process of enquiry that was first introduced to the profession by Churchman and Ackoff.

The methodological project is supported by the affordance of hierarchical process modelling (HPM). Breaking down the distinction between Hard and Soft modelling without paradigmatic wrangling is an intrinsic property of this approach. The example presented in Section 10.6 of the process industry case, where the contribution of quantitative models (in this case of vibrating structural elements[12] i.e. the *how*) is seamlessly combined with the conceptual strategising model (i.e. the *why*) to bring about the overall business transformation, supports the argument made by Checkland (1989b, p. 40) that any phenomenological process of enquiry can encompass quantitative modelling. The broad disciplines of engineering and medicine are likely to provide endless cases for where hard methods, i.e. involving quantitative modelling, need to be integrated into problem structuring in a formal arrangement.[13] This processual interpretation provides a simpler way of thinking about multimethodology.

At the same time as PSMs were emerging as a distinct project within OR (see Chapter 4), there was also a messy entanglement with systems ideas that were also forming into distinctive strands – see Paucar-Caceres and Espinosa (2011, p. 1605). Looking back at the critical period of the 1970s with the benefit of hindsight, it seems that there was a vibrancy of ideas that created a plurality of methods/methodologies that have found their way to today, in various states of health and modification, whilst at the same time spawning the new systems approaches critical system heuristics (CSH) (Ulrich, 2003), Total Systems Intervention (TSI) (Midgley, 2000) and critical systems thinking/practice (CST/P) (Jackson, 2019a), and showing some examples of synthesis across the broad swathe of systems and OR landscape (e.g. (Midgley, Munlo, & Brown, 1998)) and of integrated co-existence (Ulrich, 2012a,b). Given the degree of overlap between systems thinking and problem structuring it might seem surprising to relegate a discussion until now, especially since the overlaps, relative influence, and cross

12 By exploiting the notion of delayed reification, it is only necessary to identify this at the process level as '*predicting failure probabilities of vibrating elements*', which would only be *enacted* at some point in the future. There could in fact be a range of possible specific enactments e.g. arising from the results of an internal research project, outsourcing research to a university, buying-in expert consultancy from a specialist contractor etc.

13 There is some relationship here with Integration and Implementation Science (Bammer, 2005; Bammer et al., 2020). Also see Monks (2016), who considers OR as implementation science.

Table 16.4 Notes on Figure 16.4.

Methodology/ method/ approach	Notes
VSM	The use of VSM as a PSM is explored extensively by Lowe, Espinosa and Yearworth (2020).
CLD/QSD	I consider Qualitative System Dynamics to be roughly equivalent to the use of causal loop diagrams (CLDs) – see Wolstenholme (1999) for a discussion about the need for both quantitative and qualitative system dynamics to support management decision-making. See also Powell and Mustafee (2017)
SODA project	The evolution of SODA through JourneyMaking and Strategyfinder is a complex story (see Chapter 5) and I find it easier to refer to the '*SODA project*.' Eden and Ackermann provided an excellent overview of the developments in their '*SODA then and now*' presentation to the 62nd Conference of the United Kingdom Operational Research Society (OR62) (Eden et al., 2020). In comparison, SCA has remained relatively unchanged over the years.
Non-codified PSMs	The project was first articulated by Yearworth and White (2014) and continued by (Reynolds, 2021, 2023a)
Callon-like	Recognisable problem structuring behaviours (to a practitioner working in Soft OR; e.g. see Yearworth (2016)) but from a different research tradition. See the collection of articles in Callon, Lascoumes and Barthe (2009b) and examples of practice such as Lane et al. (2011).
HPM	Apart from Davis, MacDonald and White (2010) and the use of HPM by Lowe, Martingale and Yearworth (2016); Lowe, Espinosa and Yearworth (2020) there is little knowledge of HPM in the wider Operational Research community.

contributions have long been discussed (Checkland, 1983; Mingers & White, 2010). The core PSMs are clearly systemic in nature, obviously so in the case of SSM. The following Venn diagram can be viewed as an alternative rendering of the SOSM, decoupled from the Burrell and Morgan framework, and instead grouped by communities of practice and informed by the complementary[14] Degree Apprenticeship (DA) specification for the Systems Thinking Specialist (Table 16.4) (Institute for Apprenticeships & Technical Education, 2023).[15]

14 To the Operational Research Specialist Degree Apprenticeship specification (Institute for Apprenticeships & Technical Education, 2021).
15 Why the SODA project does not appear within the Systems Thinking set (according to the specification) is something of a mystery. Reynolds and Holwell (2010) clearly show SODA is a systems approach to managing change.

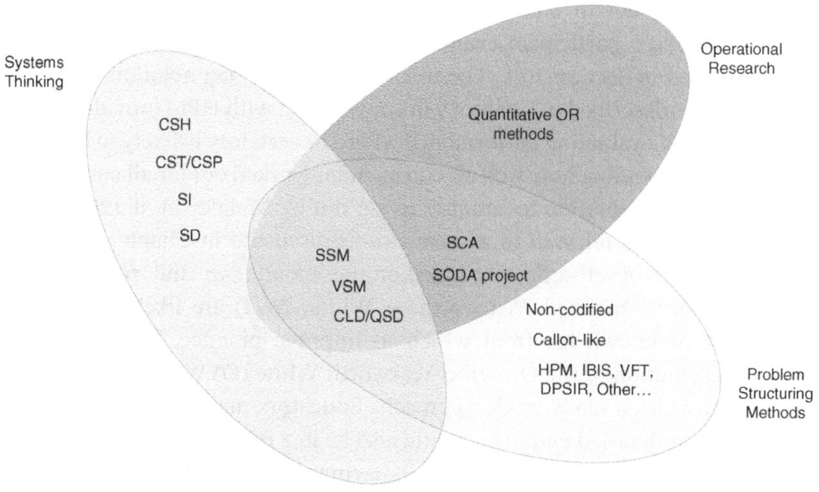

Figure 16.4 Methodologies, methods, and approaches mapped to broad communities of practice.

The relevance of this work to the knowledge requirement '*How to select and apply, a range of problem structuring methods to understand complex problems*' in the OR Specialist Degree Apprenticeship specification is clear. Assessment of meeting this requirement on such a programme should be straightforward as it will be professionally oriented and focussed on practitioners in their workplace. The situated learning environment will be similar to the EngD in Systems Programme that provided an empirical basis for much of the material in Section III. The lessons from Dewey, Schön, Checkland, Ormerod... on the process of enquiry that is central to any problem structuring aligns well with the notion in education of *authentic enquiry*, although there is a dearth of literature on the subject that is specific to professional development outside of the field of training teachers. Programmes at M/Level (or equivalents outside the UK) in science, technology, engineering and mathematics (STEM) that are *not* oriented to providing situated learning opportunities for their students will need to engage in some considerable development work with their industry/business/community/organisation partners for this practice-oriented approach to deliver its full benefits.

Whilst assessment should be straightforward where the intended learning outcomes (ILOs) and requisite knowledge, skills and behaviours (KSBs) are clearly aligned with a student-led process of enquiry into a problematic situation, the casting of methodologies themselves into process models (e.g. using HPM) would then make it possible to evaluate the performance of an intervention – either as self-evaluation or as scored by the participants. Lowe, Espinosa and Yearworth (2020) provide an example of self-evaluation in three case studies using the

Viable Systems Model in consulting engagements. Yearworth, Schien, et al. (2015, p. 9) present a participant evaluation of using HPM as a PSM in the case studies presented in Section 10.1. The use of the Italian Flag notation based on interval numbers described in Chapter 9 in conjunction with HPM provide a more nuanced means of evaluating performance where uncertainty is likely to be a key element of understanding how well an engagement worked out for all concerned.

Of course, PSM evaluation techniques as set out by Midgley et al. (2013) and White (2006) are also relevant to assessment, as discussed in Chapter 12. More elaborate processes of self-reflection based on the '*zooming-in*' and '*zooming-out*' of practice theory (Ormerod, Yearworth, & White, 2023) are likely to provide a rich source of knowledge from which to improve practice. The simplified process representation of the Ormerod Yearworth White (OYW) framework (see Table 13.1) provides a ready-made approach. Educators managing a substantial programme of student-led enquiries would also be in a position to conduct experiments along the lines set out in Section 6.3, having, of course, sought appropriate ethical approval. Chapter 14 provides suggestions for how group support system (GSS) can be used in an education setting, especially with regard to addressing scale-up to large class sizes and providing a ready means for assessment (i.e. the platform is self-documenting as well).

Methodological learning is a key element of the Generic Constitutive Definition of PSMs (Yearworth & White, 2014) and the constitutive rules in the CR/E/A framework set out in Chapter 5. This action learning cycle naturally operates at different scales simultaneously, from the individual practitioner learning and improving from engagement to engagement (Ormerod, Yearworth, & White, 2023) through to the evolution of the methodologies themselves. For example, Eden et al. (2020) present an in-depth description of the evolution of SODA over many years. Checkland (2000b) provides his own reflection on 30 years of development of SSM.[16] HPM has also exhibited a similar process of evolution since the early 1990s (see Chapter 9). Keeping reflective logs as a matter of good professional practice would enable students to engage at depth in the process of '*selecting and applying a range of problem structuring methods to understanding complex problems*' and feel that their training was actively contributing to their field rather than being treated as passive recipients of knowledge.

16.3 Introducing a Processual Turn in OR

The outline of a processual turn in engineering (in Section 10.8) supported by a new modelling approach has opened up a possible path to reimagining OR

16 Also see Figure 5.3.

practice. However, thinking in terms of '*everything is a process*' is somewhat daunting. On the one hand, in casting OR practice as a process with the broad purpose of improving human endeavours, the process of formulating problems or problematising would merely position purposeful processes from Soft OR (such as strategising, planning, transforming…) and from Hard OR (such as optimising, controlling, queueing, routing, scheduling, locating, bidding, packing, cutting, timetabling, gaming, forecasting, '*frontiering*'…) at different positions within the how/why questioning dialectic of purposeful modelling; but not fundamentally, paradigmatically, divided in any way. It would always be possible to fill-in the detail between these positions with however much process detail is required to make sense of the linkage and deliver an intervention. Also, in the same way that the processes of strategising, planning, and transforming using the core PSMs have seamlessly become coupled to the wider transformational processes highlighted in Section 13.14 (such as policy making, coordinating, ensuring, contracting…), it is not too difficult to see, *processually*, how the whole of OR as a discipline fits within the wider messiness of the world. Whilst for any given engagement the consultant and their client are likely to anchor their problematising at some initial purpose, the use of *why* questioning will always open up broader problem contexts, with the possibility of revealing opportunities for synergy and meeting shared purpose (see Section 9.3), and also possible reasons for pursuing other problematisations (different explorations of *how*). Moral choices enter into this process; OR practice is not the unquestioned implementation of solutions in response to situational logics, *why* questioning is ever present. This how/why questioning around matters of purpose would be more in keeping with the original ideals of OR as re-examined by Dyson, O'Brien and Shah (2021). Even when perfecting practice (see Chapter 6) and evaluating practice (see Chapter 12), we can easily delve into further processual accounts of activity such that we see these additionally revealed processes as further refinement of achieving the wider purpose of improving human endeavours.

How far would we need to explore the questioning of purpose through *why* questioning before we would start to encounter existential risks of the sort introduced in the motivation for this book? Existing PSMs, i.e. with well-defined epistemologies and affordance of modelling approach in the CR/E/A framework (see Chapter 5), can be re-interpreted processually to fit with this transformation; SODA with little effort (look how straightforward it was to implement HPM in Strategyfinder), SSM without too much effort (start with the model of the system to use SSM in Figure 5.3), whereas SCA would be a little harder (perhaps starting with the generic context/fit process in Table 5.6). We can take this processual view of OR as far as we and our clients want with further *how* answers to deciding the meaning of *improving* by including the processes needed to measure and control our interventions in our modelling (HPM and SSM do this as a matter of course).

On the other hand, transforming the discipline of OR towards a completely diachronic view, coupled with moving away from a substance ontology through delayed reification runs the risk of practitioners losing any firm anchor and being cast adrift in an ocean of processes. There is the real possibility of getting lost in the metaphysics of process philosophy. Soft OR has perhaps suffered from too much wrangling over paradigms and theoretical foundations; the detailed debates (summarised in Chapter 4) have served to erect barriers within the OR profession rather than enlighten. There is a warning here. Introducing a new idea, especially one as radical as a complete processual rethinking of OR is more than daunting. An immediate conceptual challenge also presents itself – reconciling a process view with the success of systems thinking, especially with regard to explanations of dynamic behaviour based on feedback. Whilst the core PSMs have an affinity with a process view, such that the modelling in HPM can stand-in for the affordances of the modelling approaches of the core methods, CLDs appear to stand apart on this matter. This is not an insurmountable obstacle to a process view, as some within the discipline of biology have shown, but there is much work to be done to take process thinking forward.

However, when everything is modelled as a process i.e. once we have a mono-category, then category errors and paradigm incommensurabilities (see Chapter 2) are rendered meaningless. The analytical process of zooming-in and zooming-out of practice (see Chapters 12 and 13) represents nothing more than moving down and up in the hierarchy of the how/why modelling dialectic. The difficult circularity that exists when problematising wicked/messy/swampy problems melts away when we view a generic process of enquiry in OR as coexisting with the processes with which we mean to intervene in the world as practitioners. In modelling the world processually we can also model our interventions within the same model (e.g. see Section 10.7). Or put more simply, *problematic situations are processes as are the means of intervention*.

Finally, returning to pragmatist philosophy, our work as OR practitioners is about getting our hands dirty with the real problems of the world and this needs to come first. Our purpose is not the search for knowledge for its own sake or realising some higher purpose but in addressing problems – we work in a problem driven, client-oriented discipline. To keep us to this purpose, there is always the grounding of problematising the next engagement.

Go and get some dirt under your fingernails.

Grant and Project Funding

Much of the work presented in this book has been the result of research collaborations funded by a variety of sources. I would like to acknowledge financial support from the following grants or projects as either Principal Investigator or Co-Investigator.

1. Systems Thinking for Comprehensive City Efficient Energy Planning (STEEP) (EU Specific Programme '*Cooperation*': Energy Grant agreement ID: 314277 €2,197,398 funded 1/10/2013–30/9/2015)
2. Renaissance of Places with Innovative Citizenship and Technology (REPLICATE) (EU SOCIETAL CHALLENGES – Secure, clean and efficient energy Grant agreement ID: 691735 €24,965,263 funded 1/2/2016–31/1/2021)
3. Resilient Economy and Society by Integrated Systems modelling (RESIST) (UK NERC, Project Reference: NE/N012143/1 £487,437 funded 1/2016–1/2020)
4. International Centre for Infrastructure Futures (ICIF) (UK EPSRC, Project Reference: EP/K012347/1 £3,444,604 funded 6/2013–4/2017)
5. Healthy Resilient Cities: Building a Business Case for Adaptation (UK NERC, Project Reference: NE/N007638/1 £20,223 funded 9/2015–9/2016)
6. Tools for assessing the systemic impact of technology deployments on energy use and climate emissions (SYMPACT) (UK EPSRC, Project Reference: EP/I000151/1 £332,548 funded)
7. Systems thinking and system dynamics modelling to support policy development using waste prevention and recycling as case studies (ST0201) (UK DEFRA, Project Id 18386 £51,110 funded 2012–2013)

The following grant associated with the establishment of the Industrial Doctorate Centre (IDC) in Systems at the University of Bristol is also acknowledged.

1. Industrial Doctorate Centre: Systems (UK EPSRC, Reference: EP/G037353/1 £4,956,495 funded 10/2009–3/2018)

Selected Bookshelf

Ackermann, F., & Eden, C. (2011). *Making strategy : mapping out strategic success* (2nd ed.). Sage: London

Baert, P. (2005). *Philosophy of the social sciences : towards pragmatism*. Polity: Cambridge

Callon, M., Lascoumes, P., & Barthe, Y. (2009). *Acting in an uncertain world : an essay on technical democracy*. MIT: Cambridge, MA

Checkland, P. (1999). *Systems thinking, systems practice : Includes a 30-year retrospective*. John Wiley & Sons: Chichester

Checkland, P., & Poulter, J. (2006). *Learning for action : a short definitive account of soft systems methodology, and its use for practitioner, teachers and students*. John Wiley & Sons: Chichester

Checkland, P., & Scholes, J. (1999). *Soft systems methodology in action : including a 30-year retrospective*. John Wiley & Sons: Chichester

Conklin, J. (2006). *Dialogue mapping : building shared understanding of wicked problems*. John Wiley & Sons: Chichester

Eden, C., Jones, S., & Sims, D. (1983). *Messing about in problems : an informal structured approach to their identification and management*. Pergamon: Oxford

Friend, J., & Hickling, A. (2005). *Planning under pressure : the strategic choice approach* (3rd ed.). Elsevier Butterworth-Heinemann: Oxford

Jackson, M. C. (2019). *Critical Systems Thinking and the Management of Complexity*. Wiley-Blackwell: Chichester

Keys, P. (1995). *Understanding the process of operational research : collected readings*. John Wiley & Sons: Chichester

Kilgour, D. M., & Eden, C. (2020). *Handbook of group decision and negotiation*. Springer International Publishing: Cham. doi:10.1007/978-3-030-12051-1

Midgley, G. (2000). *Systemic intervention : philosophy, methodology, and practice*. Kluwer Academic/Plenum: New York

Morecroft, J. (2007). *Strategic modelling and business dynamics : a feedback systems approach*. John Wiley & Sons: Chichester

Problem Structuring: Methodology in Practice, First Edition. Mike Yearworth.

Pidd, M. (2004). *Systems modelling : theory and practice.* John Wiley & Sons: Chichester

Pidd, M. (2009). *Tools for thinking : modelling in management science* (3rd ed.). John Wiley & Sons: Chichester

Rosenhead, J. (1989). *Rational analysis for a problematic world : problem structuring methods for complexity, uncertainty and conflict*. John Wiley & Sons: Chichester

Rosenhead, J., & Mingers, J. (2001). *Rational analysis for a problematic world revisited : problem structuring methods for complexity, uncertainty and conflict* (2nd ed.). John Wiley & Sons: Chichester

Salhi, S., & Boylan, J. (Eds.). (2022). *The palgrave handbook of operations research.* Palgrave Macmillan Cham. doi:10.1007/978-3-030-96935-6.

Senge, P. M. (1990). *The fifth discipline : the art and practice of the learning organization.* Random House: London

Sterman, J. D. (2000). *Business dynamics: systems thinking and modeling for a complex world.* Irwin/McGraw-Hill: Boston, MA

Vennix, J. (1996). *Group model building : facilitating team learning using system dynamics.* Wiley: Chichester

References

Abbott, A. (1988). *The system of professions: an essay on the division of expert labor*. University of Chicago Press: Chicago.

Abbott, A. (2001). *Chaos of disciplines*. University of Chicago Press: Chicago.

Ackermann, F. (1996). Participants' perceptions on the role of facilitators using group decision support systems. *Group Decision and Negotiation*, *5*(1), 93–112. https://doi.org/10.1007/BF02404178.

Ackermann, F. (2019). PSMs are dead; long live PSMs. *Journal of the Operational Research Society*, *70*(8), 1396–1397. https://doi.org/10.1080/01605682.2018.1502630.

Ackermann, F. (2020). Group support systems: Past, present, and future. In D. M. Kilgour & C. Eden (Eds.), *Handbook of Group Decision and Negotiation* (pp. 1–29). Springer International Publishing: Cham. https://doi.org/10.1007/978-3-030-12051-1_47-1.

Ackermann, F. (2024). Managing grand challenges: Extending the scope of problem structuring methods and behavioural operational research. *European Journal of Operational Research*. https://doi.org/10.1016/j.ejor.2024.01.024.

Ackermann, F., & Eden, C. (1998). *Making strategy: the journey of strategic management*. Sage Publications: London.

Ackermann, F., & Eden, C. (2001). Contrasting single user and networked group decision support systems for strategy making. *Group Decision and Negotiation*, *10*(1), 47–66. https://doi.org/10.1023/A:1008708912048.

Ackermann, F., & Eden, C. (2011a). *Making strategy: mapping out strategic success* (2nd ed.). Sage Publications: London.

Ackermann, F., & Eden, C. (2011b). Strategic management of stakeholders: Theory and practice. *Long Range Planning*, *44*(3), 179–196. https://doi.org/10.1016/j.lrp.2010.08.001.

Ackermann, F., Eden, C., & Cropper, S. (1990). Cognitive Mapping – a user's guide. *University of Strathclyde, Management Science: Theory, Method and Practice*

Problem Structuring: Methodology in Practice, First Edition. Mike Yearworth.

Working Paper Series. https://banxia.com/pdf/de/GettingStartedWithCogMapping.pdf.

Ackermann, F., Eden, C., & Williams, T. (1997). Modeling for litigation: Mixing qualitative and quantitative approaches. *Interfaces*, *27*(2), 48–65. https://doi.org/10.1287/inte.27.2.48.

Ackermann, F., Franco, L. A., Gallupe, B., & Parent, M. (2005). GSS for multi-organizational collaboration: Reflections on process and content. *Group Decision and Negotiation*, *14*(4), 307–331. https://doi.org/10.1007/s10726-005-0317-4.

Ackermann, F., Bawden, R., Bosch, O., Brocklesby, J., Bryant, J., Buede, D., Carreras, A., Checkland, P., Cushman, M., Daellenbach, H., Davies, J., Deguchi, H., Eden, C., Edwards, J., Elias, A., Franco, L. A., Friend, J., Georgiou, I., Vargas, F. G., Greenwood, A., Groessler, A., Gu, J., Hindle, G., Hipel, K., Inohara, T., Jackson, M. C., Kijima, K. J., Kilgour, M., Mabin, V., Midgley, G., Mingers, J., Montibeller, G., Mulej, M., Obeidi, A., O'Brien, F., Phillips, L., Pidd, M., Putro, U., Richardson, K., Rouwette, E., Sato, R., Shaw, D., Shiba, N., Takahashi, S., Vennix, J., Vidal, V., Westcomb, M., & White, L. (2009). The Case for Soft O.R. *ORMS Today*. https://doi.org/10.1287/orms.2009.02.09.

Ackermann, F., Franco, L. A., Rouwette, E., & White, L. (2014). Special issue on problem structuring research and practice. *EURO Journal on Decision Processes*, *2*(3), 165–172. https://doi.org/10.1007/s40070-014-0037-6.

Ackermann, F., Alexander, J., Stephens, A., & Pincombe, B. (2020). In defence of Soft OR: Reflections on teaching Soft OR. *Journal of the Operational Research Society*, *71*(1), 1–15. https://doi.org/10.1080/01605682.2018.1542960.

Ackoff, R. L. (1974a). Beyond problem solving. *General Systems*, *XIX*, 237–239.

Ackoff, R. L. (1974b). *Redesigning the future: a systems approach to societal problems*. Wiley-Interscience: New York; London.

Ackoff, R. L. (1974c). The social responsibility of operational research. *Operational Research Quarterly*, *25*(3), 361–371. https://doi.org/10.1057/jors.1974.71.

Ackoff, R. L. (1977). Optimization + objectivity = opt out. *European Journal of Operational Research*, *1*(1), 1–7. https://doi.org/10.1016/S0377-2217(77)81003-5.

Ackoff, R. L. (1979a). The future of operational-research is past. *Journal of the Operational Research Society*, *30*(2), 93–104. https://doi.org/10.2307/3009290.

Ackoff, R. L. (1979b). Resurrecting the future of operational research. *Journal of the Operational Research Society*, *30*(3), 189–199. https://doi.org/10.1057/jors.1979.41.

Ackoff, R. L. (1981). The art and science of mess management. *Interfaces*, *11*(1), 20–26. https://doi.org/10.1287/inte.11.1.20.

Ackoff, R. L. (1987). Presidents' symposium: OR a post-mortem. *Operations Research*, *35*(3), 471–474. https://doi.org/10.1287/opre.35.3.471.

Adamides, E. D., & Karacapilidis, N. (2006). A knowledge centred framework for collaborative business process modelling. *Business Process Management Journal*, *12*(5), 557–575. https://doi.org/10.1108/14637150610690993.

Adams, V., Murphy, M., & Clarke, A. E. (2009). Anticipation: Technoscience, life, affect, temporality. *Subjectivity, 28*, 246–265. https://doi.org/10.1057/sub.2009.18.

Akerlof, G. A. (1970). Market for lemons - quality uncertainty and market mechanism. *Quarterly Journal of Economics, 84*(3), 488–500. https://doi.org/10.2307/1879431.

Alford, J., & Head, B. W. (2017). Wicked and less wicked problems: A typology and a contingency framework. *Policy and Society, 36*(3), 397–413. https://doi.org/10.1080/14494035.2017.1361634.

Allender, S., Owen, B., Kuhlberg, J., Lowe, J., Nagorcka-Smith, P., Whelan, J., & Bell, C. (2015). A community based systems diagram of obesity causes. *PLoS One, 10*(7), Article e0129683. https://doi.org/10.1371/journal.pone.0129683.

Alvesson, M., & Sandberg, J. (2011). Generating research questions through problematization. *Academy of Management Review, 36*(2), 247–271. https://doi.org/10.5465/amr.2009.0188.

Andersen, D., Vennix, J., Richardson, G., & Rouwette, E. (2007). Group model building: Problem structuring, policy simulation and decision support. *Journal of the Operational Research Society, 58*(5), 691–694. https://doi.org/10.1057/palgrave.jors.2602339.

Andersson, C., & Törnberg, P. (2018). Wickedness and the anatomy of complexity. *Futures, 95*, 118–138. https://doi.org/10.1016/j.futures.2017.11.001.

Assad, A. (2011). C. West Churchman. In A. Assad & S. Gass (Eds.), *Profiles in Operations Research: Pioneers and Innovators* (pp. 171–200). Springer: New York; London. https://doi.org/10.1007/978-1-4419-6281-2_11.

Assad, A., & Gass, S. (2011). *Profiles in operations research: pioneers and innovators*. Springer: New York; London. https://doi.org/10.1007/978-1-4419-6281-2.

Atkins, J. P., Gregory, A. J., Burdon, D., & Elliott, M. (2011). Managing the marine environment: Is the DPSIR framework holistic enough? *Systems Research and Behavioral Science, 28*(5), 497–508. https://doi.org/10.1002/sres.1111.

Aubert, A. H., & Lienert, J. (2024). Operational Research for, with, and by citizens: An overview. *European Journal of Operational Research*. https://doi.org/10.1016/j.ejor.2023.10.037.

Baert, P. (2005). *Philosophy of the social sciences: towards pragmatism*. Polity: Cambridge.

Bammer, G. (2005). Integration and implementation sciences: Building a new specialization. *Ecology and Society, 10*(2), 25. https://doi.org/10.5751/ES-01360-100206.

Bammer, G., O'Rourke, M., O'Connell, D., Neuhauser, L., Midgley, G., Klein, J. T., Grigg, N. J., Gadlin, H., Elsum, I. R., Bursztyn, M., Fulton, E. A., Pohl, C., Smithson, M., Vilsmaier, U., Bergmann, M., Jaeger, J., Merkx, F., Vienni Baptista, B., Burgman, M. A., Walker, D. H., Young, J., Bradbury, H., Crawford, L., Haryanto, B., Pachanee, C. A., Polk, M., & Richardson, G. P. (2020). Expertise in research integration and implementation for tackling complex problems: When is it needed, where can it be found and how can it be strengthened? *Palgrave Communications*, *6*(1), Article 5. https://doi.org/10.1057/s41599-019-0380-0.

Beagon, U., & Bowe, B. (2023). Understanding professional skills in engineering education: A phenomenographic study of faculty conceptions. *Journal of Engineering Education*, *112*(4), 1109–1144. https://doi.org/10.1002/jee.20556.

Bell, S. (2012). DPSIR = A problem structuring method? An exploration from the "Imagine" approach. *European Journal of Operational Research*, *222*(2), 350–360. https://doi.org/10.1016/j.ejor.2012.04.029.

Bellenguez, O., Brauner, N., & Tsoukiàs, A. (2023). Is there an ethical operational research practice? And what this implies for our research? *EURO Journal on Decision Processes*, *11*, Article 100029. https://doi.org/10.1016/j.ejdp.2023.100029.

Bender, E. M., Gebru, T., McMillan-Major, A., & Shmitchell, S. (2021). On the dangers of stochastic parrots: Can language models be too big? FAccT 2021 - Proceedings of the 2021 ACM Conference on Fairness, Accountability, and Transparency. https://doi.org/10.1145/3442188.3445922.

Biglan, A. (1973a). The characteristics of subject matter in different academic areas. *Journal of Applied Psychology*, *57*(3), 195–203. https://doi.org/10.1037/h0034701.

Biglan, A. (1973b). Relationships between subject matter characteristics and the structure and output of university departments. *Journal of Applied Psychology*, *57*(3), 204–213. https://doi.org/10.1037/h0034699.

Bird, A. (1998). *Philosophy of science*. UCL Press: London.

Blei, D. M. (2012). Probabilistic topic models. *Communications of the ACM*, *55*(4), 77–84. https://doi.org/10.1145/2133806.2133826.

Blei, D. M., Ng, A. Y., & Jordan, M. I. (2003). Latent Dirichlet allocation. *Journal of Machine Learning Research*, *3*(4-5), 993–1022.

Blockley, D. (2010). The importance of being process. *Civil Engineering and Environmental Systems*, *27*(3), 189–199. https://doi.org/10.1080/10286608.2010.482658.

Blockley, D., & Godfrey, P. (2000). *Doing it differently: Systems for rethinking construction*. Thomas Telford: London.

Blockley, D., & Godfrey, P. (2017). *Doing it Differently: Systems for rethinking infrastructure* (2nd ed.). https://doi.org/10.1680/didse.60821.

Blomqvist, K. (1997). The many faces of trust. *Scandinavian Journal of Management*, *13*(3), 271–286. https://doi.org/10.1016/S0956-5221(97)84644-1.

Bourgoin, A., & Harvey, J.-F. (2018). How Consultants Project Expertise and Learn at the Same Time. *Harvard Business Review* (July-August). https://hbr.org/2018/07/how-consultants-project-expertise-and-learn-at-the-same-time.

British Academy. (2016). Crossing Paths: Interdisciplinary Institutions, Careers, Education and Applications. https://www.thebritishacademy.ac.uk/documents/213/crossing-paths.pdf.

Brown, S., Bessant, J., & Jia, F. (2018). *Strategic Operations Management* (4th ed.). Routledge.

Bryant, J., Darwin, J., & Booth, C. (2011). Strategy making with the whole organisation: OR and the art of the possible. *Journal of the Operational Research Society, 62*(5), 840–854. https://doi.org/10.1057/jors.2010.71.

Bryson, J. M., & Anderson, S. R. (2000). Applying large-group interaction methods in the planning and implementation of major change efforts. *Public Administration Review, 60*(2), 143–162. https://doi.org/10.1111/0033-3352.00073.

Bryson, J. M., Ackermann, F., Eden, C., & Finn, C. (2004). *Visible thinking: unlocking causal mapping for practical business results*. John Wiley & Sons: Chichester.

Buckingham Shum, S. (2008). Cohere: Towards web 2.0 argumentation. 2nd International Conference on Computational Models of Argument, Toulouse, France.

Buckingham Shum, S., Selvin, A. M., Sierhuis, M., Conklin, J., Haley, C. B., & Bashar Nuseibeh, B. (2006). Hypermedia support for argumentation-based rationale. In A. H. Dutoit, McCall, R., Mistrík, I., Paech, B. (Eds.), *Rationale Management in Software Engineering*. Springer-Verlag: Berlin. https://doi.org/10.1007/978-3-540-30998-7_5.

Buckle-Henning, P., Wilmshurst, J., & Yearworth, M. (2012). Understanding systems thinking: An agenda for applied research in industry. 56th Meeting of the International Society for the Systems Sciences, San Jose, CA, USA. http://journals.isss.org/index.php/proceedings56th/article/view/1909.

Bunker, B. B., & Alban, B. T. (1996). *Large group interventions: engaging the whole system for rapid change*. Jossey Bass: Indianapolis, IN.

Burger, K., White, L., & Yearworth, M. (2018). Why so serious? Theorising playful model-driven group decision support with situated affectivity. *Group Decision and Negotiation, 27*, 789–810. https://doi.org/10.1007/s10726-018-9559-9.

Burger, K., White, L., & Yearworth, M. (2019). Understanding front-end project workshops with Social Practice Theory. *International Journal of Project Management, 37*(1), 161–175. https://doi.org/10.1016/j.ijproman.2018.11.003.

Burrell, G., & Morgan, G. (1979). *Sociological paradigms and organisational analysis: elements of the sociology of corporate life*. Ashgate.

Calhoun, C. (2017). Integrating the social sciences: Area studies, quantitative methods, and problem-oriented research. In R. Frodeman, J. T. Klein, & R. C. S. Pacheco (Eds.), *The Oxford handbook of interdisciplinarity* (2nd ed., pp. 117–130). Oxford University Press: Oxford.

Callon, M. (1980). Struggles and negotiations to define what is problematic and what is not. In K. D. Knorr, R. Krohn, & R. Whitley (Eds.), *The Social Process of Scientific Investigation* (pp. 197–219). Springer Netherlands: Dordrecht. https://doi.org/10.1007/978-94-009-9109-5_8.

Callon, M. (1984). Some elements of a sociology of translation: Domestication of the scallops and the fishermen of St Brieuc Bay. *Sociological Review*, *32*(S1), 196–233. https://doi.org/10.1111/j.1467-954X.1984.tb00113.x.

Callon, M. (1999). The role of lay people in the production and dissemination of scientific knowledge. *Science, Technology & Society*, *4*(1), 81–94. https://doi.org/10.1177/097172189900400106.

Callon, M., Lascoumes, P., & Barthe, Y. (2009a). *Acting in an uncertain world: an essay on technical democracy*. MIT: Cambridge, MA.

Callon, M., Lascoumes, P., & Barthe, Y. (2009b). Hybrid forums. In M. Callon, P. Lascoumes, & Y. Barthe (Eds.), *Acting in an uncertain world: an essay on technical democracy* (pp. 13–36). MIT: Cambridge, MA.

Callon, M., Lascoumes, P., & Barthe, Y. (2009c). The organization of hybrid forums. In M. Callon, P. Lascoumes, & Y. Barthe (Eds.), *Acting in an uncertain world: an essay on technical democracy* (pp. 153–189). MIT: Cambridge, MA; London.

Callon, M., Lascoumes, P., & Barthe, Y. (2009d). Secluded research. In M. Callon, P. Lascoumes, & Y. Barthe (Eds.), *Acting in an uncertain world: an essay on technical democracy* (pp. 37–70). MIT: Cambridge, MA; London.

Carhart, N., & Rosenberg, G. (2016). A framework for characterising infrastructure interdependencies. *International Journal of Complexity in Applied Science and Technology*, *1*(1), 35–60. https://doi.org/10.1504/IJCAST.2016.10002359.

Carhart, N., & Yearworth, M. (2010). The use of group model building for analysing event causality within the nuclear industry. 28th International Conference of the System Dynamics Society, Seoul, Korea. https://proceedings.systemdynamics.org/2010/proceed/papers/P1112.pdf.

Castaldo, S., Premazzi, K., & Zerbini, F. (2010). The meaning(s) of trust. A content analysis on the diverse conceptualizations of trust in scholarly research on business relationships. *Journal of Business Ethics*, *96*(4), 657–668. https://doi.org/10.1007/s10551-010-0491-4.

Cech, E. (2012). Great problems of grand challenges: Problematizing engineering's understandings of its role in society. *International Journal of Engineering, Social Justice, and Peace*, *1*(2), 85–94.

Centre for Systems Studies. (2023, 18th May 2023). Thought Piece from Mike Jackson: The Past and Future of Critical Systems Thinking Practice.

Charmaz, K. (2006). *Constructing grounded theory*. Sage Publications: London.

Charnes, A., Cooper, W. W., & Rhodes, E. (1978). Measuring the efficiency of decision making units. *European Journal of Operational Research*, *2*(6), 429–444. https://doi.org/10.1016/0377-2217(78)90138-8.

Checkland, P. (1972). Towards a systems-based methodology for real-world problem solving. *Journal of Systems Engineering*, *3*(2), 87–116.

Checkland, P. (1976). Science and the systems paradigm. *International Journal of General Systems*, *3*(2), 127–134. https://doi.org/10.1080/03081077608934748.

Checkland, P. (1979). Techniques in 'soft' systems practice - 1. Systems diagrams - some tentative guidelines. *Journal of Applied Systems Analysis, 6*, 33–49.

Checkland, P. (1981). *Systems thinking, systems practice*. John Wiley & Sons: Chichester.

Checkland, P. (1983). O.R. And the systems movement: Mappings and conflicts. *Journal of the Operational Research Society, 34*(8), 661–675. https://doi.org/10.1057/jors.1983.160.

Checkland, P. (1989a). An application of soft systems methodology. In J. Rosenhead (Ed.), *Rational analysis for a problematic world: problem structuring methods for complexity, uncertainty, and conflict* (pp. 101–119). John Wiley & Sons: Chichester.

Checkland, P. (1989b). OR and social science: Fundamental thoughts. In M. C. Jackson, P. Keys, & S. A. Cropper (Eds.), *Operational Research and the Social Sciences* (pp. 35–41). Springer US: Boston, MA. https://doi.org/10.1007/978-1-4613-0789-1_5.

Checkland, P. (1989c). Soft systems methodology. In J. Rosenhead (Ed.), *Rational Analysis for a Problematic World: Problem Structuring Methods for Complexity, Uncertainty, and Conflict* (pp. 71–100). John Wiley & Sons: Chichester.

Checkland, P. (1999). *Systems thinking, systems practice: Includes a 30-year retrospective*. John Wiley & Sons: Chichester.

Checkland, P. (2000a). The emergent properties of SSM in use: A symposium by reflective practitioners. *Systemic Practice and Action Research, 13*(6), 799–823. https://doi.org/10.1023/A:1026431613200.

Checkland, P. (2000b). Soft systems methodology: A thirty year retrospective. *Systems Research and Behavioral Science, 17*, S11–S58. https://doi.org/10.1002/1099-1743(200011)17:1+<::aid-sres374>3.0.co;2-o.

Checkland, P. (2010). Researching real-life: Reflections on 30 years of action research. *Systems Research and Behavioral Science, 27*(2), 129–132. https://doi.org/10.1002/sres.1019.

Checkland, P. (2012). Peter Checkland on the origins of SSM. Retrieved 19th February 2018, from https://www.youtube.com/watch?v=XA2i1n-o9L0.

Checkland, P., & Casar, A. (1986). Vickers' concept of an appreciative system. *Journal of Applied Systems Analysis, 13*, 3–17.

Checkland, P., & Holwell, S. (2004). "Classic" OR and "soft" OR—an asymmetric complementarity. In M. Pidd (Ed.), *Systems Modelling: Theory and Practice*. John Wiley & Sons: Chichester.

Checkland, P., & Poulter, J. (2006). *Learning for action: a short definitive account of soft systems methodology, and its use for practitioner, teachers and students*. John Wiley & Sons: Chichester.

Checkland, P., & Scholes, J. (1990). *Soft systems methodology in action*. John Wiley & Sons: Chichester.

Checkland, P., & Scholes, J. (1999). *Soft systems methodology in action: including a 30-year retrospective.* John Wiley & Sons: Chichester.

Checkland, P., Pidd, M., & Morecroft, J. (2004). Working ideas, insights for systems modelling: The broader community of systems thinkers. 22nd International Conference of the System Dynamics Society. Retrieved 10th July 2023, from https://proceedings.systemdynamics.org/2004/SDS_2004/PAPERS/092CHECK.pdf.

Chester, L. (2010). Conceptualising energy security and making explicit its polysemic nature. *Energy Policy, 38*(2), 887–895. https://doi.org/10.1016/j.enpol.2009.10.039.

Churchman, C. W. (1967). Wicked problems. *Management Science, 14*(4), B-141–B-146. https://doi.org/10.1287/mnsc.14.4.B141.

Churchman, C. W. (1970). Operations research as a profession. *Management Science, 17*(2), B-37–B-53. https://doi.org/10.1287/mnsc.17.2.B37.

Churchman, C. W. (1971). *The design of inquiring systems: basic concepts of systems and organization.* Basic Books: New York.

Churchman, C. W., Ackoff, R. L., & Arnoff, E. L. (1957). *Introduction to operations research.* John Wiley & Sons: New York.

Coelho, D., Antunes, C. H., & Martins, A. G. (2010). Using SSM for structuring decision support in urban energy planning. *Technological and Economic Development of Economy, 16*(4), 641–653. https://doi.org/10.3846/tede.2010.39.

Collins, B., Fuentes, R., Larner, W., Pollard, S. J., Harou, J. J., Deakin Crick, R. E., Taylor, C. A., Nightingale, P., Fujiyama, T., Medda, F., Dimitriou, H., Richards, D. J., Ramchurn, S. D., Brady, T. M., Godfrey, P., Jude, S. R., Davies, A. C., Bahaj, A. S., Edkins, A. J., Tyler, N., Jennings, N., Yearworth, M., Jeffrey, P., & Varga, L. E. (2013). *International Centre for Infrastructure Futures (ICIF)* [Grant]. https://gtr.ukri.org/projects?ref=EP%2FK012347%2F1.

Conklin, J. (2003). Dialogue mapping: Reflections on an industrial strength case study. In P. A. Kirschner, Buckingham Shum, S., Carr, C. (Eds.), *Visualizing Argumentation* (pp. 117–136). Springer-Verlag: London.

Conklin, J. (2006). *Dialogue mapping: building shared understanding of wicked problems.* John Wiley & Sons: Chichester.

Contag, M., Li, G., Pawlowski, A., Domke, F., Levchenko, K., Holz, T., & Savage, S. (2017). How they did it: An analysis of emission defeat devices in modern automobiles. 2017 IEEE Symposium on Security and Privacy, SP 2017, https://doi.org/10.1109/SP.2017.66.

Coyle, R. G., Crawshay, R., & Sutton, L. (1994). Futures assessment by field anomaly relaxation. A review and appraisal. *Futures, 26*(1), 25–43. https://doi.org/10.1016/0016-3287(94)90088-4.

Crawford, L., Costello, K., Pollack, J., & Bentley, L. (2003). Managing soft change projects in the public sector. *International Journal of Project Management, 21*(6), 443–448. https://doi.org/10.1016/S0263-7863(02)00099-6.

Crick, R. D., Barr, S., Green, H., & Pedder, D. (2017). Evaluating the wider outcomes of schools: Complex systems modelling for leadership decisioning. *Educational Management Administration & Leadership*, *45*(4), 719–743. https://doi.org/10.1177/1741143215597233.

Crielaard, L., Quax, R., Sawyer, A. D. M., Vasconcelos, V. V., Nicolaou, M., Stronks, K., & Sloot, P. M. A. (2023). Using network analysis to identify leverage points based on causal loop diagrams leads to false inference. *Scientific Reports*, *13*(1), 21046. https://doi.org/10.1038/s41598-023-46531-z.

Croidieu, G., & Kim, P. H. (2017). Labor of love: Amateurs and lay-expertise legitimation in the early U.S. radio field. *Administrative Science Quarterly*, *63*(1), 1–42. https://doi.org/10.1177/0001839216686531.

Cronin, K., Midgley, G., & Jackson, L. S. (2014). Issues mapping: A problem structuring method for addressing science and technology conflicts. *European Journal of Operational Research*, *233*(1), 145–158. https://doi.org/10.1016/j.ejor.2013.08.012.

Crossley, S. A., Dascalu, M., & McNamara, D. S. (2017). How important is size? An investigation of corpus size and meaning in both latent semantic analysis and latent Dirichlet allocation. 30th International Florida Artificial Intelligence Research Society Conference (FLAIRS 2017), FL, USA.

Cummings, S., Bridgman, T., Hassard, J., & Rowlinson, M. (2017). *A new history of management*. Cambridge University Press. https://doi.org/10.1017/9781316481202.

Cunha, A., & Morais, D. (2019). Problem structuring methods in group decision making: A comparative study of their application. *Operational Research*, *19*(4), 1081–1100. https://doi.org/10.1007/s12351-017-0310-0.

Cunico, G., Zimmermann, N., & Videira, N. (2024). Playing the new devil's advocate role in facilitated modelling processes to address group homogeneity. *Journal of the Operational Research Society*, 75(8), 1587–1610. https://doi.org/10.1080/01605682.2023.2263101.

Curto, R., Norese, M. F., & Rolando, D. (2022). A multicriteria decision aid perspective that guides an incremental development of knowledge and fosters relationships and decisions. *EURO Journal on Decision Processes*, *10*, Article 100023. https://doi.org/10.1016/j.ejdp.2022.100023.

Dalkey, N., & Helmer, O. (1963). An experimental application of the delphi method to the use of experts. *Management Science*, *9*(3), 458–467. https://doi.org/10.1287/mnsc.9.3.458.

Dando, M., & Bennett, P. (1981). A kuhnian crisis in management science? *Journal of the Operational Research Society*, *32*(2), 91–103. https://doi.org/10.1057/jors.1981.22.

Dando, M., & Eden, C. (1977). Reflections on operational research. A report from the Euro II Congress. *Omega*, *5*(3), 255–270. https://doi.org/10.1016/0305-0483(77)90109-8.

Davis, J. P., & Hall, J. W. (2003). A software-supported process for assembling evidence and handling uncertainty in decision-making. *Decision Support Systems, 35*(3), 415–433. https://doi.org/10.1016/s0167-9236(02)00117-3.

Davis, J., MacDonald, A., & Marashi, S. E. (2007). Integrated performance measurement to support strategic decision making in engineering organisations. IEEE International Conference on Industrial Engineering and Engineering Management, Singapore. https://doi.org/10.1109/IEEM.2007.4419495.

Davis, J., MacDonald, A., & White, L. (2010). Problem-structuring methods and project management: An example of stakeholder involvement using Hierarchical Process Modelling methodology. *Journal of the Operational Research Society, 61*(6), 893–904. https://doi.org/10.1057/jors.2010.12.

De Liddo, A., Buckingham Shum, S. (2010). Cohere: A prototype for contested collective intelligence workshop on collective intelligence in organizations: Toward a research agenda. ACM Conference on Computer-Supported Cooperative Work, Savannah, GA, USA.

Dell'Acqua, F., McFowland, E., Mollick, E. R., Lifshitz-Assaf, H., Kellogg, K., Rajendran, S., Krayer, L., Candelon, F., & Lakhani, K. R. (2023). Navigating the jagged technological frontier: Field experimental evidence of the effects of AI on knowledge worker productivity and quality. Retrieved 11th November 2023, from https://dx.doi.org/10.2139/ssrn.4573321.

Denis, J. L., Dompierre, G., Langley, A., & Rouleau, L. (2011). Escalating indecision: Between reification and strategic ambiguity. *Organization Science, 22*(1), 225–244. https://doi.org/10.1287/orsc.1090.0501.

Dias, W. P. S. (2013). Comparing the systems approaches of Checkland and Blockley. *Civil Engineering and Environmental Systems, 30*(3–4), 221–230. https://doi.org/10.1080/10286608.2013.853749.

Dias, W. P. S., & Blockley, D. I. (1994). The integration of product and process models for design. *Design Studies, 15*(4), 417–432. https://doi.org/10.1016/0142-694X(94)90005-1.

Diefenbach, T. (2009). New public management in public sector organizations: The dark sides of managerialistic 'enlightenment'. *Public Administration, 87*(4), 892–909. https://doi.org/10.1111/j.1467-9299.2009.01766.x.

Doloi, H. K. (2011). Understanding stakeholders' perspective of cost estimation in project management. *International Journal of Project Management, 29*(5), 622–636. https://doi.org/10.1016/j.ijproman.2010.06.001.

Donaldson, A., Lane, S., Ward, N., & Whatmore, S. (2013). Overflowing with issues: Following the political trajectories of flooding. *Environment and Planning C: Government and Policy, 31*(4), 603-618. https://doi.org/10.1068/c11230.

Doswell, B., Tierney, M. J., Yearworth, M., & Stephenson, W. (2016). Planning and assessment framework for sustainable energy access. 11th Conference on

Sustainable Developments of Energy, Water and Environment Systems,4th–9th September 2016, Lisbon.

Dunford, C. N., Yearworth, M., York, D. M., & Godfrey, P. (2013). A view of Systems Practice: Enabling quality in design. *Systems Engineering*, *16*(2), 134–151. https://doi.org/10.1002/sys.21220.

Dunn, C. E. (2007). Participatory GIS - A people's GIS? *Progress in Human Geography*, *31*(5), 616–637. https://doi.org/10.1177/0309132507081493.

Dyson, R., O'Brien, F., & Shah, D. (2021). Soft OR and practice: The contribution of the founders of operations research. *Operations Research*, *69*(3), 727–738. https://doi.org/10.1287/OPRE.2020.2051.

Economic and Social Research Council. (2019). Delivery plan. Retrieved 10th July 2023, from https://www.ukri.org/wp-content/uploads/2020/09/ESRC-250920-DeliveryPlan2019.pdf.

Eden, C. (1982). Problem construction and the influence of O.R. *Interfaces*, *12*(2), 50–60. http://www.jstor.org/stable/25060235.

Eden, C. (1988). Cognitive mapping. *European Journal of Operational Research*, *36*(1), 1–13. https://doi.org/10.1016/0377-2217(88)90002-1.

Eden, C. (1989a). Operational research as negotiation. In M. C. Jackson, P. Keys, & S. A. Cropper (Eds.), *Operational Research and the Social Sciences* (pp. 43–50). Springer US: Boston, MA. https://doi.org/10.1007/978-1-4613-0789-1_5.

Eden, C. (1989b). Using cognitive mapping for strategic options development and analysis (SODA). In J. Rosenhead (Ed.), *Rational analysis for a problematic world: problem structuring methods for complexity, uncertainty, and conflict* (pp. 21–42). John Wiley & Sons: Chichester.

Eden, C. (1995). On evaluating the performance of 'wide-band' GDSS's. *European Journal of Operational Research*, *81*(2), 302–311. https://doi.org/10.1016/0377-2217(93)E0241-O.

Eden, C., & Ackermann, F. (2020). Group support systems: Concepts to practice. In D. M. Kilgour & C. Eden (Eds.), *Handbook of Group Decision and Negotiation* (pp. 1–25). Springer International Publishing: Cham. https://doi.org/10.1007/978-3-030-12051-1_59-1.

Eden, C., & Huxham, C. (1996). Action research for management research. *British Journal of Management*, *7*(1), 75–86. https://doi.org/10.1111/j.1467-8551.1996.tb00107.x.

Eden, C., & Jones, S. (1984). Using repertory grids for problem construction. *Journal of the Operational Research Society*, *35*(9), 779–790. https://doi.org/10.1057/jors.1984.160.

Eden, C., & Simpson, P. (1989). SODA and cognitive mapping in practice. In J. Rosenhead (Ed.), *Rational analysis for a problematic world: problem structuring methods for complexity, uncertainty, and conflict* (pp. 43–70). John Wiley & Sons: Chichester.

Eden, C., & Sims, D. (1979). On the nature of problems in consulting practice. *Omega, 7*(2), 119–127. https://doi.org/10.1016/0305-0483(79)90099-9.

Eden, C., Jones, S., Sims, D., & Gunton, H. (1979). Images into models. The subjective world of the policy maker. *Futures, 11*(1), 56–62. https://doi.org/10.1016/0016-3287(79)90069-7.

Eden, C., Sims, D., & Jones, S. (1979). Policy analysis and organizational politics. *European Journal of Operational Research, 3*(3), 207–215. https://doi.org/10.1016/0377-2217(79)90140-1.

Eden, C., Jones, S., & Sims, D. (1983). *Messing about in problems: an informal structured approach to their identification and management*. Pergamon: Oxford.

Eden, C., Ackermann, F., & Cropper, S. (1992). The analysis of cause maps. *Journal of Management Studies, 29*(3), 309–324. https://doi.org/10.1111/j.1467-6486.1992.tb00667.x.

Eden, C., Ackermann, F., Paucar-Caceres, A., & Abuabara, L. (2020). The practice of SODA methodologies then and now. 62nd Conference of the UK Operational Research Society (OR62), Warwick, UK.

Egan, J. (1998). *Rethinking Construction: The Report of the Construction Task Force*. HMSO. https://constructingexcellence.org.uk/wp-content/uploads/2014/10/rethinking_construction_report.pdf.

Emes, M., Smith, S., Ward, S., Smith, A., & Ming, T. (2017). Care and flow: Using soft systems methodology to understand tensions in the patient discharge process. *Health Systems, 6*(3), 260–278. https://doi.org/10.1057/s41306-017-0027-6.

EPSRC. (2011). The EPSRC industrial doctorate centre scheme. Good Practice Guidance. Retrieved 5th August 2023, from https://www.aengd.org.uk/files/6415/2344/1271/IDCGoodPracticeGuidelines.pdf.

Farsari, I., Butler, R. W., & Szivas, E. (2010). The use of cognitive mapping in analysing sustainable tourism policy: Methodological implications. *Tourism Recreation Research, 35*(2), 145–160. https://doi.org/10.1080/02508281.2010.11081630.

Faust, M. (2012). Sociological perspectives on management consulting. In M. Kipping & T. Clark (Eds.), *The Oxford Handbook of Management Consulting* (pp. 139–163). OUP: Oxford, UK.

Finnveden, G., Hauschild, M. Z., Ekvall, T., Guinée, J., Heijungs, R., Hellweg, S., Koehler, A., Pennington, D., & Suh, S. (2009). Recent developments in life cycle assessment. *Journal of Environmental Management, 91*(1), 1–21. https://doi.org/10.1016/j.jenvman.2009.06.018.

Fiorino, D. J. (1990). Citizen participation and environmental risk: A survey of institutional mechanisms. *Science, Technology & Human Values, 15*(2), 226–243. https://doi.org/10.1177/016224399001500204.

Fisher, R., & Ury, W. (1981). *Getting to yes: negotiating agreement without giving in*. Houghton Mifflin: Boston, MA.

Fletcher, A., & Davis, J. P. (2002). Decision-making with incomplete evidence. Proceedings SPE Asia Pacific Oil and Gas Conference and Exhibition, 725–739. https://doi.org/10.2118/77914-ms.

Flyvbjerg, B. (1998). *Rationality and power: democracy in practice.* University of Chicago Press: Chicago.

Flyvbjerg, B., Garbuio, M., & Lovallo, D. (2009). Delusion and deception in large infrastructure projects. *California Management Review, 51*(2), 170–194. https://doi.org/10.2307/411664.

Fomento de San Sebastian, City Council of San Sebastian, Acciona Infraestructuras, Tecnalia, Bristol City Council, University of Bristol, Arup, Municipality of Florence, SPES Consulting, & Ataf. (2013). *Systems Thinking for Comprehensive City Efficient Energy Planning (314277)* [Grant]. https://cordis.europa.eu/project/id/314277.

Forbes, G. A. (2023). *Systems Thinking and Process Philosophy: Convergent Evolution.* UK OR Society Problem Structuring Methods (PSM) Special Interest Group (SIG), 19th October 2023, https://www.youtube.com/watch?v=-5ZC1GumC2U.

Forrester, J. W. (1958). Industrial dynamics - A major breakthrough for decision makers. *Harvard Business Review, 36*(4), 37–66.

Forrester, J. W. (1961). *Industrial Dynamics.* MIT Press: Cambridge, MA.

Forrester, J. W. (1994). System dynamics, systems thinking, and soft OR. *System Dynamics Review, 10*(2-3), 245–256. https://doi.org/10.1002/sdr.4260100211.

Francis, R., & Bekera, B. (2014). A metric and frameworks for resilience analysis of engineered and infrastructure systems. *Reliability Engineering and System Safety, 121*, 90–103. https://doi.org/10.1016/j.ress.2013.07.004.

Franco, L. A. (2006). Forms of conversation and problem structuring methods: A conceptual development. *Journal of the Operational Research Society, 57*(7), 813–821. https://doi.org/10.1057/palgrave.jors.2602169.

Franco, L. A. (2013). Rethinking Soft OR interventions: Models as boundary objects. *European Journal of Operational Research, 231*(3), 720–733. https://doi.org/10.1016/j.ejor.2013.06.033.

Franco, L. A., & Greiffenhagen, C. (2018). Making OR practice visible: Using ethnomethodology to analyse facilitated modelling workshops. *European Journal of Operational Research, 265*(2), 673–684. https://doi.org/10.1016/j.ejor.2017.08.016.

Franco, L. A., & Hämäläinen, R. (2016). Behavioural operational research: Returning to the roots of the OR profession. *European Journal of Operational Research, 249*(3), 791–795. https://doi.org/10.1016/j.ejor.2015.10.034.

Franco, L. A., & Nielsen, M. F. (2018). Examining group facilitation in situ: The use of formulations in facilitation practice. *Group Decision and Negotiation, 27*, 735–756. https://doi.org/10.1007/s10726-018-9577-7.

Franco, L. A., & Rouwette, E. A. J. A. (2022). Problem structuring methods: Taking stock and looking ahead. In S. Salhi & J. Boylan (Eds.), *The Palgrave Handbook of*

Operations Research (pp. 735–780). Springer International Publishing: Cham. https://doi.org/10.1007/978-3-030-96935-6_23.

Franco, L. A., Cushman, M., & Rosenhead, J. (2004). Project review and learning in the construction industry: Embedding a problem structuring method within a partnership context. *European Journal of Operational Research, 152*(3), 586–601. https://doi.org/10.1016/S0377-2217(03)00059-6.

Franco, L. A., Shaw, D., & Westcombe, M. (2007). Taking problem structuring methods forward. *Journal of the Operational Research Society, 58*(5), 545–546. https://doi.org/10.1057/palgrave.jors.2602366.

Françozo, R. V., & Belderrain, M. C. N. (2022). A problem structuring method framework for value-focused thinking. *EURO Journal on Decision Processes, 10*, 100014. https://doi.org/10.1016/j.ejdp.2022.100014.

Frank, D. J., & Meyer, J. W. (2007). University expansion and the knowledge society. *Theory and Society, 36*(4), 287–311. https://doi.org/10.1007/s11186-007-9035-z.

Frankopan, P. (2023). *The earth transformed: an untold history*. Bloomsbury Publishing: London.

Freeman, R. E. (1984). *Strategic management: a stakeholder approach*. Pitman: Boston, MA.

Freeman, R., & Yearworth, M. (2017). Climate change and cities: Problem structuring methods and critical perspectives on low-carbon districts. *Energy Research and Social Science, 25*, 48–64. https://doi.org/10.1016/j.erss.2016.11.009.

Fregonara, E., Curto, R., Grosso, M., Mellano, P., Rolando, D., & Tulliani, J. (2013). Environmental technology, materials science, architectural design, and real estate market evaluation: A multidisciplinary approach for energy-efficient buildings. *Journal of Urban Technology, 20*(4), 57–80. https://doi.org/10.1080/10630732.2013.855512.

Fregonara, E., Giordano, R., Rolando, D., & Tulliani, J. M. (2016). Integrating environmental and economic sustainability in new building construction and retrofits. *Journal of Urban Technology, 23*(4), 3–28. https://doi.org/10.1080/10630732.2016.1157941.

French, S. (2021). From soft to hard elicitation. *Journal of the Operational Research Society, 73*(6), 1181–1197. https://doi.org/10.1080/01605682.2021.1907244.

Friel, S., Pescud, M., Malbon, E., Lee, A., Carter, R., Greenfield, J., Cobcroft, M., Potter, J., Rychetnik, L., & Meertens, B. (2017). Using systems science to understand the determinants of inequities in healthy eating. *PLoS One, 12*(11), Article e0188872. https://doi.org/10.1371/journal.pone.0188872.

Friend, J. (1989). The strategic choice approach. In J. Rosenhead (Ed.), *Rational analysis for a problematic world: problem structuring methods for complexity, uncertainty, and conflict* (pp. 121–158). John Wiley & Sons: Chichester.

Friend, J. (1992). New directions in software for strategic choice. *European Journal of Operational Research, 61*(1-2), 154–164. https://doi.org/10.1016/0377-2217(92)90277-G.

Friend, J. (1993). Searching for appropriate theory and practice in multi-organizational fields. *Journal of the Operational Research Society, 44*(6), 585–598. https://doi.org/10.1057/jors.1993.100.

Friend, J. (1995). Supporting developmental decision processes: The evolution of an OR approach. *International Transactions in Operational Research, 2*(3), 225–232. https://doi.org/10.1016/0969-6016(95)00006-S.

Friend, J. (2001). The strategic choice approach. In J. Rosenhead & J. Mingers (Eds.), *Rational Analysis for Problematic World Revisited* (pp. 115–149). John Wiley & Sons: Chichester.

Friend, J., & Hickling, A. (2005). *Planning under pressure: the strategic choice approach* (3rd ed.). Elsevier Butterworth-Heinemann: Oxford.

Friend, J., & Stradspan Limited. (2002). *STRAD the STRategic ADviser User's Manual v2.3*. Stradspan Limited.

Friend, J., Power, J., & Yewlett, C. (1974). *Public planning: the inter-corporate dimension*. Tavistock Publications: London.

Friend, J., Norris, M., & Stringer, J. (1988). The institute for operational research: An initiative to extend the scope of OR. *Journal of the Operational Research Society, 39*(8), 705–713. https://doi.org/10.1057/jors.1988.125.

Frodeman, R., Klein, J. T., & Pacheco, R. C. S. (2017). *The Oxford handbook of interdisciplinarity* (2nd ed.). Oxford University Press: Oxford.

Fuller, S. (2003). *Kuhn vs Popper: the struggle for the soul of science*. Icon: Thriplow.

Garfinkel, H. (1996). Ethnomethodology's program. *Social Psychology Quarterly, 59*(1), 5–21. https://doi.org/10.2307/2787116.

Geels, F. W. (2005). Processes and patterns in transitions and system innovations: Refining the co-evolutionary multi-level perspective. *Technological Forecasting and Social Change, 72*(6), 681–696. https://doi.org/10.1016/j.techfore.2004.08.014.

Geissdoerfer, M., Savaget, P., Bocken, N. M. P., & Hultink, E. J. (2017). The Circular Economy–A new sustainability paradigm? *Journal of Cleaner Production, 143*, 757–768. https://doi.org/10.1016/j.jclepro.2016.12.048.

Georgiou, I. (2009). A graph-theoretic perspective on the links-to-concepts ratio expected in cognitive maps. *European Journal of Operational Research, 197*(2), 834–836. https://doi.org/10.1016/j.ejor.2008.07.030.

Georgiou, I. (2012). Messing about in transformations: Structured systemic planning for systemic solutions to systemic problems. *European Journal of Operational Research, 223*(2), 392–406. https://doi.org/10.1016/j.ejor.2012.06.010.

Georgiou, I., & Heck, J. (2021). The emergence of problem structuring methods, 1950s–1989: An atlas of the journal literature. *Systems Research and Behavioral Science, 38*(6), 756–796. https://doi.org/10.1002/sres.2737.

Ghosh, S., & Troutt, M. D. (2012). Complex compound option models - Can practitioners truly operationalize them? *European Journal of Operational Research, 222*(3), 542–552. https://doi.org/10.1016/j.ejor.2012.05.007.

Gibson, J. J. (1977). The theory of affordances. In R. Shaw & J. Bransford (Eds.), *Perceiving, Acting, and Knowing*. John Wiley & Sons: New York.

Giordano, R., Passarella, G., Uricchio, V. F., & Vurro, M. (2007). Integrating conflict analysis and consensus reaching in a decision support system for water resource management. *Journal of Environmental Management, 84*(2), 213–228. https://doi.org/10.1016/j.jenvman.2006.05.006.

Glaser, B. G. (1978). *Theoretical sensitivity: advances in the methodology of grounded theory*. Sociology Press: Mill Valley, CA.

Godfrey, P. (2009). *Industrial Doctorate Centre: Systems* (EPSRC EP/G037353/1) [Grant]. University of Bristol. https://gow.epsrc.ukri.org/NGBOViewGrant.aspx?GrantRef=EP/G037353/1.

Godfrey, P. (2012). The Engineering Doctorate (EngD): Developing Leaders for Tomorrow with Industry. *CLAIU – EU (Council of Association of long-cycle Engineers, of a university or higher school of engineering of the European Union).* Retrieved 10th July 2023, from http://claiu.fabi.be/home/wp-content/uploads/2011/12/The-Engineering-Doctorate-final-Godfrey.pdf.

Goh, Y. M., Brown, H., & Spickett, J. (2010). Applying systems thinking concepts in the analysis of major incidents and safety culture. *Safety Science, 48*(3), 302–309. https://doi.org/10.1016/j.ssci.2009.11.006.

Goh, Y. M., Love, P. E. D., Stagbouer, G., & Annesley, C. (2012). Dynamics of safety performance and culture: A group model building approach. *Accident Analysis and Prevention, 48*, 118–125. https://doi.org/10.1016/j.aap.2011.05.010.

Gomes Júnior, A., & Schramm, V. (2021). Problem structuring methods: A review of advances over the last decade. *Systemic Practice and Action Research.* https://doi.org/10.1007/s11213-021-09560-1.

Goodeve, C. (1948a). Methods of operational research. *Nature, 161*(4089), 379–381. https://doi.org/10.1038/161379a0.

Goodeve, C. (1948b). Operational research. *Nature, 161*(4089), 377–378. https://doi.org/10.1038/161377a0.

Goodeve, C. (1948c). Present and future of operational research. *Nature, 161*(4089), 381–384. https://doi.org/10.1038/161381a0.

de Gooyert, V., Rouwette, E., van Kranenburg, H., & Freeman, E. (2017). Reviewing the role of stakeholders in Operational Research: A stakeholder theory perspective. *European Journal of Operational Research, 262*(2), 402–410. https://doi.org/10.1016/j.ejor.2017.03.079.

Government Office for Science. (2004). Future flooding. https://www.gov.uk/government/publications/future-flooding.

Government Office for Science. (2012). High impact low probability risks: Blackett review. Retrieved 28th May 2014, from https://www.gov.uk/government/publications/high-impact-low-probability-risks-blackett-review.

Green, S. D. (1999). A participative research strategy for propagating soft methodologies in value management practice. *Construction Management and Economics, 17*(3), 329–340. https://doi.org/10.1080/014461999371538.

Gregory, A. J., Atkins, J. P., Burdon, D., & Elliott, M. (2013). A problem structuring method for ecosystem-based management: The DPSIR modelling process. *European Journal of Operational Research, 227*(3), 558–569. https://doi.org/10.1016/j.ejor.2012.11.020.

Gregory, A. J., Atkins, J. P., Midgley, G., & Hodgson, A. M. (2020). Stakeholder identification and engagement in problem structuring interventions. *European Journal of Operational Research, 283*(1), 321–340. https://doi.org/10.1016/j.ejor.2019.10.044.

Guarnieri, P., e Silva, L. C., & Levino, N. A. (2016). Analysis of electronic waste reverse logistics decisions using Strategic Options Development Analysis methodology: A Brazilian case. *Journal of Cleaner Production, 133*, 1105–1117. https://doi.org/10.1016/j.jclepro.2016.06.025.

Guide Jr, V. D. R., & Van Wassenhove, L. N. (2007). Dancing with the devil: Partnering with industry but publishing in Academia: Commentary. *Decision Sciences, 38*(4), 531–546. https://doi.org/10.1111/j.1540-5915.2007.00169.x.

Hall, J. (2003). Condition monitoring and asset management of Flood and Coastal Defence systems. Retrieved 10th July 2023, from https://assets.publishing.service.gov.uk/government/uploads/system/uploads/attachment_data/file/290721/scho1205bkav-e-e.pdf.

Hall, J., Blockley, D., & Davis, J. (1998). Uncertain inference using interval probability theory. *International Journal of Approximate Reasoning, 19*(3–4), 247–264. https://doi.org/10.1016/S0888-613X(98)10010-5.

Hall, J., Davis, J., & Taylor, C. (1999). *Condition Monitoring and Asset Management for Complex Infrastructure Systems* (EPSRC GR/M53073/01) [Grant]. University of Bristol. https://gow.epsrc.ukri.org/NGBOViewGrant.aspx?GrantRef=GR/M53073/01.

Hall, J. W., Le Masurier, J. W., Baker-Langman, E. A., Davis, J. P., & Taylor, C. A. (2004). A decision-support methodology for performance-based asset management. *Civil Engineering and Environmental Systems, 21*(1), 51–75. https://doi.org/10.1080/1028660031000135086.

Hannigan, T. R., Haan, R. F. J., Vakili, K., Tchalian, H., Glaser, V. L., Wang, M. S., Kaplan, S., & Jennings, P. D. (2019). Topic modeling in management research: Rendering new theory from textual data. *Academy of Management Annals, 13*(2), 586–632. https://doi.org/10.5465/annals.2017.0099.

Harper, A., Mustafee, N., & Yearworth, M. (2021). Facets of trust in simulation studies. *European Journal of Operational Research, 289*(1), 197–213. https://doi.org/10.1016/j.ejor.2020.06.043.

Harrison, M. I., Koppel, R., & Bar-Lev, S. (2007). Unintended consequences of information technologies in health care-an interactive sociotechnical analysis. *Journal of the American Medical Informatics Association, 14*(5), 542–549. https://doi.org/10.1197/jamia.M2384.

Harrop, N., Wood-Harper, T., & Gillies, A. (2006). Neglected user perspectives in the design of an online hospital bed-state system: Implications for the National Programme for IT in the NHS. *Health Informatics Journal, 12*(4), 293–303. https://doi.org/10.1177/1460458206069759.

Hartz, R. (2023). Demeaning, depriving, and isolating the academic subject: A case study of the degradation of professional autonomy. *Management Learning*, 13505076231207938. https://doi.org/10.1177/13505076231207938.

Harwood, S. (2019). Whither is problem structuring methods (PSMs)? *Journal of the Operational Research Society, 70*(8), 1391–1392. https://doi.org/10.1080/01605682.2018.1502628.

Head, B. W. (2022). *Wicked Problems in Public Policy*. Palgrave Macmillan Cham.

Hewlett-Packard. (2007). HP Halo Collaboration Meeting Room. Retrieved 1st May 2023, from https://www.hp.com/hpinfo/newsroom/press_kits/2008/halo/ds_halo-meetingroom.pdf.

Hickling, A. (1989). The strategic choice approach. In J. Rosenhead (Ed.), *Rational analysis for a problematic world: problem structuring methods for complexity, uncertainty, and conflict* (pp. 159–192). John Wiley & Sons: Chichester.

Hickling, A. (2001). Gambling with frozen fire? In J. Rosenhead & J. Mingers (Eds.), *Rational Analysis for Problematic World Revisited* (pp. 151–180). John Wiley & Sons: Chichester.

Hindle, G. A. (2011). Case Article—Teaching soft systems methodology and a blueprint for a module. *INFORMS Transactions on Education, 12*(1), 31–42. https://doi.org/10.1287/ited.1110.0068ca.

Hjortsø, C. N. (2004). Enhancing public participation in natural resource management using Soft OR - An application of strategic option development and analysis in tactical forest planning. *European Journal of Operational Research, 152*(3), 667–683. https://doi.org/10.1016/S0377-2217(03)00065-1.

HM Treasury. (2015). Valuing Infrastructure Spend: Supplementary Guidance to the Green Book. http://www.gov.uk/government/uploads/system/uploads/attachment_data/file/417822/PU1798_Valuing_Infrastructure_Spend_-_lastest_draft.pdf.

Hood, C. (1991). A public management for all seasons? *Public Administration, 69*(1), 3–19. https://doi.org/10.1111/j.1467-9299.1991.tb00779.x.

Hopkinson, P., Zils, M., Hawkins, P., & Roper, S. (2018). Managing a complex global circular economy business model: Opportunities and challenges. *California Management Review*, *60*(3), 71–94. https://doi.org/10.1177/0008125618764692.

House of Commons Committee of Public Accounts. (2013). *The dismantled National Programme for IT in the NHS.* (HC294). The Stationery Office Limited: London Retrieved from http://www.publications.parliament.uk/pa/cm201314/cmselect/cmpubacc/294/294.pdf.

Hovmand, P. S. (2014). *Community based system dynamics* (Vol. 9781461487630). Springer: New York. https://doi.org/10.1007/978-1-4614-8763-0.

Inam, A., Adamowski, J., Halbe, J., & Prasher, S. (2015). Using causal loop diagrams for the initialization of stakeholder engagement in soil salinity management in agricultural watersheds in developing countries: A case study in the Rechna Doab watershed, Pakistan. *Journal of Environmental Management*, *152*, 251–267. https://doi.org/10.1016/j.jenvman.2015.01.052.

Institute for Apprenticeships & Technical Education. (2021). *Operational Research Specialist (ST0884)*. Retrieved 10th July 2023 from https://www.instituteforapprenticeships.org/apprenticeship-standards/operational-research-specialist-v1-0.

Institute for Apprenticeships & Technical Education. (2023). *Systems Thinking Practitioner (ST0787)*. Retrieved 22nd July 2023 from https://www.instituteforapprenticeships.org/apprenticeship-standards/systems-thinking-practitioner-v1-1.

International Federation for Systems Research. (2014). Systems thinking: New directions in theory, practice and application. Proceedings of the Seventeenth IFSR Conversation, 27th April - 2nd May 2014, Linz, Austria. http://archive-ifsr.org/wp-content/uploads/2015/05/IFSR-Conversation-2014-Proceedings.pdf.

International Society for the Systems Sciences, & International Council on Systems Engineering. (2011). Relationship Agreement. 3. Retrieved 2nd December 2019, from https://docs.google.com/viewer?a=v&pid=sites&srcid=ZGVmYXVsdGRvbWFpbnxzeXNzY2l3Z3xneDo0NjFlNGE1MWNmOWYxN2Q1.

International Society for the Systems Sciences, & International Council on Systems Engineering. (2017). Memorandum of Understanding. 6. Retrieved 2nd December 2019, from https://docs.google.com/viewer?a=v&pid=sites&srcid=ZGVmYXVsdGRvbWFpbnxzeXNzY2l3Z3xneDo3ZDUwNzQwNDRmMmFjZjY2.

Jackson, M. C. (1982). Nature of 'soft' systems thinking: The work of Churchman, Ackoff and Checkland. *Journal of Applied Systems Analysis*, *9*, 17–39.

Jackson, M. C. (1993). The system of systems methodologies - A guide to researchers. *Journal of the Operational Research Society*, *44*(2), 208–209. https://doi.org/10.1057/jors.1993.42.

Jackson, M. C. (2000). *A systems approach to management*. Kluwer: New York.

Jackson, M. C. (2001). Critical systems thinking and practice. *European Journal of Operational Research, 128*(2), 233–244. https://doi.org/10.1016/s0377-2217(00)00067-9.

Jackson, M. C. (2003). *Systems thinking: creative holism for managers*. John Wiley & Sons: Chichester.

Jackson, M. C. (2006). Beyond problem structuring methods: Reinventing the future of OR/MS. *Journal of the Operational Research Society, 57*(7), 868–878. https://doi.org/10.1057/palgrave.jors.2602093.

Jackson, M. C. (2019a). *Critical systems thinking and the management of complexity*. Wiley-Blackwell: Chichester.

Jackson, M. C. (2019b). The future of OR is present. UK OR Society Beale Lecture, February 2019, https://www.youtube.com/watch?v=1W8HqmOC7Q4.

Jackson, M. C. (2023). Rebooting the systems approach by applying the thinking of Bogdanov and the pragmatists. *Systems Research and Behavioral Science, 40*(2), 349-365. https://doi.org/10.1002/sres.2908

Jackson, M. C., & Keys, P. (1984). Towards a system of systems methodologies. *Journal of the Operational Research Society, 35*(6), 473–486. https://doi.org/10.1057/jors.1984.101.

Jackson, M. C., Keys, P., & Cropper, S. (1989). *Operational research and the social sciences: Specialized conference: Papers*. Plenum Press: New York.

Jalali, M. S., & Beaulieu, E. (2023). Strengthening a weak link: Transparency of causal loop diagrams — current state and recommendations. *System Dynamics Review*. https://doi.org/10.1002/sdr.1753.

Jenkins, G. M. (1969). The systems approach. *Journal of Systems Engineering, 1*(1), 3–49.

Jenkins, K., McCauley, D., Heffron, R., Stephan, H., & Rehner, R. (2016). Energy justice: A conceptual review. *Energy Research and Social Science, 11*, 174–182. https://doi.org/10.1016/j.erss.2015.10.004.

Johnson, M., Midgley, G., Wright, J., & Chichirau, G. (2018). Community Operational Research: Innovations, internationalization and agenda-setting applications. *European Journal of Operational Research, 268*(3), 761–770. https://doi.org/10.1016/j.ejor.2018.03.004.

Jones, C. (2003). As if business ethics were possible, 'within such limits'. *Organization, 10*(2), 223–248. https://doi.org/10.1177/1350508403010002003.

Jones, S., & Eden, C. (1981). O.R. in the community. *Journal of the Operational Research Society, 32*(5), 335–345. https://doi.org/10.1057/jors.1981.71.

Jones, W., Kotiadis, K., & O'Hanley, J. R. (2021). Maximising stakeholder learning by looping again through the simulation life-cycle: A case study in public transport. *Journal of the Operational Research Society*. https://doi.org/10.1080/01605682.2021.2007806.

Josimović, B., Cvjetić, A., & Furundžić, D. (2021). Strategic Environmental Assessment and the precautionary principle in the spatial planning of wind farms – European experience in Serbia. *Renewable and Sustainable Energy Reviews, 136*, 110459. https://doi.org/10.1016/j.rser.2020.110459.

Kamari, A., Corrao, R., & Kirkegaard, P. H. (2017). Sustainability focused decision-making in building renovation. *International Journal of Sustainable Built Environment, 6*(2), 330–350. https://doi.org/10.1016/j.ijsbe.2017.05.001.

Kamari, A., Jensen, S. R., Corrao, R., & Kirkegaard, P. H. (2019). A holistic multi-methodology for sustainable renovation. *International Journal of Strategic Property Management, 23*(1), 50–64. https://doi.org/10.3846/ijspm.2019.6375.

Kammeier, H. D. (1998). A computrer-aided strategic approach to decision-making in urban planning: An exploratory case study in Thailand. *Cities, 15*(2), 105–119. https://doi.org/10.1016/S0264-2751(97)10017-8.

Kane, L., & Del Mistro, R. (2003). Changes in transport planning policy: Changes in transport planning methodology? *Transportation, 30*(2), 113–131. https://doi.org/10.1023/A:1022562125856.

Keeney, R. L. (1996). Value-focused thinking: Identifying decision opportunities and creating alternatives. *European Journal of Operational Research, 92*(3), 537–549. https://doi.org/10.1016/0377-2217(96)00004-5.

Kelly, G. A. (1955). *The psychology of personal constructs.* Norton: New York.

Keys, P. (1991). *Operational research and systems: the systemic nature of operational research.* Plenum: New York.

Keys, P. (1995). *Understanding the process of operational research: collected readings.* John Wiley & Sons: Chichester.

Keys, P. (1998). OR as technology revisited. *Journal of the Operational Research Society, 49*(2), 99–108. https://doi.org/10.1057/palgrave.jors.2600502.

Keys, P. (2006). On becoming expert in the use of problem structuring methods. *Journal of the Operational Research Society, 57*(7), 822–829. https://doi.org/10.1057/palgrave.jors.2602194.

Khakee, A., & Strömberg, K. (1993). Applying futures studies and the strategic choice approach in urban planning. *Journal of the Operational Research Society, 44*(3), 213–224. https://doi.org/10.1057/jors.1993.46.

Khisty, C. J. (1995). Soft-systems methodology as learning and management tool. *Journal of Urban Planning and Development, 121*(3), 91–107. https://doi.org/10.1061/(ASCE)0733-9488(1995)121:3(91).

Kilgour, D. M., & Eden, C. (2020). *Handbook of group decision and negotiation.* Springer International Publishing: Cham. https://doi.org/10.1007/978-3-030-12051-1.

Kirby, M. W. (2011). Charles Frederick Goodeve. In A. Assad & S. Gass (Eds.), *Profiles in Operations Research: Pioneers and Innovators* (pp. 83–94). Springer: New York; London. https://doi.org/10.1007/978-1-4419-6281-2_5.

Kirby, M. W., & Rosenhead, J. (2011). Patrick Blackett. In A. Assad & S. Gass (Eds.), *Profiles in Operations Research: Pioneers and Innovators* (pp. 1–29). Springer: New York; London. https://doi.org/10.1007/978-1-4419-6281-2_1.

Kittel, C. (1947). The nature and development of operations research. *Science, 105*(2719), 150–153. https://doi.org/10.1126/science.105.2719.150.

Kopainsky, B., & Luna-Reyes, L. F. (2008). Closing the loop: Promoting synergies with other theory building approaches to improve system dynamics practice. *Systems Research and Behavioral Science, 25*, 471–486. https://doi.org/10.1002/sres.913.

Kotiadis, K., & Mingers, J. (2006). Combining PSMs with hard OR methods: The philosophical and practical challenges. *Journal of the Operational Research Society, 57*(7), 856–867. https://doi.org/10.1057/palgrave.jors.2602147.

Kotiadis, K., Tako, A. A., Rouwette, E. A. J. A., Vasilakis, C., Brennan, J., Gandhi, P., Wegstapel, H., Sagias, F., & Webb, P. (2013). Using a model of the performance measures in Soft Systems Methodology (SSM) to take action: A case study in health care. *Journal of the Operational Research Society, 64*(1), 125–137. https://doi.org/10.1057/jors.2012.21.

Kuhn, T. S. (1962). *The structure of scientific revolutions*. University of Chicago Press: Chicago.

Kunc, M. (2021). Soft operational research (OR). *Journal of Operational Research Society*. https://www.tandfonline.com/journals/tjor20/collections/SoftOperationalResearch(OR).

Kunz, W., & Rittel, H. W. J. (1970). Issues as Elements of Information Systems. (Working Paper No. 131). Retrieved 20th November 2013, from http://magrawal.myweb.usf.edu/phd/articles/ibis_wp_70.pdf.

Kurtz, C. F., & Snowden, D. J. (2003). The new dynamics of strategy: Sense-making in a complex and complicated world. *IBM Systems Journal, 42*(3), 462–483. https://doi.org/10.1147/sj.423.0462.

Kutsch, E., Maylor, H., Weyer, B., & Lupson, J. (2011). Performers, trackers, lemmings and the lost: Sustained false optimism in forecasting project outcomes - Evidence from a quasi-experiment. *International Journal of Project Management, 29*(8), 1070–1081. https://doi.org/10.1016/j.ijproman.2011.01.010.

Kysar, D. A., & McGarity, T. O. (2006). Did NEPA drown New Orleans? The levees, the blame game, and the hazards of hindsight. *Duke Law Journal, 56*(1), 179–235.

Lami, I. M., & Todella, E. (2022). A multi-methodological combination of the strategic choice approach and the analytic network process: From facts to values and vice versa. *European Journal of Operational Research*. https://doi.org/10.1016/j.ejor.2022.10.029.

Lami, I. M., & White, L. (2022). Are PSMs relevant in a digital age? Towards an ethical dimension. In S. Salhi & J. Boylan (Eds.), *The Palgrave Handbook of Operations Research* (pp. 781–801). Springer International Publishing: Cham. https://doi.org/10.1007/978-3-030-96935-6_24.

Lami, I. M., Abastante, F., Bottero, M., Masala, E., & Pensa, S. (2014). Integrating multicriteria evaluation and data visualization as a problem structuring approach to support territorial transformation projects. *EURO Journal on Decision Processes, 2*(3-4), 281–312. https://doi.org/10.1007/s40070-014-0033-x.

Landry, M. (1995). A note on the concept of 'problem'. *Organization Studies, 16*(2), 315–343. https://doi.org/10.1177/017084069501600206.

Landström, C., Whatmore, S. J., Lane, S. N., Odoni, N. A., Ward, N., & Bradley, S. (2011). Coproducing flood risk knowledge: Redistributing expertise in critical 'participatory modelling'. *Environment and Planning A, 43*(7), 1617–1633. https://doi.org/10.1068/a43482.

Lane, D. C. (1999). Social theory and system dynamics practice. *European Journal of Operational Research, 113*(3), 501–527. https://doi.org/10.1016/S0377-2217(98)00192-1.

Lane, D. C. (2000). Diagramming conventions in system dynamics. *Journal of the Operational Research Society, 51*(2), 241–245. https://doi.org/10.1057/palgrave.jors.2600864.

Lane, D. C. (2008). The emergence and use of diagramming in system dynamics: A critical account. *Systems Research and Behavioral Science, 25*(1), 3–23. https://doi.org/10.1002/Sres.826.

Lane, D. C. (2022). Fons et origo: Reflections on the 60th anniversary of Industrial Dynamics. *System Dynamics Review, 38*(3), 292–324. https://doi.org/10.1002/sdr.1717.

Lane, D. C., & Oliva, R. (1998). The greater whole: Towards a synthesis of system dynamics and soft systems methodology. *European Journal of Operational Research, 107*(1), 214–235. https://doi.org/10.1016/S0377-2217(97)00205-1.

Lane, S. N., Odoni, N., Landstrom, C., Whatmore, S. J., Ward, N., & Bradley, S. (2011). Doing flood risk science differently: An experiment in radical scientific method. *Transactions of the Institute of British Geographers, 36*(1), 15–36. https://doi.org/10.1111/j.1475-5661.2010.00410.x.

Lane, D. C., Munro, E., & Husemann, E. (2016). Blending systems thinking approaches for organisational analysis: Reviewing child protection in England. *European Journal of Operational Research, 251*(2), 613–623. https://doi.org/10.1016/j.ejor.2015.10.041.

Laouris, Y., & Michaelides, M. (2018). Structured Democratic Dialogue: An application of a mathematical problem structuring method to facilitate reforms with local authorities in Cyprus. *European Journal of Operational Research, 268*(3), 918–931. https://doi.org/10.1016/j.ejor.2017.04.039.

Latour, B. (1987). *Science in action: how to follow scientists and engineers through society*. Open University Press: Milton Keynes.

Lawrence, J. R. (1966a). Editorial. *Journal of the Operational Research Society, 17*(4), 331–334. https://doi.org/10.1057/jors.1966.69.

Lawrence, J. R. (1966b). *Operational research and the social sciences*. Tavistock: London.

Lawson, H. (2011). Systems Thinking and its Application: An Introduction. 9. Retrieved 2nd December 2019, from https://docs.google.com/viewer?a=v&pid=sites&srcid=ZGVmYXVsdGRvbWFpbnxzeXNzY2l3Z3xneDo3MzRlMWQyMDZmODdjMTA1.

Lazarus, R. J. (2009). Super wicked problems and climate change: Restraining the present to liberate the future. *Cornell Law Review*, *94*(5), 1153–1234. http://cornelllawreview.org/articles/super-wicked-problems-and-climate-change-restraining-the-present-to-liberate-the-future/.

Lehaney, B., & Paul, R. J. (1996). The use of soft systems methodology in the development of a simulation of out-patient services at Watford General Hospital. *Journal of the Operational Research Society*, *47*(7), 864–870. https://doi.org/10.1057/jors.1996.112.

Leong, C., Soon, D. W. X., Ong, C., & Howlett, M. (2023). Problems in applying Soft OR methods to climate actions: Lessons from two cases of governmental use. *npj Climate Action*, *2*(1), 1. https://doi.org/10.1038/s44168-023-00037-6.

Lesso, W. (2009). The essence of what attracted me to O.R. *ORMS Today*, *42*(3). https://doi.org/10.1287/orms.2009.03.09.

Little, D. (2011). Causal mechanisms in the social realm. In P. M. Illari, F. Russo, & J. Williamson (Eds.), *Causality in the Sciences* (pp. 273–316). Oxford University Press: Oxford.

Lombardi, P. (2023). Environmental, social and inter-generational justice at territorial level: The contribution of planning evaluation and multicriteria decision theory. In G. Napoli, G. Mondini, A. Oppio, P. Rosato, & S. Barbaro (Eds.), *Values, Cities and Migrations. Green Energy and Technology* (pp. 9–17): Springer Science and Business Media Deutschland GmbH.

Lowe, D. (2020). *SCOOPs model pathologies*. [Personal Communication] 11th September 2020.

Lowe, D., & Yearworth, M. (2019). Response to viewpoint: Whither problem structuring methods (PSMs)? *Journal of the Operational Research Society*, *70*(8), 1393–1395. https://doi.org/10.1080/01605682.2018.1502629.

Lowe, D., Martingale, L., & Yearworth, M. (2016). Guiding interventions in a multi-organisational context: Combining the Viable System Model and Hierarchical Process Modelling for use as a Problem Structuring Method. *Journal of the Operational Research Society*, *67*(12), 1481–1495. https://doi.org/10.1057/jors.2016.6.

Lowe, D., Espinosa, A., & Yearworth, M. (2020). Constitutive rules for guiding the use of the viable system model: Reflections on practice. *European Journal of Operational Research*, *287*(3), 1014–1035. https://doi.org/10.1016/j.ejor.2020.05.030.

Luna-Reyes, L. F., & Andersen, D. L. (2003). Collecting and analyzing qualitative data for system dynamics: Methods and models. *System Dynamics Review, 19*(4), 271–296. https://doi.org/10.1002/Sdr.280.

MacDonald, A. D. J. (2010). *On the foundations of performance management* [PhD, Civil Engineering, University of Bristol].

Mantere, S., & Ketokivi, M. (2013). Reasoning in organization science. *Academy of Management Review, 38*(1), 70–89. https://doi.org/10.5465/amr.2011.0188.

Marashi, S. E. (2006). *Managing Discourse and Uncertainty for Decision-making in Civil and Infrastructure Engineering Systems* [PhD, Civil Engineering, University of Bristol].

Marashi, S. E., & Davis, J. (2006). An argumentation-based method for managing complex issues in design of infrastructural systems. *Reliability Engineering and System Safety, 91*(12), 1535–1545. https://doi.org/10.1016/j.ress.2006.01.013.

Marashi, S. E., & Davis, J. (2007). A systems-based approach for supporting discourse in decision making. *Computer-Aided Civil and Infrastructure Engineering, 22*, 511–526.

Marashi, S. E., Davis, J., & Hall, J. (2008). Combination methods and conflict handling in evidential theories. *International Journal of Uncertainty Fuzziness and Knowledge-Based Systems, 16*(3), 337–369. https://doi.org/10.1142/s0218488508005315.

Martin, J. (2011). Joint Workshop of INCOSE & ISSS. 16. Retrieved 2nd December 2019, from https://docs.google.com/viewer?a=v&pid=sites&srcid=ZGVmYXVsdGRvbWFpbnxzeXNzY2l3Z3xneDozYTc5NDU0MjQ4ZTU4NmQ2.

Marttunen, M., Lienert, J., & Belton, V. (2017). Structuring problems for Multi-Criteria Decision Analysis in practice: A literature review of method combinations. *European Journal of Operational Research, 263*(1), 1–17. https://doi.org/10.1016/j.ejor.2017.04.041.

Maskrey, S. A., Mount, N. J., Thorne, C. R., & Dryden, I. (2016). Participatory modelling for stakeholder involvement in the development of flood risk management intervention options. *Environmental Modelling and Software, 82*, 275–294. https://doi.org/10.1016/j.envsoft.2016.04.027.

McCardle-Keurentjes, M., & Rouwette, E. A. J. A. (2018). Asking questions: A sine qua non of facilitation in decision support? *Group Decision and Negotiation*. https://doi.org/10.1007/s10726-018-9573-y.

McCardle-Keurentjes, M. H. F., Rouwette, E. A. J. A., Vennix, J. A. M., & Jacobs, E. (2018). Potential benefits of model use in group model building: Insights from an experimental investigation. *System Dynamics Review, 34*(1–2), 354–384. https://doi.org/10.1002/sdr.1603.

Meadows, D. (2008). *Thinking in systems: a primer*. Chelsea Green Publishing.

Menand, L. (2001). *The metaphysical club*. Farrar, Straus & Giroux: New York.

Mendoza, G. A., & Prabhu, R. (2006). Participatory modeling and analysis for sustainable forest management: Overview of soft system dynamics models and applications. *Forest Policy and Economics*, *9*, 179–196. https://doi.org/10.1016/j.forpol.2005.06.006.

Metcalfe, M. (2008). Pragmatic inquiry. *Journal of the Operational Research Society*, *59*(8), 1091–1099. https://doi.org/10.1057/palgrave.jors.2602443.

Midgley, G. (2000). *Systemic intervention: philosophy, methodology, and practice*. Kluwer Academic/Plenum: New York.

Midgley, G., & Ochoa-Arias, A. E. (2004). *Community operational research: OR and systems thinking for community development*. Kluwer Academic/Plenum: New York.

Midgley, G., Munlo, I., & Brown, M. (1998). The theory and practice of boundary critique: Developing housing services for older people. *Journal of the Operational Research Society*, *49*(5), 467–478. https://doi.org/10.1057/palgrave.jors.2600531.

Midgley, G., Cavana, R., Brocklesby, J., Foote, J., Wood, D., & Ahuriri-Driscoll, A. (2013). Towards a new framework for evaluating systemic problem structuring methods. *European Journal of Operational Research*, *229*(1), 143–154. https://doi.org/10.1016/j.ejor.2013.01.047.

Miller, G. A. (1956). The magical number seven, plus or minus two: Some limits on our capacity for processing information. *Psychological Review*, *63*, 81–97. https://doi.org/10.1037/h0043158.

Mingers, J. (2000). An idea ahead of its time: The history and development of soft systems methodology. *Systemic Practice and Action Research*, *13*(6), 733–755. https://doi.org/10.1023/A:1026475428221.

Mingers, J. (2001). Multimethodology - Mixing and matching methods. In J. Rosenhead & J. Mingers (Eds.), *Rational Analysis for Problematic World Revisited* (pp. 289–310). John Wiley & Sons: Chichester.

Mingers, J. (2011). Soft OR comes of age-but not everywhere! *Omega*, *39*(6), 729–741. https://doi.org/10.1016/j.omega.2011.01.005.

Mingers, J., & Brocklesby, J. (1997). Multimethodology: Towards a framework for mixing methodologies. *Omega*, *25*(5), 489–509. https://doi.org/10.1016/s0305-0483(97)00018-2.

Mingers, J., & Gill, A. (1997). *Multimethodology: the theory and practice of integrating management science methodologies*. John Wiley & Sons: Chichester.

Mingers, J., & Rosenhead, J. (2004). Problem structuring methods in action. *European Journal of Operational Research*, *152*(3), 530–554. https://doi.org/10.1016/s0377-2217(03)00056-0.

Mingers, J., & White, L. (2010). A review of the recent contribution of systems thinking to operational research and management science. *European Journal of Operational Research*, *207*(3), 1147–1161. https://doi.org/10.1016/j.ejor.2009.12.019.

Mitchell, R. K., Agle, B. R., & Wood, D. J. (1997). Toward a theory of stakeholder identification and salience: Defining the principle of who and what really counts. *Academy of Management Review, 22*(4), 853–886. https://doi.org/10.5465/AMR.1997.9711022105.

Moallemi, E. A., Zare, F., Hebinck, A., Szetey, K., Molina-Perez, E., Zyngier, R. L., Hadjikakou, M., Kwakkel, J., Haasnoot, M., Miller, K. K., Groves, D. G., Leith, P., & Bryan, B. A. (2023). Knowledge co-production for decision-making in human-natural systems under uncertainty. *Global Environmental Change, 82*, 102727. https://doi.org/10.1016/j.gloenvcha.2023.102727.

Monks, T. (2016). Operational research as implementation science: Definitions, challenges and research priorities. *Implementation Science, 11*(1), Article 81. https://doi.org/10.1186/S13012-016-0444-0.

Morecroft, J. (2007). *Strategic modelling and business dynamics: a feedback systems approach*. John Wiley & Sons: Chichester.

Morozov, E. (2013). *To save everything, click here: technology, solutionism and the urge to fix problems that don't exist*. Allen Lane: London.

Morozov, E. (2023). The problem with artificial intelligence? It's neither artificial nor intelligent. *The Guardian*.

Morrison, J. B. (2003). *Co-evolution of process and content in organizational change: explaining the dynamics of start and fizzle* [PhD, Sloan School of Management, Massachusetts Institute of Technology]. http://dspace.mit.edu/handle/1721.1/29592.

Morton, A., Ackermann, F., & Belton, V. (2007). Problem structuring without workshops? Experiences with distributed interaction within a PSM process. *Journal of the Operational Research Society, 58*(5), 547–556. https://doi.org/10.1057/palgrave.jors.2602210.

Müller, M. O., Groesser, S. N., & Ulli-Beer, S. (2012). How do we know who to include in collaborative research? Toward a method for the identification of experts. *European Journal of Operational Research, 216*(2), 495–502. https://doi.org/10.1016/j.ejor.2011.07.014.

Mural. (2023). *2023 Collaboration Trends Report*.

National Academy of Engineering. (2017). NAE Grand Challenges for Engineering. Retrieved 10th January 2022, from https://www.nae.edu/File.aspx?id=187214.

Naughton, J. (1977). *The Checkland methodology: A reader's guide*. Open University: Milton Keynes.

Naugle, A., Langarudi, S., & Clancy, T. (2023). What is System Dynamics? Defining Characteristics and the Opportunities they Create. Retrieved 26th July 2023, from https://arxiv.org/abs/2307.11801.

Neely, A. (2009). Exploring the financial consequences of the servitization of manufacturing. *Operations Management Research, 1*(2), 103–118. https://doi.org/10.1007/s12063-009-0015-5.

Nelson, L. K. (2020). Computational grounded theory: A methodological framework. *Sociological Methods and Research, 49*(1), 3–42. https://doi.org/10.1177/0049124117729703.

Neves, L. P., Dias, L. C., Antunes, C. H., & Martins, A. G. (2009). Structuring an MCDA model using SSM: A case study in energy efficiency. *European Journal of Operational Research, 199*(3), 834–845. https://doi.org/10.1016/j.ejor.2009.01.053.

Newberry, P., & Carhart, N. (2023). Constructing causal loop diagrams from large interview data sets. *System Dynamics Review*. https://doi.org/10.1002/sdr.1745.

Nicholson, D. J., & Dupré, J. (2018). A manifesto for a processual philosophy of biology. In D. J. Nicholson & J. Dupré (Eds.), *Everything Flows: Towards a Processual Philosophy of Biology* (pp. 3–45). OUP: Oxford. https://doi.org/10.1093/oso/9780198779636.003.0001.

Nicolini, D. (2013). *Practice theory, work, and organization: an introduction* (1st ed.). Oxford University Press: Oxford.

Nikolova, N., & Devinney, T. (2012). The nature of client-consultant interaction: A critical review. In M. Kipping & T. Clark (Eds.), *The Oxford Handbook of Management Consulting* (pp. 389–409). OUP: Oxford, UK.

O'Brien, F., & Meadows, M. (2007). Developing a visioning methodology: Visioning Choices for the future of operational research. *Journal of the Operational Research Society, 58*(5), 557–575. https://doi.org/10.1057/palgrave.jors.2602259.

Oliveira, D. F. D., & Souza, R. G. D. (2023). Life cycle sustainability impact categories for sustainable procurement. *Journal of Cleaner Production, 383*, Article 135448. https://doi.org/10.1016/j.jclepro.2022.135448.

OpenAI. (2023). GPT-4 Technical Report. Retrieved 15th April 2023, from https://arxiv.org/abs/2303.08774.

Ormerod, R. J. (1997). The design of organisational intervention: Choosing the approach. *Omega, 25*(4), 415–435. https://doi.org/10.1016/S0305-0483(97)00006-6.

Ormerod, R. J. (2005). Putting soft or methods to work: The case of IS strategy development for the UK Parliament. *Journal of the Operational Research Society, 56*(12), 1379–1398. https://doi.org/10.1057/palgrave.jors.2601963.

Ormerod, R. J. (2006). The history and ideas of pragmatism. *Journal of the Operational Research Society, 57*(8), 892–909. https://doi.org/10.1057/palgrave.jors.2602065.

Ormerod, R. J. (2008). The transformation competence perspective. *Journal of the Operational Research Society, 59*(11), 1435–1448. https://doi.org/10.1057/palgrave.jors.2602482.

Ormerod, R. J. (2014). OR competences: The demands of problem structuring methods. *EURO Journal on Decision Processes, 2*(3–4), 313–340. https://doi.org/10.1007/s40070-013-0021-6.

Ormerod, R. J. (2020a). The history and ideas of sociological functionalism: Talcott Parsons, modern sociological theory, and the relevance for OR. *Journal of the*

Operational Research Society, *71*(12), 1873–1899. https://doi.org/10.1080/01605682.2019.1640590.

Ormerod, R. J. (2020b). The pragmatic logic of OR consulting practice: Towards a foundational view. *Journal of the Operational Research Society*, *71*(11), 1691–1709. https://doi.org/10.1080/01605682.2019.1630331.

Ormerod, R. J. (2022). The economic logic of OR consulting practice: Towards a foundational view. *Systems Research and Behavioral Science*, *39*(4), 685–707. https://doi.org/10.1002/sres.2816.

Ormerod, R. J. (2023). The logic of semiotics applied to mathematical and social interaction in operational research consulting practice: Towards a foundational view. *Systems Research and Behavioral Science*, *40*(1), 16–42. https://doi.org/10.1002/sres.2828.

Ormerod, R. J., & Ulrich, W. (2013). Operational research and ethics: A literature review. *European Journal of Operational Research*, *228*(2), 291–307. https://doi.org/10.1016/j.ejor.2012.11.048.

Ormerod, R. J., Yearworth, M., & White, L. (2023). Understanding participant actions in OR interventions using practice theories: A research agenda. *European Journal of Operational Research*, *306*(2), 810–827. https://doi.org/10.1016/j.ejor.2022.08.030.

Oteros-Rozas, E., Martín-López, B., Daw, T. M., Bohensky, E. L., Butler, J. R. A., Hill, R., Martin-Ortega, J., Quinlan, A., Ravera, F., Ruiz-Mallén, I., Thyresson, M., Mistry, J., Palomo, I., Peterson, G. D., Plieninger, T., Waylen, K. A., Beach, D. M., Bohnet, I. C., Hamann, M., Hanspach, J., Hubacek, K., Lavorel, S., & Vilardy, S. P. (2015). Participatory scenario planning in place-based social-ecological research: Insights and experiences from 23 case studies. *Ecology and Society*, *20*(4), Article 32. https://doi.org/10.5751/ES-07985-200432.

Paay, J., Kjeldskov, J., & O'Hara, K. (2011). BISi: A blended interaction space. Conference on Human Factors in Computing Systems (CHI 2011), Vancouver, BC, Canada. https://doi.org/10.1145/1979742.1979644.

Pahl-Wostl, C. (2007). The implications of complexity for integrated resources management. *Environmental Modelling and Software*, *22*(5), 561–569. https://doi.org/10.1016/j.envsoft.2005.12.024.

Pamplona, D. A., & Alves, C. J. P. (2020). An overview of air delay: A case study of the Brazilian scenario. *Transportation Research Interdisciplinary Perspectives*, *7*, Article 100189. https://doi.org/10.1016/j.trip.2020.100189.

Parry, R., & Mingers, J. (1991). Community operational research: Its context and its future. *Omega*, *19*(6), 577–586. https://doi.org/10.1016/0305-0483(91)90008-H.

Paucar-Caceres, A. (2010). Mapping the changes in management science: A review of 'soft' OR/MS articles published in Omega (1973–2008). *Omega*, *38*(1), 46–56. https://doi.org/10.1016/j.omega.2009.04.001.

Paucar-Caceres, A., & Espinosa, A. (2011). Management science methodologies in environmental management and sustainability: Discourses and applications.

Journal of the Operational Research Society, 62(9), 1601–1620. https://doi.org/10.1057/jors.2010.110.

Pawson, R., & Tilley, N. (1997). *Realistic evaluation*. Sage Publications: London.

Perrone, A., Inam, A., Albano, R., Adamowski, J., & Sole, A. (2020). A participatory system dynamics modeling approach to facilitate collaborative flood risk management: A case study in the Bradano River (Italy). *Journal of Hydrology, 580*, Article 124354. https://doi.org/10.1016/j.jhydrol.2019.124354.

Petropoulos, F., Laporte, G., Aktas, E., Alumur, S. A., Archetti, C., Ayhan, H., Battarra, M., Bennell, J. A., Bourjolly, J.-M., Boylan, J., Breton, M., Canca, D., Charlin, L., Chen, B., Cicek, C. T., Cox, L. A. Jr., Currie, C. S. M., Demeulemeester, E., Ding, L., Disney, S. M., Ehrgott, M., Eppler, M. J., Erdoğan, G., Fortz, B., Franco, L. A., Frische, J., Greco, S., Gregory, A. J., Hämäläinen, R. P., Herroelen, W., Hewitt, M., Holmström, J., Hooker, J. N., Işık, T., Johnes, J., Kara, B. Y., Karsu, Ö., Kent, K., Köhler, C., Kunc, M., Kuo, Y.-H., Letchford, A. N., Leung, J., Li, D., Li, H., Lienert, J., Ljubić, I., Lodi, A., Lozano, S., Lurkin, V., Martello, S., McHale, I. G., Midgley, G., Morecroft, J., Mutha, A., Oğuz, C., Petrovic, S., Pferschy, U., Psaraftis, H. N., Rose, S., Saarinen, L., Salhi, S., Song, J.-S., Sotiros, D., Stecke, K. E., Strauss, A. K., Tarhan, İ., Thielen, C., Toth, P., Vanden Berghe, G., Vasilakis, C., Vaze, V., Vigo, D., Virtanen, K., Wang, X., Weron, R., White, L., Van Woensel, T., Yearworth, M., Alper Yıldırım, E., Zaccour, G., & Zhao, X. (2024). Operational research: Methods and applications. *Journal of the Operational Research Society, 75*(3), 423–617. https://doi.org/10.1080/01605682.2023.2253852.

Pidd, M. (2004). *Systems modelling: theory and practice*. John Wiley & Sons: Chichester.

Pidd, M. (2009). *Tools for thinking: modelling in management science* (3rd ed.). John Wiley & Sons: Chichester.

Pluchinotta, I., Salvia, G., & Zimmermann, N. (2022). The importance of eliciting stakeholders' system boundary perceptions for problem structuring and decision-making. *European Journal of Operational Research, 302*(1), 280–293. https://doi.org/10.1016/j.ejor.2021.12.029.

Pollack, J. (2009). Multimethodology in series and parallel: Strategic planning using hard and soft OR. *Journal of the Operational Research Society, 60*(2), 156–167. https://doi.org/10.1057/palgrave.jors.2602538.

Poole, M. S., & van de Ven, A. H. (1989). Using paradox to build management and organization theories. *Academy of Management Review, 14*(4), 562–578. https://doi.org/10.2307/258559.

Popper, K. (1963). *Conjectures and Refutations. The growth of scientific knowledge. (Essays and Lectures)*. Routledge & Kegan Paul: London.

Powell, J. H., & Coyle, R. G. (1997). A network-based futures method for strategic business planning. *Journal of the Operational Research Society, 48*(8), 793–803. https://doi.org/10.1057/palgrave.jors.2600437.

Powell, J. H., & Mustafee, N. (2017). Widening requirements capture with soft methods: An investigation of hybrid M&S studies in health care. *Journal of the Operational Research Society*, *68*(10), 1211–1222. https://doi.org/10.1057/s41274-016-0147-6.

Rabinow, P. (2011). Dewey and Foucault: What's the problem? *Foucault Studies*, (11), 11–19. https://doi.org/10.22439/fs.v0i11.3202.

Raitt, R. A. (1979). O.R. and science. *Journal of the Operational Research Society*, *30*(9), 835–836. https://doi.org/10.1057/jors.1979.195.

Rand, G. K. (2024). From Operational Research Quarterly to Journal of the Operational Research Society: The early editors. *Journal of the Operational Research Society*, *75*, 3–12. https://doi.org/10.1080/01605682.2023.2244339.

Ranyard, J., Fildes, R., & Hu, T.-I. (2015). Reassessing the scope of OR practice: The influences of problem structuring methods and the analytics movement. *European Journal of Operational Research*, *245*(1), 1–13. https://doi.org/10.1016/j.ejor.2015.01.058.

Rauschmayer, F., Kavathatzopoulos, I., Kunsch, P. L., & Le Menestrel, M. (2009). Why good practice of OR is not enough—Ethical challenges for the OR practitioner. *Omega*, *37*(6), 1089–1099. https://doi.org/10.1016/j.omega.2008.12.005.

REPLICATE Project Consortium. (2015). *H2020-SCC-2015 Smart Cities and Communities Innovation Action (IA) Project no. 691735 - REnaissance of Places with Innovative Citizens (REPLICATE). Project Description of Work* [Grant]. https://cordis.europa.eu/project/id/691735.

Reynolds, E. (2021). *Discovering Everyday Problem Structuring* [PhD, Warwick Business School, University of Warwick]. https://wrap.warwick.ac.uk/178300.

Reynolds, E. (2023a). PS as a commonplace everyday activity. 65th Conference of the UK Operational Research Society (OR65), University of Bath, UK.

Reynolds, E. (2023b). When our backs are turned: Exploring how people structure problems within an everyday workplace. UK OR Society Problem Structuring Methods (PSM) Special Interest Group (SIG),October 2023, https://www.youtube.com/watch?v=RTpZ46duuZI.

Reynolds, M., & Holwell, S. (2010). *Systems approaches to managing change: a practical guide*. Springer: London.

Rhodes, C. (2016). Democratic business ethics: Volkswagen's emissions scandal and the disruption of corporate sovereignty. *Organization Studies*, *37*(10), 1501–1518. https://doi.org/10.1177/0170840616641984.

Rhyne, R. (1981). Whole-pattern futures projection, using field anomaly relaxation. *Technological Forecasting and Social Change*, *19*(4), 331–360. https://doi.org/10.1016/0040-1625(81)90005-6.

Rhyne, R. (1994). Futures assessment by field anomaly relaxation. A review and appraisal. *Futures*, *26*(1), 25–43. https://doi.org/10.1016/0016-3287(94)90088-4.

Rhyne, R. (1995). Field anomaly relaxation - The arts of usage. *Futures*, *27*(6), 657–674. https://doi.org/10.1016/0016-3287(95)00032-R.

Riley, D. (2012). We've been framed! Ends, means, and the ethics of the grand(iose) challenges. *International Journal of Engineering, Social Justice, and Peace*, *1*(2), 123–136. http://library.queensu.ca/ojs/index.php/IJESJP/article/view/4307.

Ritchey, T. (2006). Problem structuring using computer-aided morphological analysis. *Journal of the Operational Research Society*, *57*(7), 792–801. https://doi.org/10.1057/palgrave.jors.2602177.

Rittel, H., & Webber, M. (1973). Dilemmas in a general theory of planning. *Policy Sciences*, *4*(2), 155–169. https://doi.org/10.1007/BF01405730.

Rivero, N. P. P., Morais, D. C., & De Sousa Pereira, L. (2020). Assessment of actions to tackle the shortages of water in La Paz, Bolivia. *Water Policy*, *22*(2), 177–192. https://doi.org/10.2166/wp.2020.087.

Rooksby, J., Sommerville, I., & Pidd, M. (2006). A hybrid approach to upstream requirements: IBIS and cognitive mapping. In *Rationale Management in Software Engineering* (pp. 137–154). https://doi.org/10.1007/978-3-540-30998-7_6.

Rose, M. E., & Kitchin, J. R. (2019). pybliometrics: Scriptable bibliometrics using a Python interface to Scopus. *SoftwareX*, *10*. https://doi.org/10.1016/j.softx.2019.100263.

Rosenberg, G. (2003). *Systems-based approaches for managing environmental risk* [PhD, Civil Engineering, University of Bristol].

Rosenberg, G., Carhart, N., Edkins, A. J., & Ward, J. A. (2014). Development of a Proposed Interdependency Planning and Management Framework. https://doi.org/10.14324/20141455020.

Rosenhead, J. (1978). Operational research in health services planning. *European Journal of Operational Research*, *2*(2), 75–85. https://doi.org/10.1016/0377-2217(78)90103-0.

Rosenhead, J. (1986). Custom and practice. *Journal of the Operational Research Society*, *37*(4), 335–343. https://doi.org/10.1057/jors.1986.61.

Rosenhead, J. (1989a). Diverse untiy: The principles and prospects for problem structuring methods. In J. Rosenhead (Ed.), *Rational analysis for a problematic world: problems structuring methods for complexity, uncertainty and conflict* (pp. 341–358). John Wiley & Sons: Chichester.

Rosenhead, J. (1989b). Introduction: Old and new paradigms of analysis. In J. Rosenhead (Ed.), *Rational analysis for a problematic world: problems structuring methods for complexity, uncertainty and conflict* (pp. 1–20). John Wiley & Sons: Chichester.

Rosenhead, J. (1989c). *Rational analysis for a problematic world: problem structuring methods for complexity, uncertainty and conflict*. John Wiley & Sons: Chichester.

Rosenhead, J. (1992). Into the swamp - The analysis of social-issues. *Journal of the Operational Research Society*, *43*(4), 293–305. https://doi.org/10.1057/jors.1992.44.

Rosenhead, J. (1996). What's the problem? An introduction to problem structuring methods. *Interfaces*, *26*(6), 117–131. https://doi.org/10.1287/inte.26.6.117.

Rosenhead, J., & Mingers, J. (2001a). A new paradigm for analysis. In J. Rosenhead & J. Mingers (Eds.), *Rational analysis for a problematic world revisited: problem structuring methods for complexity, uncertainty and conflict* (2nd ed., pp. 1–19). John Wiley & Sons: Chichester.

Rosenhead, J., & Mingers, J. (2001b). *Rational analysis for a problematic world revisited: problem structuring methods for complexity, uncertainty and conflict* (2nd ed.). John Wiley & Sons: Chichester.

Rousseau, D. (2019). A vision for advancing systems science as a foundation for the systems engineering and systems practice of the future. *Systems Research and Behavioral Science*, *36*(5), 621-634. https://doi.org/10.1002/sres.2629.

Rouwette, E., Bastings, I., & Blokker, H. (2011). A comparison of group model building and strategic options development and analysis. *Group Decision and Negotiation*, *20*(6), 781–803. https://doi.org/10.1007/s10726-010-9207-5.

Rowe, G., & Wright, G. (2011). The Delphi technique: Past, present, and future prospects — Introduction to the special issue. *Technological Forecasting and Social Change*, *78*(9), 1487–1490. https://doi.org/10.1016/j.techfore.2011.09.002.

Rowlinson, M. (2023). *Problematising management research*. [Personal Communication] 19th July 2023.

Sage, A. P. (1977). *Methodology for large-scale systems*. McGraw-Hill: New York; London.

Salhi, S., & Boylan, J. (Eds.). (2022). *The palgrave handbook of operations research*. Palgrave Macmillan: Cham. https://doi.org/10.1007/978-3-030-96935-6.

Schad, J., & Bansal, P. (2018). Seeing the forest and the trees: How a systems perspective informs paradox research. *Journal of Management Studies*, *55*(8), 1490–1506. https://doi.org/10.1111/joms.12398.

Schaffernicht, M. (2010). Causal loop diagrams between structure and behaviour: A critical analysis of the relationship between polarity, behaviour and events. *Systems Research and Behavioral Science*, *27*(6), 653–666. https://doi.org/10.1002/sres.1018.

Schmidt, J. C. (2011). What is a problem? *Poiesis und Praxis*, *7*(4), 249–274. https://doi.org/10.1007/s10202-011-0091-0.

Schön, D. A. (1983). *The reflective practitioner*. Basic Books.

Schön, D. A. (1987). *Educating the reflective practitioner*. Jossey-Bass: San Francisco, CA.

Schön, D. A. (1992). The Theory of Inquiry: Dewey's Legacy to Education. *Curriculum Inquiry*, *22*(2), 119–139. https://doi.org/10.1080/03626784.1992.11076093.

Schramm, V. B., & Schramm, F. (2018). An approach for supporting problem structuring in water resources management and planning. *Water Resources Management*, *32*(9), 2955–2968. https://doi.org/10.1007/s11269-018-1966-9.

Schutz, A. (1967). *The phenomenology of the social world Translated by Walsh, G. & Lehnert, F.* Northwestern University Press.

Searle, J. R. (1995). *The construction of social reality*. Allen Lane: London.

Senge, P. M. (1990). *The fifth discipline: the art and practice of the learning organization*. Random House: London.

SERC. (1990). *The Engineering Doctorate. A SERC Working Party report to the Engineering Board of the Science and Engineering Research Council.*

Seyfang, G., & Haxeltine, A. (2012). Growing grassroots innovations: Exploring the role of community-based initiatives in governing sustainable energy transitions. *Environment and Planning C: Government and Policy*, *30*(3), 381–400. https://doi.org/10.1068/c10222.

Sfard, A. (1991). On the dual nature of mathematical conceptions: Reflections on processes and objects as different sides of the same coin. *Educational Studies in Mathematics*, *22*(1), 1–36. https://doi.org/10.1007/BF00302715.

Shaw, D., Westcombe, M., Hodgkin, J., & Montibeller, G. (2004). Problem structuring methods for large group interventions. *Journal of the Operational Research Society*, *55*(5), 453–463. https://doi.org/10.1057/palgrave.jors.2601712.

Shaw, D., Franco, L. A., & Westcombe, M. (2006). Problem structuring methods: New directions in a problematic world. *Journal of the Operational Research Society*, *57*(7), 757–758. https://doi.org/10.1057/palgrave.jors.2602193.

Siau, K., & Tan, X. (2005). Improving the quality of conceptual modeling using cognitive mapping techniques. *Data and Knowledge Engineering*, *55*(3), 343–365. https://doi.org/10.1016/j.datak.2004.12.006.

Sievert, C., & Shirley, K. (2014). LDAvis: A method for visualizing and interpreting topics. Association for Computational Linguistics Workshop on Interactive Language Learning, Visualization, and Interfaces (ACL2014), Baltimore, MA. USA.

Silva, W. D. O., & Fontana, M. E. (2021). Integrative multi-attribute negotiation model to define stakeholders' responsibilities in the reverse flow channel. *Journal of Cleaner Production*, *279*, Article 123752. https://doi.org/10.1016/j.jclepro.2020.123752.

Simchi-Levi, D. (2022). From the Editor. *Operations Research*, *69*(1), 1–6. https://doi.org/10.1287/mnsc.2022.4644.

Simpson, B., & den Hond, F. (2022). The contemporary resonances of classical pragmatism for studying organization and organizing. *Organization Studies*, *43*(1), 127–146. https://doi.org/10.1177/0170840621991689.

Singer, J., Sillitto, H., Bendz, J., Chroust, G., Hybertson, D., Lawson, H. W., Martin, J., Martin, R., Singer, M., & Takaku, T. (2012). The systems praxis framework. 16th IFSR Conversation – Systems and Science at Crossroads, Linz, Austria.

Slack, N., & Brandon-Jones, A. (2019). *Operations management* (9th ed.). Pearson.

Smith, G. F. (1988). Towards a heuristic theory of problem structuring. *Management Science*, *34*(12), 1489–1506. http://www.jstor.org/stable/2632039.

Smith, C., & Shaw, D. (2019). The characteristics of problem structuring methods: A literature review. *European Journal of Operational Research, 274*(2), 403–416. https://doi.org/10.1016/j.ejor.2018.05.003.

Snowden, D. J., & Boone, M. E. (2007). A leaders framework for decision making. *Harvard Business Review, 85*(11), 69–76. https://hbr.org/2007/11/a-leaders-framework-for-decision-making.

Sorby, S., Fortenberry, N. L., & Bertoline, G. (2021). Stuck in 1955, Engineering Education Needs a Revolution. *Issues in Science and Technology*. Retrieved 29th September 2023, from https://issues.org/engineering-education-change-sorby-fortenberry-bertoline/.

Sørensen, C. G., Fountas, S., Nash, E., Pesonen, L., Bochtis, D., Pedersen, S. M., Basso, B., & Blackmore, S. B. (2010). Conceptual model of a future farm management information system. *Computers and Electronics in Agriculture, 72*(1), 37–47. https://doi.org/10.1016/j.compag.2010.02.003.

de Sousa Pereira, L., & Costa Morais, D. (2020). The strategic choice approach to the maintenance management of a water distribution system. *Urban Water Journal, 17*(1), 23–31. https://doi.org/10.1080/1573062X.2020.1734945.

Sowa, J. (1991). *Principles of semantic networks: explorations in the representation of knowledge*. Morgan Kaufmann.

SPES Consulting, & Comune di Firenze. (2017). *REPLICATE Deliverable D4.2 Pilot action measures advancement sheets V1* (EU H2020-SCC-2015 (691735) project REPLICATE (REnaissance of Places with Innovative Citizenship and TEchnology).

STEEP. (2011). *Systems Thinking for Comprehensive City Efficient Energy Planning (Proposal)* (STEEP PROJECT (314277) - Systems Thinking for Comprehensive City Efficient Energy Planning.

Sterman, J. D. (2000). *Business dynamics: systems thinking and modeling for a complex world*. Irwin/McGraw-Hill: Boston, MA.

Stoecker, J. L. (1993). The Biglan classification revisited. *Research in Higher Education, 34*(4), 451–464. https://doi.org/10.1007/BF00991854.

Strauss, A. L., & Corbin, J. (1998). *Basics of qualitative research: techniques and procedures for developing grounded theory* (2nd ed.). Sage Publications: Thousand Oaks, CA.

Strijbos, S. (2017). Systems thinking. In R. Frodeman, J. T. Klein, & R. C. S. Pacheco (Eds.), *The Oxford Handbook of Interdisciplinarity* (2nd ed., pp. 291–302). Oxford University Press: Oxford.

Sydelko, P., Midgley, G., & Espinosa, A. (2021). Designing interagency responses to wicked problems: Creating a common, cross-agency understanding. *European Journal of Operational Research, 294*(1), 250–263. https://doi.org/10.1016/j.ejor.2020.11.045.

Taket, A. (2002). Facilitation: Some contributions to theorising the practice of operational research. *Journal of the Operational Research Society*, *53*(2), 126–136. https://doi.org/10.1057/palgrave.jors.2601209.

Tako, A., & Kotiadis, K. (2015). PartiSim: A multi-methodology framework to support facilitated simulation modelling in healthcare. *European Journal of Operational Research*, *244*(2), 555–564. https://doi.org/10.1016/j.ejor.2015.01.046.

Tako, A. A., Kotiadis, K., & Vasilakis, C. (2010). A conceptual modelling framework for stakeholder participation in simulation studies. Proceedings of the 2010 Operational Research Society Simulation Workshop, SW 2010.

Talisse, R. (2013). *Robert Talisse on Pragmatism* [Interview]. https://fivebooks.com/best-books/robert-talisse-on-pragmatism/.

Tavella, E., & Lami, I. (2019). Negotiating perspectives and values through soft OR in the context of urban renewal. *Journal of the Operational Research Society*, *70*(1), 136–161. https://doi.org/10.1080/01605682.2018.1427433.

Tavella, E., & Papadopoulos, T. (2015a). Expert and novice facilitated modelling: A case of a Viable System Model workshop in a local food network. *Journal of the Operational Research Society*, *66*(2), 247–264. https://doi.org/10.1057/jors.2013.187.

Tavella, E., & Papadopoulos, T. (2015b). Novice facilitators and the use of scripts for managing facilitated modelling workshops. *Journal of the Operational Research Society*, *66*(12), 1967–1988. https://doi.org/10.1057/jors.2015.7.

Taylor, R., & Tofts, C. (2008). *Taking a RaSP to Enterprise Stakeholder Dissonance*. https://www.hpl.hp.com/techreports/2008/HPL-2008-82.html.

Taylor, R., Tofts, C., & Yearworth, M. (2004). *Open Analytics*. https://www.hpl.hp.com/techreports/2004/HPL-2004-138R1.pdf.

Taylor, R. H., van Wijk, L. G. A., May, J. H. M., & Carhart, N. J. (2015). A study of the precursors leading to 'organisational' accidents in complex industrial settings. *Process Safety and Environmental Protection*, *93*, 50–67. https://doi.org/10.1016/j.psep.2014.06.010.

The Operational Research Society. (2023). *About OR*. Retrieved 10th October 2023 from https://www.theorsociety.com/about-or/.

The Royal Academy of Engineering. (2010). *Philosophy of Engineering: Volume 1 of the proceedings of a series of seminars held at The Royal Academy of Engineering*.

Tippin, M., Kalbach, J., & Chin, D. (2018). *The definitive guide to facilitating remote workshops: Insights, tools, and case studies from digital-first companies and expert facilitators*.

Todella, E., Lami, I. M., & Armando, A. (2018). Experimental use of strategic choice approach (SCA) by individuals as an architectural design tool. *Group Decision and Negotiation*. https://doi.org/10.1007/s10726-018-9567-9.

Tully, P., White, L., & Yearworth, M. (2019). The value paradox of problem structuring methods. *Systems Research and Behavioral Science*, *36*(4), 424–444. https://doi.org/10.1002/sres.2557.

Turner, B. L., & Goodman, M. (2023). Capturing the science behind the craft: A reporting framework to improve quality and confidence in nonsimulated models. *System Dynamics Review*. https://doi.org/10.1002/sdr.1752.

Tutton, R. (2017). Wicked futures: Meaning, matter and the sociology of the future. *Sociological Review*, *65*(3), 478–492. https://doi.org/10.1111/1467-954X.12443.

Ulrich, W. (2003). Beyond methodology choice: Critical systems thinking as critically systemic discourse. *Journal of the Operational Research Society*, *54*(4), 325–342. https://doi.org/10.1057/palgrave.jors.2601518.

Ulrich, W. (2012a). Operational research and critical systems thinking-an integrated perspective Part 1: OR as applied systems thinking. *Journal of the Operational Research Society*, *63*(9), 1228–1247. https://doi.org/10.1057/jors.2011.141.

Ulrich, W. (2012b). Operational research and critical systems thinking-an integrated perspective Part 2: OR as argumentative practice. *Journal of the Operational Research Society*, *63*(9), 1307–1322. https://doi.org/10.1057/jors.2011.145.

University of Bristol, & EPSRC. (2020). Systems Centre: Industrial Doctorate Centre in Systems Impact Report. Retrieved 18th January 2020, from https://uob-my.sharepoint.com/personal/fesac_bristol_ac_uk/Documents/IDC%20in%20Systems%20Impact%20Report.pdf.

Vaughan, D. (1996). *The Challenger launch decision: risky technology, culture, and deviance at NASA*. University of Chicago Press: Chicago; London.

Vaughan, D. (2006). NASA revisited: Theory, analogy, and public sociology. *American Journal of Sociology*, *112*(2), 353–393. https://doi.org/10.1086/506413.

Vaz, A. L. A., Ferreira, F. A. F., Pereira, L. F., Correia, R. J. C., & Banaitis, A. (2022). Strategic visualization: The (real) usefulness of cognitive mapping in smart city conceptualization. *Management Decision*, *60*(4), 916–939. https://doi.org/10.1108/MD-11-2020-1512.

de Ven, A. H. V., & Poole, M. S. (2005). Alternative approaches for studying organizational change. *Organization Studies*, *26*(9), 1377–1404. https://doi.org/10.1177/0170840605056907.

Vennix, J. (1996). *Group model building: facilitating team learning using system dynamics*. John Wiley & Sons: Chichester.

Videira, N., Schneider, F., Sekulova, F., & Kallis, G. (2014). Improving understanding on degrowth pathways: An exploratory study using collaborative causal models. *Futures*, *55*, 58–77. https://doi.org/10.1016/j.futures.2013.11.001.

Vidoni, M. (2022). Beyond Hard and Soft OR: Operational research from a software engineering perspective. *Journal of the Operational Research Society*, *73*(4), 693–715. https://doi.org/10.1080/01605682.2020.1865848.

Voinov, A., Seppelt, R., Reis, S., Nabel, J. E. M. S., & Shokravi, S. (2014). Values in socio-environmental modelling: Persuasion for action or excuse for inaction. *Environmental Modelling & Software*, *53*, 207–212. https://doi.org/10.1016/j.envsoft.2013.12.005.

Wagner, H. (1975). *Principles of operations research: with applications to managerial decisions* (2nd ed.). Prentice-Hall: Englewood Cliffs, NJ; London.

Wang, W., Liu, W., & Mingers, J. (2015). A systemic method for organisational stakeholder identification and analysis using Soft Systems Methodology (SSM). *European Journal of Operational Research, 246*(2), 562–574. https://doi.org/10.1016/j.ejor.2015.05.014.

Warfield, J. N. (1973). Binary matrices in system modeling. *IEEE Transactions on Systems Man and Cybernetics, SMC3*(5), 441–449. https://doi.org/10.1109/TSMC.1973.4309270.

Warfield, J. N. (1974a). Developing interconnection matrices in structural modeling. *IEEE Transactions on Systems, Man and Cybernetics, SMC-4*(1), 81–87. https://doi.org/10.1109/TSMC.1974.5408524.

Warfield, J. N. (1974b). Toward interpretation of complex structural models. *IEEE Transactions on Systems, Man and Cybernetics, 4*(5), 405–417. https://doi.org/10.1109/TSMC.1974.4309336.

Watson, S. (2003). *A Generic Process for Assessing Climate Change Impacts on the Electricity Supply Industry and Utilities* (UKERC GR/S18915/01) [Grant]. Loughborough University.

Watzlawick, P., Weakland, J. H., & Fisch, R. (1974). *Change: principles of problem formation and problem resolutions*. Norton: New York.

de Weck, O. (2016). MIT Engineering Systems Division R. I. P. Eulogy for a successful experiment 1998–2015. *MIT Faculty Newsletter, XXVIII*(4). Retrieved 10th July 2023, from http://web.mit.edu/fnl/volume/284/deweck.html.

de Weck, O. L., Roos, D., & Magee, C. L. (2011). *Engineering systems: meeting human needs in a complex technological world*. MIT Press: Cambridge, MA; London.

White, L. (2002). Size matters: Large group methods and the process of operational research. *Journal of the Operational Research Society, 53*(2), 149–160. https://doi.org/10.1057/palgrave.jors.2601298.

White, L. (2006). Evaluating problem-structuring methods: Developing an approach to show the value and effectiveness of PSMs. *Journal of the Operational Research Society, 57*(7), 842–855. https://doi.org/10.1057/palgrave.jors.2602149.

White, L. (2009). Understanding problem structuring methods interventions. *European Journal of Operational Research, 199*(3), 823–833. https://doi.org/10.1016/j.ejor.2009.01.066.

White, L. (2016). Behavioural operational research: Towards a framework for understanding behaviour in OR interventions. *European Journal of Operational Research, 249*(3), 827–841. https://doi.org/10.1016/j.ejor.2015.07.032.

White, L. (2025). Fifty years of Soft OR. *European Journal of Operational Research*.

White, L., & Bourne, H. (2007). Voices and values: Linking values with participation in OR/MS in public policy making. *Omega, 35*(5), 588–603. https://doi.org/10.1016/j.omega.2005.11.002.

White, L., Yearworth, M., & Burger, K. (2015). Understanding PSM interventions through sense-making and the mangle of practice lens. In B. Kamiński, G. Kersten, & T. Szapiro (Eds.), *15th International Conference on Group Decision and Negotiation (GDN 2015) Warsaw, Poland* (pp. 13–27). Springer: Cham. https://doi.org/10.1007/978-3-319-19515-5_2.

White, L., Burger, K., & Yearworth, M. (2016). Understanding behaviour in problem structuring methods interventions with activity theory. *European Journal of Operational Research*, *249*(3), 983–1004. https://doi.org/10.1016/j.ejor.2015.07.044.

Whitehead, A. N. (1929). *Process and reality: an essay in cosmology*. Cambridge University Press.

Williams, T., Eden, C., Ackermann, F., & Tait, A. (1995). The effects of design changes and delays on project costs. *Journal of the Operational Research Society*, *46*(7), 809–818. https://doi.org/10.1057/jors.1995.114.

Wilson, B. (1984). *Systems: concepts, methodologies and applications*. John Wiley & Sons: Chichester.

Winzer, C. (2012). Conceptualizing energy security. *Energy Policy*, *46*, 36–48. https://doi.org/10.1016/j.enpol.2012.02.067.

Wittenborn, A. K., Rahmandad, H., Rick, J., & Hosseinichimeh, N. (2016). Depression as a systemic syndrome: Mapping the feedback loops of major depressive disorder. *Psychological Medicine*, *46*(3), 551–562. https://doi.org/10.1017/S0033291715002044.

Wolstenholme, E. F. (1999). Qualitative vs quantitative modelling: The evolving balance. *Journal of the Operational Research Society*, *50*(4), 422–428. https://doi.org/10.1057/palgrave.jors.2600700.

Woolley, R. N., & Pidd, M. (1981). Problem structuring—A literature review. *Journal of the Operational Research Society*, *32*(3), 197–206. https://doi.org/10.1057/jors.1981.42.

Yearworth, M. (2011). The role of the systems superviser for an engineering doctorate in systems. 55th Meeting of the International Society for the Systems Sciences, 17th–22nd July 2011, Hull, UK.

Yearworth, M. (2015a). *Healthy Resilient Cities: Building a Business Case for Adaption (NERC NE/N007638/1)* [Grant]. Bristol. https://gotw.nerc.ac.uk/list_full.asp?pcode=NE%2FN007638%2F1.

Yearworth, M. (2015b). PSMs as reliable machines for delivering client results? *OR Society PSM SIG. Dstl Seminar 16th October 2015*. Retrieved 7th January 2023, from https://www.theorsociety.com/media/3564/16thoct_4xsigs_mikeyearworth_12012016205032.pdf.

Yearworth, M. (2016). Sustainability as a 'super-wicked' problem; opportunities and limits for engineering methodology. *Intelligent Buildings International*, *8*(1), 37–47. https://doi.org/10.1080/17508975.2015.1109789.

Yearworth, M. (2020). The theoretical foundation(s) for Systems Engineering? *Systems Research and Behavioral Science*, *37*, 184–187. https://doi.org/10.1002/sres.2667.

Yearworth, M. (2021). Hierarchical process modelling (HPM) in problem structuring. 63rd Conference of the UK Operational Research Society (OR63), 16th September 2021, Southampton, UK. https://hdl.handle.net/10871/126885.

Yearworth, M. (2022). From problem formulation to problem structuring and back. 32nd European Conference on Operational Research (EURO 2022), 3rd–6th July 2022, Espoo, Finland. https://hdl.handle.net/10871/130142.

Yearworth, M. (2024). Towards the Development of a Problem Structuring Platform Supported by Machine Learning. 66th Conference of the UK Operational Research Society (OR66), 11th September 2024, Bangor University.

Yearworth, M., & Edwards, G. (2014). On the desirability of integrating research methods into overall systems approaches in the training of engineers: Analysis using SSM. *Systems Research and Behavioral Science*, *31*(1), 47–66. https://doi.org/10.1002/sres.2167.

Yearworth, M., & White, L. (2013). The uses of qualitative data in multimethodology: Developing causal loop diagrams during the coding process. *European Journal of Operational Research*, *231*(1), 151–161. https://doi.org/10.1016/j.ejor.2013.05.002.

Yearworth, M., & White, L. (2014). The non-codified use of problem structuring methods and the need for a generic constitutive definition. *European Journal of Operational Research*, *237*(3), 932–945. https://doi.org/10.1016/j.ejor.2014.02.015.

Yearworth, M., & White, L. (2015). Soft Engineering: Anticipating an increasingly messy landscape for engineering practice. 1st International Conference on Anticipation (Anticipation 2015), Trento, Italy. https://hdl.handle.net/1983/7357a5b2-272d-41e1-bb24-edfecbc3fbec.

Yearworth, M., & White, L. (2017). Demystifying facilitation: A new approach to investigating the role of facilitation in group decision support processes. In D. Bajwa, S. T. Koeszegi, & R. Vetschera (Eds.), *Group Decision and Negotiation. Theory, Empirical Evidence, and Application: 16th International Conference, GDN 2016, Bellingham, WA, USA, June 20–24, 2016, Revised Selected Papers* (pp. 69–86). Springer International Publishing: Cham. https://doi.org/10.1007/978-3-319-52624-9_6.

Yearworth, M., & White, L. (2018). Spontaneous emergence of Community OR: Self-initiating, self-organising problem structuring mediated by social media. *European Journal of Operational Research*, *268*(3), 809–824. https://doi.org/10.1016/j.ejor.2018.01.024.

Yearworth, M., & White, L. (2020). Group support systems: Experiments with an online system and implications for same-time/different-places working. In D. M. Kilgour & C. Eden (Eds.), *Handbook of Group Decision and Negotiation* (pp. 681–706). Springer International Publishing: Cham. https://doi.org/10.1007/978-3-030-12051-1_48-1.

Yearworth, M., & White, L. (2021). A problem structuring method implemented using a group support system. 31st European Conference on Operational Research (EURO 2021), Athens, Greece. http://hdl.handle.net/10871/37710.

Yearworth, M., Edwards, G., & Rosenberg, G. (2011). Systems practice in engineering: Reflections on teaching research methods and contribution to methodological development. 9th Annual Conference on Systems Engineering Research (CSER 2011), Los Angeles, USA. https://hdl.handle.net/1983/1721.

Yearworth, M., Dunford, C. N., York, D. M., & Godfrey, P. (2012). Critical awareness of worldviews in organisational change. 25th Conference on Operational Research (EURO 2012), Vilnius, Lithuania. https://hdl.handle.net/1983/90c7c97f-c371-4b8d-af8a-29c0f18d43c7.

Yearworth, M., Edwards, G., Davis, J., Burger, K., & Terry, A. (2013). Integrating problem solving and research methods teaching for systems practice in engineering. In C. J. J. Paredis, C. Bishop, & D. Bodner (Eds.), *11th Annual Conference on Systems Engineering Research (CSER 2013)* (pp. 1072–1081). Atlanta, GA. https://doi.org/10.1016/j.procs.2013.01.113.

Yearworth, M., Lowe, D. A., Schien, D., & Walworth, T. (2015). *From deciding to acting: Hierarchical process modelling for problem structuring*. Calculating and Communicating Uncertainty (CCU2015), London, UK. https://hdl.handle.net/1983/43b50dde-2ffa-452c-95b5-446d1c132be2.

Yearworth, M., Schien, D., Burger, K., Shabajee, P., & Freeman, R. (2015). STEEP Project Deliverable D2.1(R2) - Energy Master Plan Process Modelling. *STEEP PROJECT (314277) - Systems Thinking for Comprehensive City Efficient Energy Planning*, pp78. Retrieved 26th January 2023, from https://www.grounded.systems/wp-content/uploads/2023/01/01_STEEP_D2.1_Energy_Master_Plan_process_model_update_M24_DEF_sent.pdf.

Yearworth, M., White, L., & Ormerod, R. (2015). The performative idiom and PSMs. 27th European Conference on Operational Research (EURO 2015), Glasgow, Scotland. https://hdl.handle.net/1983/0a2f32e2-84e0-4fe0-a68f-a4ae22509288.

Yearworth, M., Willis Singer, J., Adcock, R., Hyberston, D., Singer, M., Chroust, G., & Kijima, K. (2015). Systems Engineering in a context of systemic cooperative praxis (SCOOPs): Development and implications. In J. Wade & R. Cloutier (Eds.), *13th Annual Conference on Systems Engineering Research (CSER 2015)* (pp. 214–223). Hoboken, NJ. https://doi.org/10.1016/j.procs.2015.03.048.

Yearworth, M., Tully, P., White, L., & Burger, K. (2018). Quantifying the value of problem structuring interventions? 28th European Conference on Operational Research (EURO 2018), Valencia, Spain. https://ore.exeter.ac.uk/repository/handle/10871/36991.

Zamfirescu-Pereira, J. D., Wong, R. Y., Hartmann, B., & Yang, Q. (2023). Why Johnny can't prompt: How non-AI experts try (and fail) to design LLM prompts.

Conference on Human Factors in Computing Systems - Proceedings, https://doi.org/10.1145/3544548.3581388.

Zimmermann, N., Pluchinotta, I., Salvia, G., Touchie, M., Stopps, H., Hamilton, I., Kesik, T., Dianati, K., & Chen, T. (2021). Moving online: Reflections from conducting system dynamics workshops in virtual settings. *System Dynamics Review*, *37*(1), 59–71. https://doi.org/10.1002/sdr.1667.

Zyphur, M. J., & Pierides, D. C. (2020). Statistics and probability have always been value-laden: An historical ontology of quantitative research methods. *Journal of Business Ethics*, *167*(1), 1–18. https://doi.org/10.1007/s10551-019-04187-8.

Index

Please note that page numbers referring to Figures are followed by the letter '*f*', while references to Tables are followed by the letter '*t*'; Page numbers followed by 'n' refers to footnotes.

Problem Structuring: Methodology in Practice, First Edition. Mike Yearworth.

i

j

k

l

m

p

q

r

Printed and bound by CPI Group (UK) Ltd, Croydon, CR0 4YY
23/09/2024
14561759-0001